The Changing Mile Revisited

The University of Arizona Press/Tucson

The Changing Mile Revisited

An Ecological Study of Vegetation Change with Time in the Lower Mile of an Arid and Semiarid Region

Raymond M. Turner

Robert H. Webb

Janice E. Bowers

James Rodney Hastings

The University of Arizona Press
© 2003 Arizona Board of Regents
All rights reserved
⊛ This book is printed on acid-free, archival-quality
paper.
Manufactured in the United States of America
First Printing

08 07 06 05 04 03   6 5 4 3 2 1

Library of Congress Cataloging-in-Publication Data
Turner, R. M. (Raymond M.)
The changing mile revisited : an ecological study of
vegetation change with time in the lower mile of an
arid and semiarid region / Raymond M. Turner,
Robert H. Webb, Janice E. Bowers, James R. Hastings.
p. cm.
Includes bibliographical references (p.).
ISBN 0-8165-2306-1 (cloth : alk. paper)
1. Desert ecology—Arizona.   2. Desert ecology—
Mexico.   3. Vegetation dynamics—Arizona.
4. Vegetation dynamics—Mexico.   5. Repeat
photography—Arizona.   6. Repeat photography—
Mexico.   I. Webb, Robert H.   II. Bowers, Janice Emily.
III. Title.
QK938.D4 T87   2003
581.7′54′097917—dc21   2002015042

British Library Cataloguing-in-Publication Data
A catalogue record for this book is available from the
British Library.

Contents

Figures

Table

Plates

Preface

More than forty years have passed since this examination of landscape change began. Following the publication of *The Changing Mile* in 1965, the thrust of the early work temporarily slowed and then shifted in geographic emphasis. Rod Hastings's death in 1974 caused a loss of momentum. However, the slack was overcome in the 1980s as the U.S. Geological Survey continued its long-term commitment to studies of landscape change at the Desert Laboratory in Tucson, Arizona. As a result, the archive of matched photographs at the Desert Laboratory continued to expand from a few hundred "matches" in 1965 to more than 5,800 today. Many of these newer paired photographs are from the Colorado Plateau and bordering areas, from Baja California, from New Mexico's "bootheel," and even from Kenya.

When this long-term study began during the 1958–62 period, the original authors did not propose a schedule for revisiting the camera stations, and no research plan was developed for reexamining the sites. In the absence of such direction, the new authors decided in the early 1990s that, after three decades, enough time had passed for detectable landscape changes to have occurred. The time had come to revisit the original sites and document recent shifts in the scenes. Colleagues in several organizations were approached with proposals for supporting a new look at the region's landscape. Ultimately, grants were made to the U.S. Geological Survey to promote a revision and an update of the original repeat photography venture. These supportive organizations were the U.S. Forest Service, Arizona Game and Fish Department, U.S. Bureau of Land Management, The Nature Conservancy, and Pima County. We gratefully add these names to the earlier list of supporters that included the University of Arizona, the Rockefeller Foundation, and the Office of Naval Research. To these organizations and the following staff members, we owe a large debt of thanks: Gerald Gottfried, Carleton Edminister, Larry Allen, Terry Johnson, Ted Cordery, John Cook, and Julia Fonseca.

With this financial backing available, Tucson photographer Dominic Oldershaw was enlisted to visit most of the sites and recapture the views on film (see Appendix 3). Photographic prints from his darkroom were then produced for the authors' use in their subsequent visits to the sites and for use as plates in this volume. Dominic's excellent work, mostly completed in 1994, has been of singular importance throughout the course of this study.

To the list of more than thirty individuals acknowledged in the preface to *The Changing Mile*, we must add names of many who gave assistance during this second examination of changes. Those who have provided support by giving advice on a number of subjects are Julio Betancourt, Conrad Bahre, David Brown, Dan Robinett, Bill Halvorson, Steven McLaughlin, David Meko, Thomas Swetnam, Justin Turner, Larry Allen, Jean Hassell, Ron Bemis, and Thomas Dudley. Peter Griffiths and Peter Unmack assisted with computer generation of maps and illustrations. Diane Boyer provided essential expertise for converting the photographic images into their digital analogs. We were assisted in the field by Toni Yocum, Jeanne Turner, Barry Spicer, Jack Boyer, Georgianne Boyer, Don Morgan, Nancy Nagle, and Ray Nagle. We thank the numerous landowners who granted permission to enter their private land during our photograph-matching field trips. Thomas E. Sheridan, David E. Brown, and two anonymous reviewers read parts or all of the manuscript and made many helpful suggestions.

In returning to the original locales, we should emphasize anew that our photographs present views of vegetation within a desert region, not merely a desert. Some photographs show truly desert vistas, and others depict vegetation from elevated areas too moist to be regarded as desert. The camera stations are located at sea level and at intervening elevations up to approximately one mile above sea level—the vertical mile in which landscape changes are described.

As a concession to most of our readers, we have chosen to use numerous endnote insertions of pertinent literature rather than in-text references to authors and publication dates. We have adhered to the earlier practice of employing common names for most plants and animals; the Latin equivalents are given as endnotes or, for plants, as entries in appendix 2. Measurements are given in English units.

The third set of photographs new to this edition were taken with 4x5 inch, large-format cameras (Crown Graphic or Horseman 45 FA), usually fitted with a 135-millimeter focal-length lens. For exact data, readers should consult the detailed records maintained at the Desert Laboratory. Sources of original photographs are given in appendix 3, along with additional data such as photographer, geographic coordinates, and elevation.

Special acknowledgment is due one in the current roster of authors. Rod Hastings initiated this study in the late 1950s following conversations with Jim McDonald, atmospheric physicist, Atmospheric Sciences Department, University of Arizona. Supported by that department, Hastings began repeating historic photographs throughout southern Arizona in an attempt to determine whether shifts in vegetation might have ties with shifts in various climatic features. Previously, his main interest had been the colonial-era history of this borderlands region. Grasping the need to diversify, he branched into the fields of climatology and biology. Joined by Ray Turner in 1959, their joint effort subsequently culminated in the first edition of this book. To Hastings, however, goes the major share of credit for conceiving the book's grasp and defining its original thrust.

Introduction

A series of dramatic changes in the natural landscape of the American Southwest began in the 1880s. Within a decade, topography and vegetation were markedly altered. By the summer of 1890, arroyo cutting completely reworked the form of most watercourses in southern Arizona.[1] The San Pedro River, which formerly wound through a marshy, largely unchanneled valley, in August 1890 began carving a steep-walled trench through which storm waters thereafter emptied rapidly and torrentially into the Gila River. Reaches that once had perennial flow became intermittent, and the new channel was dry most of the time along much of its length.[2] The same thing happened with striking synchroneity along all the major watercourses of southern Arizona: the Santa Cruz River, Cienega Creek, San Simon Creek, Babocomari Creek, Sonoita Creek, and the Sonoyta River.[3] Channel entrenchment occurred elsewhere in the West at about the same time.[4] Irrigation ditches were left high and dry. Washes adjusted their levels to those of the main streams, dissecting and redissecting the valley floors. Much of the better farm land washed away, and much of the rest was rendered unusable. For farmers and ranchers, the economic loss was severe. Malaria, a major plague among the early settlers, disappeared, but so did the marshes, beaver, and fish.[5]

Another kind of change occurred away from the streams. In 1903, David Griffiths, agriculturist with the Office of Farm Management, noted that, on the west side of the Santa Rita Mountains, "close examination of the broad, gentle, grassy slopes between the arroyos . . . reveals a very scattering growth of mesquite (*Prosopis velutina*) which is in the form of twigs 2 to 3 feet high with an occasional larger shrub in some of the more favorable localities." He suspected that the population was likely to grow.[6] Another seven years of observation enabled him to assert that "the time is coming when these foothill grassy areas, which now have only an occasional small shrub, will be as shrubby as the deserts and lower foothills . . . if not more so."[7] Burroweed, a small shrub, had also "thickened and increased perceptibly during the last five years." He thought it quite probable that "the grasses unmolested would hold their own against its encroachment, but with the grassy vegetation weakened by grazing it may increase to such an extent as to crowd out nearly all of the valuable plants."[8]

In the same year J. J. Thornber, a botanist at the University of Arizona, agreed that "the mesquite is one of our species that appears to be on the increase." He also called attention to the deterioration of the grazing ranges:

> When once perennial grasses are killed out [by overgrazing and trampling] they are indeed slow to reassert themselves. Such denuded areas are claimed by the less valuable six-weeks grasses . . . or worse yet, they are seized upon by one or more of the species of obnoxious weeds . . . unpalatable . . . for grazing purposes. On an area where practically everything else is grazed, they alone are left untouched by stock to continue reproduction, and to thus spread farther over the adjacent poorly grazed ranges. Unfortunately, too much of our once valuable grazing domain has become thus converted into unproductive weed wastes which hold the ground against valuable grazing plants.[9]

Also in 1910, an *annus mirabilis* in the perception of changing conditions, Forrest Shreve of the Desert Laboratory of the Carnegie Institution of Washington noticed that things were not right with the saguaro, or giant cactus:

> Young plants less than 1 dm [4 inches] in height are so rare, or inconspicuous, that nine botanists who have had excellent opportunities to find them report that they have never done so . . . [leading to] the conclusion that it is not maintaining itself. . . . A fuller knowledge of its germination and the behavior of its seedlings, together with a more complete knowledge of the periodicity of certain climatic elements within its range will be sure to throw light on the fall in its rate of establishment.[10]

During the next half century, burroweed and snakeweed proliferated on overgrazed ranges,[11] the predicted mesquite invasion became a fact,[12] and the failure of the saguaro to repopulate in some areas was confirmed.[13] During the past fifty years, additional studies have proven the accuracy of Shreve's observation of failed saguaro establishment at the Desert Laboratory in Tucson[14] and have described the plant's decline in some other areas.[15] Furthermore, proliferation of certain woody plants has continued to attract attention.[16] In the original *The Changing Mile*, we noted that these changes and others constituted a striking shift in the natural vegetation of the region; they are no less striking today.

In seeking to understand the magnitude and timing of these changes, we face challenging scientific and human problems. On the one hand, vegetational change is intimately involved with hydrological, climatological, and ecological processes. On the other hand, it is closely tied to human activities by origin and implication. In explaining vegetational changes, one camp holds humans solely responsible, directly or indirectly, for all the landscape changes;[17] the other points at natural factors, primarily climatic, operating independently of people.[18] This study will not settle the matter once and for all; indeed, it may raise more questions than it answers. But it does attempt to present a comprehensive view of the changes that have occurred since 1880, to sketch their historical context, to review the principal explanations that have been advanced to account for them, and to evaluate the evidence to date.

Now that more than thirty years have passed since the previous photographs were taken for *The Changing Mile*, time enough has elapsed to reexamine, with a new set of photographs in hand, the question of causation. Whether these changes are the result of climate or human influences such as fire suppression or grazing management will be examined anew. Perhaps, with the passage of time, interactions seen only dimly before will appear more clearly, and conclusions, previously tempered by restraint for want of decisive experimental testing, will clearly surface now that the insight provided by time and testing is available.

Repeat Photography as a Scientific Tool

As a technique for studying landscape change, repeat photography dates back to 1888, when it was used to document changes in glaciers.[19] One bibliography describes more than 450 such studies undertaken worldwide between 1888 and 1984.[20] The first extensive use of the technique for studying vegetation change was in Africa[21] and then in the United States.[22] Following these early works, several investigators in the American West made extensive use of repeat photography to evaluate vegetation change.[23]

The original photographs that form the basis of our study (see appendix 3) were taken mostly in the late 1800s and early 1900s by a variety of photographers for different purposes. George Roskruge (1845–1928), for example, was a surveyor who worked primarily in southeastern Arizona. Among many other tasks, he prepared maps of the Tumacacori and Calabasas land claims and performed road surveys in the Santa Catalina Mountains and the Santa Cruz River Valley. His photographs document not only survey markers and other features of interest to surveyors but also the landscapes in which he worked. Volney Spalding (1849–1918), Daniel MacDougal (1865–1958), and Forrest Shreve (1878–1950), all scientists associated with the Desert Laboratory on Tumamoc Hill in Tucson, Arizona, took many landscape photographs in connection with their studies of desert plant ecology. Spalding and Shreve documented permanent vegetation plots at the Desert Laboratory, for instance, and MacDougal captured pristine desert vegetation in the volcanic crater that bears his name. For *The Changing Mile*, the originals taken by these and other photographers were matched mostly by James Rodney Hastings in 1962. Dominic Oldershaw and Raymond M. Turner made most of the matches for *The Changing Mile Revisited* between 1984 and 2000.

Our interpretation of photographs involved comparing the originals and replicates for changes in plant biomass. For the most part, foreground trees and shrubs were readily recognizable in the photographs. We used past field notes, photographs labeled during earlier field visits, and recent field checks to identify more distant plants. For each photograph pair or triplet, we listed all identifiable trees, shrubs, and cacti and assessed whether their apparent cover had increased, decreased, or stayed the same. Altogether, we accumulated over 1,500 records—that is, individual assessments for sixty-one plant species.

Despite some technical difficulties encountered in making an acceptable match, then interpreting it,[24] repeat photography is an excellent means of documenting many aspects of biological change. Only maps of long-term permanent vegetation plots surpass it in accuracy and historical reach. Aerial photography, which has been used to study landscape changes since it first became available in the late 1920s or 1930s, is blind to earlier times when many of the changes were initiated. In addition, species composition of vegetation can be difficult to ascertain. Aerial photography is an accurate means of quantitatively assessing recent geomorphic changes, although the nineteenth-century historical context is often lost.

Another recent tool, multispectral satellite imagery, is also widely touted as a means of evaluating landscape change. Although satellite imagery has some decided advantages—namely large spatial coverage and ability to detect many different types of spectral bands—information about species composition and geomorphic change is limited to dominant species and large changes. Moreover, systematic satellite imagery was not available until the early 1970s, making the historical perspective very short.

We assert that ground-based photographs,

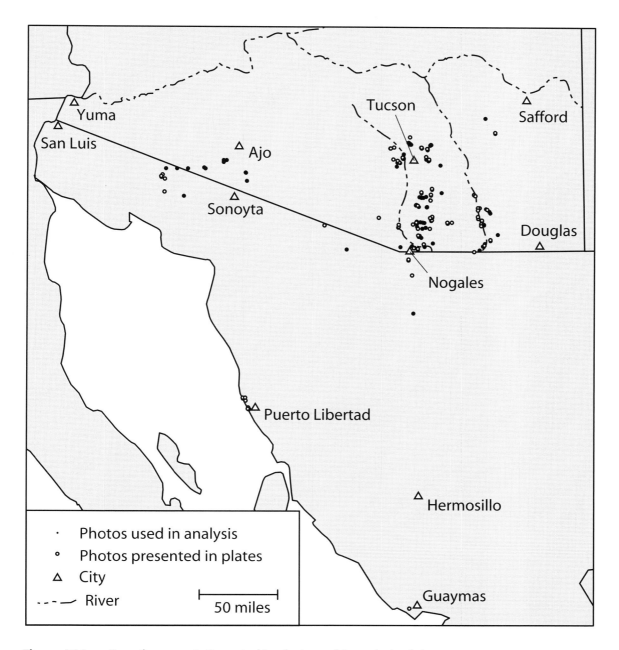

Figure I.1 Location of camera stations used in photographic analysis of change.

with their limited views, are superior to either aerial photographs or satellite imagery in the following circumstances: (1) if long-term (i.e., greater than 100 years) environmental changes are to be reconstructed; (2) if detailed information on species composition is desired; and (3) if small-scale landscape changes are to be documented. Although a single ground photograph has limited scope, it provides great detail. Multiple views from multiple stations in a large region are thus more informative than low-elevation aerial photography. Ground photography has other advantages, as well, such as ease of ground checking. Historic aerial photographs are seldom accompanied by ground-base data. Thus, when a tree of one species grows up beneath a tree of another species and

eventually replaces it, an interpreter who relied on aerial views alone would mistakenly assume that no change had taken place.

Evaluation of human impact remains one of the most elusive elements in studies of change. Some studies using repeat photography have taken place where humans have had little impact,[25] and findings from those may help tease out the effects of humans in less well controlled studies. Accurate past records of land use are mostly nonexistent, and records of a general nature are scattered. Some of our early photographs show conditions that almost certainly reflect long-term human impacts. The Desert Laboratory in Tucson, for example, lies near an earlier permanent water source and had heavy livestock use as well as other heavy use by humans.[26] These "artificial" impacts may have been of sufficient duration to place the Desert Laboratory in a special Pastoral Ecosystem category,[27] where grazing and associated forage removal have become driving forces determining the character of the ecosystem.

When only a few old photographs are available for a region, repeat photography is limited by the lack of broad coverage.[28] In our region, however, old photographs show much more terrain than can be depicted in a book of this size. We believe that, in this revision of *The Changing Mile*, we have successfully worked around this deficiency. Altogether, we examined and interpreted about 300 sets of matched photographs from 260 stations (figure I.1). Although we cannot reproduce more than a fraction of them in this book, our conclusions are based upon the much larger sample. Instead of merely describing changes, we quantified them (as described in later chapters), hence our conclusions should be more objective and quantitative than in most previous studies.

Repeat photography may have other limitations, such as a propensity for illustrating disturbed sites in many areas.[29] Our current sample includes many sites that seem relatively undisturbed, with the caveat that all but a few were almost certainly grazed, and some were doubtless cut over for fuelwood. Some photographs in the present sample were purposely selected to show disturbed sites because they illustrate important phenomena associated with disturbance. The land-use history of most of our photograph sites is largely unknown, making interpretation of causation difficult. Some would argue that causality can only be evaluated by experimentation,[30] yet how could one design an experiment that would address the all-encompassing, large-scale questions affecting change in the Sonoran Desert region?

The same criticisms can be (and have been) leveled against most other techniques for analyzing landscape change, such as repeat aerial photography and section-line descriptions from surveyors' field notes. Because our earliest photographs post-date Anglo-American settlement, we cannot assume that we can know what "pristine" conditions were. Knowledge of pre-Columbian conditions, while desirable, is not necessary to demonstrate that dramatic and consistent changes have occurred in the vegetation of our region since the late 1800s.

The Changing Mile Revisited

Chapter One The Desert Habitat

Our repeat photography spans an elevational mile of Arizona and Sonora over which annual rain increases from less than 4 inches to 20 inches. In the following pages, we will describe the setting by first characterizing the desert and then progressing upslope to the more mesic zones.

As defined by rainfall, any area that gets less than 10 inches of precipitation per year is generally considered a desert.[1] Temperature considerations are also involved, such that a cool climate with less than 10 inches of precipitation annually may support a more mesic vegetation than a hot climate with the same amount.[2] In general, two kinds of deserts can be distinguished: those deprived of an upwind source of oceanic moisture, and those that lie in a latitude chronically lacking in atmospheric lifting processes. The isolation from a moisture source may be due to great distance from a body of water, as with the Gobi Desert, surrounded on all sides by the Eurasian landmass. Moisture deprivation may be induced by the "rain-shadow" effect found to the leeward of a mountain range, as with the Mojave Desert in the shadow of the Sierra Nevada. Low-latitude west-coast deserts, the second kind, owe their existence to the arrangement of the wind systems, atmospheric pressure belts, and ocean currents. The desert along the west coast of central Baja California is a typical west-coast desert.

Arid regions cluster around the horse latitudes (about 25–35° latitude, north and south of the equator); these zones are characterized by high-pressure cells in which dry, stable air masses from aloft sink and diverge. The rainfall deficiency in these deserts does not result from lack of precipitable water overhead but from a lack of lifting in the atmosphere in some seasons of the year.[3] Humid and semiarid belts lie north and south of the horse latitudes and are characterized by converging and rising air masses. Deserts in the horse latitudes are too far from the equator to benefit much from the dynamic processes of the Intertropical Convergence Zone (ITCZ), even at the peak of its summer migration, and too near the equator to be heavily affected by frontal systems spinning off from the subpolar lows. Lying at the western edge of the continental landmasses, horse-latitude deserts receive few benefits from trade winds or east-coast hurricanes but do receive unpredictable input from dissipating west-coast tropical cyclones.

The North American Deserts

The area in North America falling within the desert region amounts to about 440,000 square miles and is shown in figure 1.1. It straddles the boundary between the United States and Mexico and includes most of the peninsula of Baja California, major parts of Arizona, Chihuahua, Coahuila, Nevada, Sonora, and Utah, and lesser parts of California, Colorado, Durango, Idaho, New Mexico, Nuevo León, Oregon, San Luis Potosí, Texas, Utah, Washington, Wyoming, and Zacatecas. As world deserts go, it is not large: it occupies about one-eighth the area of the Sahara Desert or half the area of the Arabian Desert.[4]

On the basis of climate and biology, the North American Desert contains four regional deserts. The Great Basin, the northernmost, has cold winters with frequent snowfall. Its vegetation, dominated by sagebrush, has evolved from the primitive, temperate Arcto-Tertiary forest.[5] The Mojave Desert, next southward, lies in California, Nevada, Utah, and Arizona. Most of the scanty precipitation here falls during the cool winter season.[6] The vegetation has a strong likeness to those parts of the Sonoran Desert where winter rainfall also prevails and, biologically, has been considered part of that desert.[7] The Sonoran Desert has the warmest winter temperatures of the four deserts, plus hot summers. Precipitation is biseasonal, varying from predominantly winter at the northwest to predominantly summer at the east and southeast. The Chihuahuan Desert, although hot in the summer, can be cold in the winter. Its rainfall is biseasonal, with most falling in the summer months. We examine landscape changes in the two southernmost of these three deserts, the Chihuahuan and the Sonoran.

The Chihuahuan Desert

From the Mexican state of Chihuahua, the Chihuahuan Desert extends north into New Mexico, east into Texas and Coahuila, west into Sonora and Arizona, and south into San Luis Potosí, Nuevo León, Zacatecas, and Durango. It occupies the northern end of the Mexican plateau, an elevated tableland bordered on one side by the Sierra Madre of the West and on the other by either lowland plains or the eastern Sierra Madre.

Like the Great Basin, the Chihuahuan Desert is high: nearly half lies above 4,000 feet, and its upper elevational limit is well above 6,000

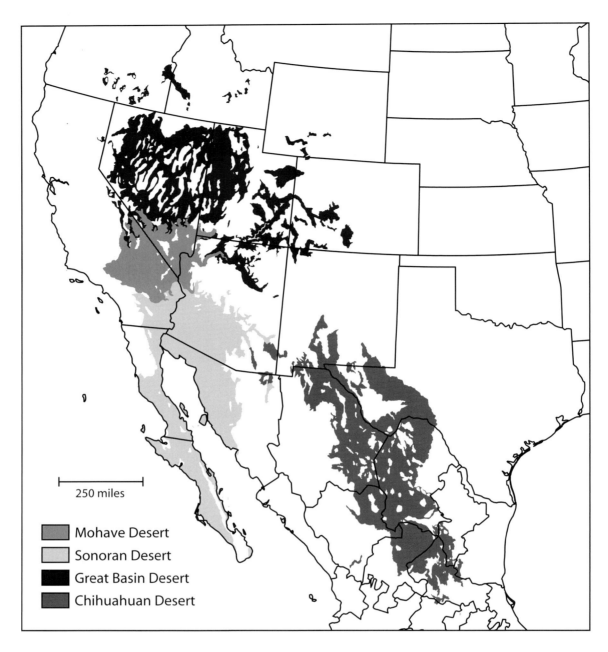

Figure 1.1 The North American Desert and its subdivisions. (After Brown et al. 1998, Reichenbacher et al. 1998.)

feet.[8] In contrast to the Mojave Desert, the major rains occur in summer. The largely shrubby perennial vegetation is more varied in appearance than the low, monotonous shrublands of the Great Basin but still less varied than the arborescent associations of the Sonoran Desert. The Chihuahuan Desert contains relatively few trees and tall cacti but has an abundance of smaller cacti, spiny shrubs, succulent-leaved century plants and yuccas, and isolated islands of grassland.

Separated from the main part of the North American Desert, the Chihuahuan Desert grades into the semiarid steppes of the Great Plains on the north and east; to the west the Continental Divide roughly separates it from the Sonoran Desert. At no point is creosote bush—the common denominator of the three southern deserts—continuous across the grassland that lies along the Divide and serves as a transition between the arid regions on either side. A large island of Chihuahuan Desert flora occurs along the San Pedro and San Simon Valleys of Arizona, well within the range of this study, and well away from the main body of the parent desert.[9] Unmistakably Chihuahuan plants such as tarbush, all thorn, mortonia, and Chihuahuan whitethorn grow abundantly on limestone soils where, in the last century, there were almost pure stands of grass.

The Sonoran Desert

The Sonoran Desert is the lowest and hottest of the North American deserts. Most of the Sonoran Desert lies in the Mexican states of Sonora, Baja California, and Baja California Sur, but a substantial section extends north into Arizona and California. The desert proper lies between sea level and about 3,000 feet, depending on latitude and local soil and microclimatic conditions. Isolated mountain masses often called "sky islands" rise above the desert. Because atmospheric moisture is present over the region (occasionally in abundance),[10] here, as in the Chihuahuan Desert, orographic lifting and rainout combine to make the mountains moist. Their tops may receive over 30 inches of precipitation per year and yet be located within 10 miles of creosote-bush plains. These steep gradients of rainfall and temperature are reflected in a sharp vertical zonation of the plant life, making the mountains an ideal place to observe the effect of climatic change on plant communities. Indeed, the higher grassland and oak woodland zones, which will be described later, are of as much interest to us as the desert.

Occupying an area of about 120,000 square miles, the Sonoran Desert is 870 miles long and 400 miles across at its widest point[11] and extends irregularly across twelve degrees of latitude, from above Needles, California, at 35° N to near San José del Cabo at the tip of Baja California at 23° N. Large-scale physiographic and climatic influences inevitably come into play over such a large area. From west to east, the land surface tilts upward, and the climate becomes increasingly continental with a larger daily and seasonal range of temperature.[12] In addition, a winter rainy season gives way to a summer rainy season, and total rainfall increases. North to south, the hours of daylight, the temperature, and the intensity and angle of sunlight all reflect the transition to lower latitudes. Several important characteristics of vegetation change largely in response to these shifting climatic parameters.

The Climate of the Region: Precipitation

The Synoptic Meteorology of the Rainfall Regimes

Arizona and Sonora are geographically placed in the path of several moisture sources. In winter, most moisture entering the Sonoran Desert comes embedded in extratropical cyclones from the North Pacific Ocean, although subtropical air from as far as the equatorial Pacific Ocean may also enter the region.[13] In summer, the dominant moisture source is the Gulf of Mexico, although the Pacific Ocean and Gulf of California also contribute. In late summer and early fall, dissipating tropical cyclones bring moisture in from the southern west coast of Mexico.[14] In all seasons, moisture recycled from other places on the continent may enter the deserts, but this source is relatively small.[15]

Moisture sources to the southwestern deserts are steered and fed by several semipermanent circulation features. In winter, the Aleutian Low, in the Gulf of Alaska, commonly spins off low-pressure systems that may move into Arizona or even Sonora. Two semipermanent high-pressure cells, known as gyres, steer moisture into Arizona. The most dependable source is the Bermuda high off the East Coast of North America, which is thought to be largely responsible for the Arizona monsoon. The Eastern Pacific high off the West Coast affects the steerage of both tropical cyclones and the subtropical jetstream, which may fuel winter frontal systems.

Winter rains are brought by low-pressure systems embedded in the westerlies, which migrate southward during the fall and northward during the spring. These systems may be either frontal in nature, and associated with large-scale ridge-trough structures in the upper atmosphere, or they may be low-pressure systems cut off from the general circulation. Frontal systems may be either cold and relatively dry, if their origin is over the continental western United States, or warm and relatively wet, if supplemented by moisture embedded in the subtropical jetstream. Cutoff lows generally peak in spring and fall but can occur at any time during the winter; they are eddies of moisture that become detached from the westerlies and wander down the Pacific Coast, exiting to the center of North America on a path that traverses Arizona and New Mexico. The Mojave and Sonoran Deserts profit marginally from these infrequent incursions of cold, rainy weather; the Chihuahuan Desert, on the lee side of the Sierra Madre, benefits less frequently when the subtropical jetstream brings in warm, wet air.

With the northward advance of the sun in late winter and early spring, the low-pressure systems cross the continent on a more northerly route. The decreased frequency of incursions heralds the onset of a dry season that becomes more severe through spring and finally is either ended or reinforced by events related, this time, to the northward migrations of the two semipermanent highs. The Sonoran and Mojave Deserts, on the western continental slope, come under the influence of the Eastern Pacific high. The Chihuahuan Desert, on the eastern watershed of the continent, falls under the influence of the Bermuda high.

In the Northern Hemisphere, the western edge of a high-pressure system is associated with converging, rising, unstable air, whereas the eastern edge has sinking air masses from diverging, stable masses aloft. Over the deserts of the western slope, the drought intensifies, giving rise to a dry, increasingly hot period appropriately called by Forrest Shreve the "arid fore-summer." Over the desert of the eastern watershed, however, the drought is dissipated by moist, southeasterly flow reaching progressively farther inland from the Gulf of Mexico as the high-pressure cell advances. May rainfall in Tamaulipas shows a sharp increase over April; June rainfall in Zacatecas, Durango, and Chihuahua, is a sharp increase over May.[16]

At the end of June, an abrupt global readjustment brings a meteorological climax to Sonoran activities[17] and provides one important exception to the statement that western coastal deserts do not benefit from easterly activity. The highs move rapidly northwestward and enlarge. The Sonoran Desert is freed from the subsident eastern end of the Pacific anticyclone, which continues, however, to inhibit rainfall over the Mojave Desert and California. At the same time, the moist tongue at the western edge of the Bermuda high extends over the Continental Divide, bringing air from the Gulf of Mexico. The Sonoran summer monsoon begins; rainfall over most of the Chihuahuan Desert continues to be abundant.

During summer months, moisture is delivered to the Southwest in three dominant types of storms. Local summer thunderstorms, fueled by hot rising air from the desert floor, brings the most predictable precipitation. Occasionally, larger systems, including the rare frontal system from the Pacific Ocean, bring what is called mesoscale convective complexes into the region.[18] These storms, which can include features such as squall lines, cover relatively large areas and can lead to large-scale flooding.[19] Finally, tropical cyclones can dissipate over land during the summer months, although they are more likely to occur in September and October. All of the storm types have caused floods in the Sonoran Desert.[20]

Global Teleconnections and Climate of the Sonoran Desert

When considering the climate of the southwestern United States, one must consider first the global climate system as a whole, because our region is sensitive to global-scale climatic processes. One of the most important aspects of this system is the trajectory of winds over the Northern Hemisphere. Referred to as the general circulation pattern, the movement of air over the Northern Hemisphere has been classified into two distinct types, with numerous variations in between.[21] In zonal flow, low-pressure systems track mostly west to east across the oceans and continents, with little north-to-south movement. In me-

ridional flow, the path is very sinuous, and waves and troughs form in the pressure field aloft. Meridional flow causes much movement of storms in a north-south direction and also serves as a mechanism for mixing of tropical and extratropical moisture sources.

The dominance of one type over the other has changed during the twentieth century.[22] Between the early 1960s and the present, meridional flow has dominated. This has caused high precipitation variability throughout the western United States because meridional flow can bring either floods or droughts, depending upon the location of the waves and troughs as well as the availability of moisture. Extremes can be pronounced, particularly if the waves and troughs become fixed in space; storms spawned in the tropical Pacific Ocean, for example, pounded California in 1997 because of a fixed meridional wave pattern that drew moisture into the state along the subtropical jetstream. Zonal flow tends to cause protracted drought to average conditions, primarily because the tropical moisture source is largely severed. Zonal flow was particularly prevalent in the 1940s and 1950s. Because of lack of data, no determination of flow patterns have been made before 1947, although one suspects that meridional flow also predominated around the turn of the twentieth century, given the wet conditions and frequent floods.

A phenomenon in the Pacific Ocean is now thought to have a profound effect on much of the climate of the southwestern United States. This phenomenon is known as the El Niño–Southern Oscillation (ENSO), the former name coming from its typical arrival off the coast of Peru around Christmas time, and the latter name referring to a back-and-forth shifting of warm water from the western to the eastern Pacific Ocean.[23] During El Niño conditions (also known as "warm ENSO conditions"), a warm water pool forms in the equatorial Pacific Ocean off of Peru and spreads northward and southward, bringing abnormally warm water as far north as California. The result is above-average winter precipitation and high potential for flooding as well as a weak tendency for failure of the summer monsoon. The opposite of El Niño conditions, referred to as La Niña conditions, results in colder-than-normal water in the western Pacific. The result is well below-average winter precipitation and a tendency for above-average monsoons.[24]

El Niño events recur every four to seven years on average but may dominate for several consecutive years. El Niño affects general circulation patterns, creating a tendency for more subtropical moisture to enter the Southwest.[25] Because of this and the fact that the warm oceanic water generates more atmospheric moisture, precipitation increases, particularly in winter and particularly in February and March. The reliability of the precipitation increases with southerly distance in Arizona[26] and is strongest in the central Sonoran Desert.

The combination of general circulation patterns and ENSO creates what has come to be known as global teleconnections, or correlations between climates of different regions.[27] For example, the probability is high that, if the Pacific Northwest is in a period of drought, the Southwest will be in a wet period; the converse is also true. Global teleconnection maps reveal the interconnectedness of climate around the Earth; periods of Australian and Sahelian drought typically are times when rainfall is high in the Southwest. Teleconnection maps also suggest that certain regions—for example, the Sonoran Desert and the southeastern United States—share a common climatic signal with other desert regions owing to the similarity in their location within the global climate system.

Hydroclimatology and Flood Frequency

One of the implications of decadal climatic variability is its effects on rivers, their channels, and the riparian vegetation growing along their banks. Arizona, and to a lesser extent Sonora, is unique in North America for the diversity of storm types that cause floods. K. K. Hirschboeck listed nine types of flood-producing storms in the Gila River basin.[28] These types can be simplified to four: summer monsoonal thunderstorms; winter frontal systems; summer and fall dissipating tropical cyclones; and spring and fall cutoff low-pressure systems. Of these four types, the most predictable—and the one least affected by long-term changes—is the type comprising local summer thunderstorms. The influence of winter frontal systems and tropical cyclones changes through time; in some cases, years to a decade or more may pass before tropical cyclones affect some areas of the Sonoran Desert.

During summer thunderstorms, rainfall intensities generally exceed soil infiltration rates, and the resulting runoff redistributes moisture from upland sites to channels. Most runoff from typical summer thunderstorms infiltrates locally, within several miles of where the rainfall occurred. The spotty nature of summer thunderstorms creates a mosaic of saturated soil upslope of relatively dry channels, there forming moist, riparian microenvironments in which the less xeric native plants can flourish. Summer thunderstorm runoff encourages these riparian microenvironments in another way, by depositing sediments over floodplains and replenishing nutrients to the riverine environment. Periods dominated by summer thunderstorms, with few, large winter frontal systems or dissipating tropical cyclones, may be the times when the incised channels, called *arroyos*, fill up with sediments.

Winter frontal systems and warm-season regional storms have different effects. Although rainfall intensities of dissipating tropical cyclones may be lower than the more typical thunderstorms, so much water is added to already wet watersheds that much of the precipitation runs off. The same is true for winter frontal systems or cutoff lows, particularly if rainfall has been high beforehand. After some extremely large storms, the resulting floods escape from the desert through the Gila, Sonora, and Yaqui Rivers. Examples of these storms are Tropical Storm Octave in 1983 and large winter storms in 1891 and 1993.[29] These floods were violent enough to sweep away riparian vegetation growing along the floodplains, resetting the successional clock for this type of vegetation in the Sonoran Desert. High flood frequency before the turn of the century may be one reason why the earliest photographs of river channels in the Sonoran Desert show little riparian vegetation and few trees.

The frequency of floods caused by some storm types has changed through time in the Sonoran Desert. For the Santa Cruz River at Tucson, the incidence of floods caused by winter frontal systems and dissipating tropical cyclones greatly increased after about 1960, while floods caused by the summer monsoon remained unchanged.[30] Changes in flood frequency, particularly when combined with some land uses (such as overgrazing by livestock), may be the reason for the initiation of arroyo cutting that greatly changed the landscapes of the Sonoran Desert before 1900 (see chapter 3).

The Relation of Desert Vegetation to Climate

Shreve's definition of a desert is topographic and climatic:

> It is essentially a region of low and unevenly distributed rainfall, low humidity, high air temperatures with great daily and seasonal ranges, very high surface soil temperatures, strong wind, soil with low organic content and high content of mineral salts, violent erosional work by water and wind, sporadic flow of streams, and poor development of normal dendritic drainage.[31]

He did not fail to note the vegetational implications of this definition:

> From the biological standpoint desert is best defined in terms of the limitations which have just been mentioned. For plants these serve to prevent the full degree of develop-

ment that would enable them to form a closed covering, attain a considerable size, maintain vegetative activity throughout the year, and meet the environmental conditions without structural features or types of physiological behavior that tend to reduce their maximum performance.[32]

In the case of the Sonoran Desert, these general characteristics have to be qualified. Some plants do attain a considerable size: the saguaro, for example, or its even bigger relative, the cardón. Something resembling a closed cover can be found in riparian communities and the forest mottes of the Foothills of Sonora.[33] At any time of year, at least a few of the members of a community may be vegetatively active.[34] Finally, near the Gulf of California, the higher moisture content of the air may be a factor in shaping distinctive shore communities. Moreover, some plants undoubtedly make use of the heavy fogs that drift inland from the Pacific across the west-central region of Baja California. Qualifications like these emphasize the special climatic and microclimatic characteristics that set the Sonoran Desert apart from others; they remind us, too, that the plant life faithfully reflects these special conditions.

The Regional Range of Annual Precipitation

Except for snowfall at higher elevations in the mountains, precipitation in the desert region falls mainly as rain. The average annual amount varies from about 1.2 inches per year at Bataques, west of the Colorado River in Baja California, to 21.8 inches per year at Oracle, Arizona, in semidesert grassland surrounded by Arizona Upland desert.[35] For all of the climate stations we analyze for this book, the average annual precipitation is 11.98 inches. At our camera stations, few of which have reliable weather records, annual precipitation probably ranges from perhaps 4 inches in the Pinacate Mountains of Sonora and at Punta de Cirio on the Gulf of California to about 18 inches at El Plomo Mine.[36]

These annual averages by no means indicate the variability in water stresses that desert plants must undergo. In 1956, Yuma recorded 0.25 inches of rain, but 11.41 inches fell in 1905. San Luís, a Mexican station on the lower Colorado River just south of Yuma, reported less than 0.01 inches in 1956, but 15.2 inches fell in 1927 and 14.93 inches fell in 1983. Perennial plants must routinely be able to survive many weeks or months of high temperature without precipitation of any kind and with soil moisture values well below the wilting point. During June, rain fell only eight times at Yuma in the 81 years from 1893 through 1974 and only nine times at Yuma Citrus Farm in the 76 years between 1921 and 1997. In

drier sections of the desert, droughts of seven to eleven months duration are not uncommon.[37]

Coefficients of Variation for Precipitation

A useful statistical measure of temporal variability in precipitation is the coefficient of variation (CV), which has been calculated for several regions of the Sonoran Desert.[38] In general, the CVs of annual precipitation vary inversely with mean annual precipitation.[39] For four of the stations lying in our region, the CVs are 77 percent at San Luis, Sonora (mean annual precipitation = 5.20 inches); 65 percent at Yuma, Arizona (mean annual precipitation = 3.26 inches); 38 percent at Guaymas, Sonora (mean annual precipitation = 9.25 inches); 30 percent at Tucson, Arizona (mean annual precipitation = 11.46 inches); and 22 percent at the San Rafael Ranch, near the Arizona-Sonora border (mean annual precipitation = 17.55 inches).

The CVs reveal the great interannual variability that desert vegetation must endure. Even if annual precipitation is 65 percent below average in only one year out of three, this still represents a major climatic stress in the lower elevation desert. Given these high CVs, it is not surprising that the population of annual plants undergoes large oscillations from year to year. During some years, few may germinate, or mature; during other years, the desert may approach lushness.

The CVs of precipitation also vary monthly and seasonally. CVs of monthly precipitation are highest in May and June (2.16 percent and 2.13 percent, respectively) and decrease with mean annual precipitation. The lowest CVs for monthly precipitation occur in July and August (0.86 percent and 0.77 percent, respectively), whereas most of the winter months have CVs between 1.1 and 1.7 percent. Contrary to what might be expected, the spatially spotty, convective storms of summer, often less than a mile in diameter, are a more dependable moisture source from one year to the next than winter storms. The implications of the seasonal precipitation variability for perennial plants are clear: precipitation is most dependable from year to year during the hot summer, when vegetative activity is optimum for many species and when water stress is the highest.

One exception to this generality is that summer rainfall along the western edge of the Sonoran Desert, and particularly over the northwestern corner, is less dependable. In Yuma, the summer CV is 94 percent, which is substantially larger than the winter CV of 75 percent. In northern and western Baja California the same is true, although in the eastern and southern parts of the peninsula the more general picture holds, and summer produces the minima.[40] Summer variability is also greatest in the low, hot sink west of the Colorado River in Mexico, the most arid region in North America and, in summer, one of the hottest. In this case particularly, where the plant life most needs dependability to mitigate aridity, it does not get it and must labor under an additional handicap. "Certain parts of [this region] have the thinnest plant covering to be found in North America."[41]

Seasonal Distribution of Precipitation

The seemingly anomalous highs in summer CVs over the western part of the region are directly related to factors governing the seasonal distribution of precipitation. The rainfall of the Sonoran Desert is transitional in its regime between the Chihuahuan, with summer rainfall, and the Mojave, which receives the bulk of its precipitation during winter months. There are, in effect, two rainy seasons over most of the Sonoran Desert, and biseasonality, along with its subtropical nature, is a major factor in shaping its plant life.

Shreve illustrates the influence of seasonality by citing three stations with nearly identical yearly amounts of precipitation but different distributions. Ensenada, on the west coast of Baja California, has a winter rainy season; Nogales, on the Arizona-Sonora border, has the Sonoran double-maximum, one coming in December–January, the other in July–August; Monclova, in eastern Coahuila, is characterized by summer rainfall. "The vegetation of the three localities is respectively chaparral, evergreen oak woodland and arid bushland. It is doubtful if a single native plant is common to the floras of Ensenada and Monclova."[42]

The Implications of Biseasonality

The principal synoptic events that control the seasonal distribution of precipitation account for many of the differences among the three "creosote-bush deserts." The bimodal rainfall distribution that prevails at our photographic sites has been characterized by R. A. Bryson, who appends to his description a provocative and pertinent notion:

> Just north of Arizona, then, winter rainfall is more abundant than summer, but the double-maximum is dominant; just south of Arizona the double-maximum weakens and summer rainfall is dominant, just west of Arizona the summer rainfall disappears and winter rains dominate the southern California area, and just east of Arizona the winter rainfall loses *relative* importance and the summer peak dominates the annual march. Little wonder that through the centuries the ecol-

ogy of the area, strongly controlled by moisture and its pattern of availability, has been apparently unstable.[43]

Much the same picture arises from a simpler analysis involving the ratio of winter to total annual rainfall, summer being defined as the six hot months (May–October) and winter as the six cold months (November–April). In figure 1.2, isograms for this datum have been plotted for Arizona. Their values, ranging from 30 percent to 70 percent in the desert region and the adjacent highlands, show not only the general biseasonal pattern but also the lack of homogeneity within it. Two prominent gradients appear: one extends from west to east, with summer rainfall becoming increasingly dominant; the other shows a similar trend from north to south.

The east-west trend is probably of critical importance to the survival of such Sonoran Desert dominants as the saguaro and foothill paloverde, whose ranges suggest that regular summer moisture is a requirement.[44] The north-south trend may be of equal importance. Many plants find their northern or southernmost limits in this area of rapidly changing gradients, and the shifting patterns of seasonal precipitation may well play a part in terminating their ranges.

The north-south gradients are also of interest in connection with a problem closely related to vegetation change: the arroyo cutting that began along the streams of Arizona about 1890. The peculiar topography of the southeastern part of the state makes it one of the few regions in the United States that drains northward, and the valleys of both the San Pedro and the Santa Cruz Rivers sharply intersect the gradients of seasonal precipitation. In the course of 100 miles, summer precipitation in these valleys drops from over 70 percent of the annual total to 50 percent. At least one group of observers has suggested that shifts in the seasonal distribution may be responsible for the onset of arroyo cutting,[45] and, though the tangible evidence is slight, the possibility nevertheless exists.

Effective Precipitation

Effective precipitation is the fraction of total precipitation that reaches the roots of desert plants. In the hydrologic cycle, some water is intercepted in plant canopies before it reaches the ground; some of this reaches plant roots via stemflow. Water in the canopies and very near the soil surface mostly evaporates with little benefit to the plants. In very wet periods, some water percolates below the rooting depth of most desert plants and recharges the ground water. Effective precipitation becomes soil water that is drawn from depths of between about 2 and 36 inches, al-

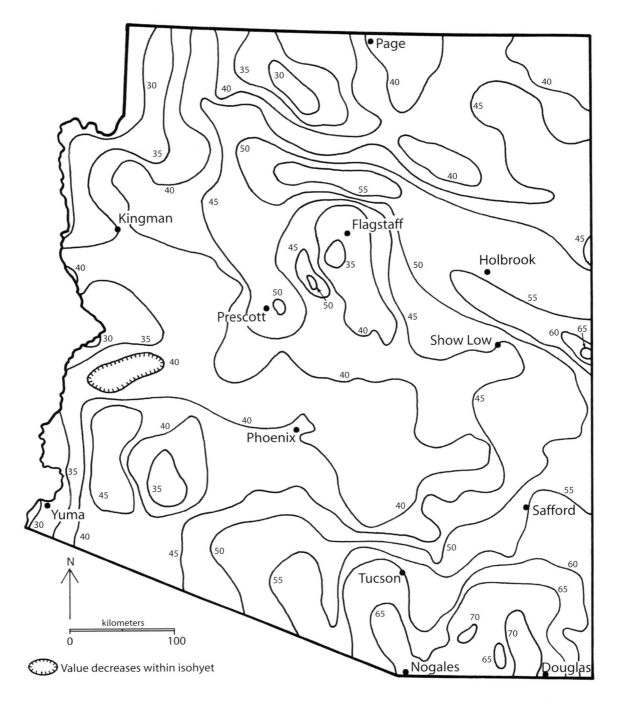

Figure 1.2 Percentage of annual precipitation falling in the four summer months (June through September). (Based upon ARIS 1975.)

though values within that range decrease with increasing aridity and are affected by soil calcic horizons (*caliche*).

The amount of precipitation that is effective varies seasonally and with storm type. During gentle winter storms, nearly all precipitation is effective. During wet years with large, regional storms, the soil column becomes saturated, and both runoff and ground-water recharge occur. A threshold exists where additional winter precipitation does not benefit plants. In these winters,

plant growth is related more to the number and timing of storms through the winter than to the total rainfall.

Most summer thunderstorms have intensities that greatly exceed the infiltration capacity of the soil.[46] As a result, much of the precipitation lost to upland plants via runoff is gained by riparian ecosystems downstream. Much of the summer precipitation—particularly late season precipitation when soils are wetter—is ineffective to desert plants. One exception is when moisture from dissipating tropical cyclones enters the region. Rainfall from this moisture source often has lower intensities and longer durations, resulting in saturated soils and, occasionally, in large floods.

The Efficient Use of Precipitation

Because most summer rain evaporates, ecologists have tended to underestimate its importance to perennial plants. Actually, the shallow-rooted succulents of the Sonoran Desert may make considerable use of summer moisture. Experiments on the grounds of the Carnegie Institution's old Desert Laboratory indicate that the efficiency of the saguaro in absorbing water varies directly with temperature[47] and that, given identical amounts of rainfall, the cactus is capable of more rapidly picking up and storing a given volume in summer than in winter. During the former season, rehydration may be detected by stem swelling within twenty-four hours following significant showers.[48] Many desert plants are capable of producing "rain roots" that elongate quickly following small amounts of rainfall. In some succulents studied, rain roots are visible six to eight hours after droughted plants are given water.[49] The abundance of succulents in a hot desert with biseasonal precipitation, as well as their relative scarcity in a winter-rain desert (e.g., the Mojave Desert), may rest in part upon their ability to take immediate advantage of even small amounts of summer rainfall.[50]

The ability of succulents to effectively capture water during the hot summer months via specialized rain roots is just one part of their multifaceted adaptation to aridity. Little advantage would be gained if water were taken up only to be immediately lost through epidermis and stomates, the plant surface and its pores. In most plants, stomates open during the day and close at night, permitting inward diffusion of CO_2, which is required for photosynthesis. Oxygen and water meanwhile diffuse outward and are lost to the atmosphere. Such water loss, a necessary if undesirable side effect of photosynthesis, is a disadvantage to plants in hot, dry climates.

Succulents such as cacti and agaves have structural and physiologic traits that greatly diminish water losses. Their low surface:volume ratio provides maximum water storage capacity coupled with minimum exposure of water-losing surfaces. In addition, the epidermis is covered by an especially thick, waxy cuticle that severely limits passage of water molecules. Coupled with these water-conserving anatomical features are physiologic adaptations that permit the stomates to remain closed during daylight hours and to be opened at night. The series of chemical steps making this possible is called crassulacean acid metabolism, or CAM. During the night, when their stomates are open and gas exchange occurs freely, CAM plants incorporate atmospheric CO_2 into a compound called malate. When the sun comes up, this internal source of CO_2 is released, the stomates close, and photosynthesis begins anew. Because temperatures are relatively cool at night, CAM plants lose much less water to the atmosphere than typical non-CAM plants, which must conduct photosynthesis under the higher temperatures of daytime. CAM plants thus have high gains of photosynthate per unit of water lost.

A further adaptation made by arid region plants in their struggle to conserve water under conditions of water scarcity involves the internal biochemical pathway of the carbon in the CO_2 molecules that are exchanged between the plant tissue and the atmosphere. About half of all grass species and nearly all other plants employ a biochemical process whereby CO_2 is converted into a three-carbon compound in an early step leading to the manufacture of the complex carbohydrate products of photosynthesis. These so-called C_3 plants contrast with another group that includes the remainder of the grasses and a scattering of nongrass plants. Here, the biochemical pathway leads to a four-carbon product as an early step in photosynthesis. Members of this group of C_4 plants use water more efficiently and have greater photosynthetic capacity at high temperatures than do C_3 plants, ideal adaptations for desert living. Except for the saltbushes, however, most of the woody plants seen in our photographs from desert to woodland are C_3. While only half of all grasses are C_4, this physiologic group is greatly overrepresented in the grasslands of our region, where perhaps 95 percent of the production is by grasses with the C_4 pathway.[51] Thus, the significance of carbon pathway differences will more likely apply to grasslands than to deserts.[52]

The Climate of the Region: Temperature

The Influence of Temperature

Plant responses to temperature are no less important than their responses to rainfall in determining which species can tolerate desert conditions; if anything, the ways in which a plant may react to temperature are more numerous and complex. Some of these ways are as follows:

- The temperature at which seeds germinate varies widely from one species to another. With desert annuals, this special requirement is important in determining both their spatial distribution and whether they occur as winter or summer annuals, in response to winter or summer rainfall.[53] Similarly, seeds of perennial plants are often responsive to rainfall during but one of the two contrasting rainy seasons in our region. Because of high temperature requirements for germination, seeds of velvet mesquite, saguaro, foothill paloverde, and ocotillo germinate only during the summer months. Creosote bush, regardless of its location in Chihuahuan, Sonoran, or Mojave Deserts, germinates mostly in late summer and fall. Seeds of triangleleaf bursage respond to midwinter rains, and brittlebush seeds respond to cool-season rains falling between October and April.[54]

- In some cases, a critical physiological function requires a specific temperature range. One example is the saguaro and its inability to pick up and store water in cold weather. A similar limitation has been noted for the staghorn cholla, whose root growth is slow at temperatures below 68°F and optimum at 93°F. Its range is necessarily limited to places where a rainy season coincides with hot weather.[55] Roots are relatively intolerant of high temperatures, a constraint that is probably responsible for the paucity of roots in the upper few inches of the soil.[56]

- Within a single plant, some processes may be most efficiently carried on at one temperature, other processes at another, so that appropriate diurnal and seasonal variations in temperature are necessary if the plant is to carry on all of its functions.

- Unusually cold temperatures may kill some of the native plants at the cold margin of their range and may thereby alter the structure and species composition of biotic communities in such areas.[57] Of more frequent occurrence are catastrophic freezes that merely injure plants without lasting effect on the structure and composition of communities.[58] Although a given occurrence may be rare in the weather records, the probability is high that, in the many thousands of years over which the native vegetation has evolved, it has already experienced and withstood similar extremes. Nevertheless, the paleobotanical record makes it clear that long-term shifts in climate can displace plant distributions geographically and drastically alter the composition of the vegetation.[59]

The Regional Range of Temperature

Because temperatures are determined primarily by large-scale factors like solar radiation and the earth's lag response, curves showing the annual distribution of average monthly temperatures look much the same for all stations in the Sonoran Desert. A well-defined yearly maximum occurs at most places in July; a minimum occurs in January. Inland, the range between the two is largest and the extremes of the curve tend to be most sharply defined; that is, the climate at these inland stations is more continental than that of the coastal stations.[60] As a result of their proximity to the ocean, certain stations—La Paz, Ensenada, Santa Rosalía, and Puerto Peñasco, for example—tend to have temperature maxima that are delayed into August.

The driest part of the region—that lying around the head of the Gulf of California—is also the hottest. Average July temperatures at Yuma, Gila Bend, San Luís, Bataques, and Mexicali all exceed 90°F. Average January temperatures for the same group cluster around 55°F.[61] Of the places where we have photographic stations, the Pinacate Mountain region and Guaymas are probably the hottest. In the former, the range of average monthly temperatures is from about 55° to 95°F; at the latter, the range is from 65° to somewhat under 90°F.[62]

But, as with precipitation, means in temperature do not indicate the real stresses to which plants are subject, particularly when standardized measurements are made well above the ground. On a June day in Tucson, a relatively temperate desert station, a temperature of 161°F was recorded at a depth of 0.16 inches in the soil. At the same time, a thermometer exposed in a standard shelter read only 108.5°F.[63] Although the lower temperature measured above the ground may have more significance for a large plant, the higher temperature just beneath the surface is clearly more important to the germination of seeds and to the survival of seedlings. The subsurface temperature, or the even harsher temperature at the soil surface, most nearly approaches the limiting condition for plant life.

The Distribution of Freezing Weather

In light of the stereotypes that emphasize deserts as hot regions, the opposite extremes—those of cold—are even more surprising. Occasional winter cyclones penetrate as far into Mexico as the 20th parallel, bringing cold, arctic air with them.[64] Although Shreve and coworkers flatly state that "the entire area of the Sonoran Desert is subject to occasional frosts,"[65] in fact there may be a few coastal stations that are completely frost-free.

The effects of frost vary considerably in magnitude and frequency over the Sonoran Desert. The frequency of freezing temperatures is highest at higher elevations, although higher elevations may not have the lowest minimum temperatures. Relatively warm ocean water buffers freezing temperatures near coastlines, resulting in very few periods of freezing weather in coastal Baja California or Sonora. Cold-air drainage affects some valleys that have high topographic relief, particularly internally drained valleys that have playas. Valleys such as the Tucson basin—at about 2,200 feet and surrounded by mountains 7,000–9,000 feet high—have a mosaic of environments that differ greatly in their frost occurrence because of cold-air drainage.

Freezing temperatures can cause relatively minor damage to leaves and tender shoots, or the entire above-ground biomass may be killed. The amount of damage to frost-sensitive plants is a function of the lowest temperature reached as well as the amount of time below freezing. Minor damage can also result from radiative cooling, where air temperatures are greater than freezing but temperatures on leaf and stem surfaces fall below freezing. Damage can be greater if the plants were actively growing before the freeze (e.g., as in December 1978).

During the spectacular cold spell of January 1937, thermometers recorded freezing temperatures in southern Sonora during 1937: 30°F at Navajoa, 32° at Ciudad Obregon, and 29° at Cedros. During that same period, temperatures reached 33°F at Libertad and 34° at Empalme. Observers reported frost damage as far south as Cedros, near the Sonora-Sinaloa boundary, where injury occurred not only to tender, subtropical plants but also to hardy types like canyon ragweed and soapberry, which range northward into the United States.[66]

Near Cedros, the desert grades into thornscrub forest, and Shreve has observed that the boundary coincides with the beginning of the frost-free zone. He proposes implicitly, and W. V. Turnage and A. L. Hinckley do so explicitly, that the line where freezing weather no longer occurs be considered the southern climatological limit for the Sonoran Desert.[67]

If one accepts this view, then the mere occurrence of freezing can be dismissed as a factor making for diversification among desert plant communities, because it is a factor to which all of them are subject. But the frequency, intensity, and duration of freezing spells still remain of interest, and it seems likely that at least one of them may control the vertical zonation of plants on the mountain ranges within the desert.

The Frequency and Intensity of Freezing Weather

For the northern part of the region, frequency is relatively easy to examine, because Arizona data are available for the average number of days per year that have minimum temperatures below 32°F. Plotting these values against elevation for eighteen stations in the southeastern part of the state, one gets the scatter shown in figure 1.3a. There is, in fact, almost no relation to elevation above sea level. Bisbee at 5,350 feet has fewer than one-third as many freezing days as Willcox at 4,200 feet; fewer than half as many freezing days occur at Benson at 3,635 feet.[68]

Values for "record low," when used as a rough approximation to intensity, show a similar lack of dependence (figure 1.3b). San Simon at 3,608 feet has experienced a low of −5°F; Fort Grant at 4,880 feet, +2°F; Bisbee, 5,350 feet, +6°F. It is clear that neither the number of freezing days nor the intensity of cold bears a wholly consistent relation to elevation or to vegetation zonation.

A valuable study of freezing temperatures at three stations near Tucson sheds some light on the lack of correlation between low temperature and elevation. For a period of five years, thermographs were maintained at two locations near the Desert Laboratory of the Carnegie Institution: the "hill" station on the shoulder of Tumamoc Hill and the "garden" station at the foot. Only half a mile of distance and 330 feet of elevation separated the instruments, yet the garden station on some nights recorded temperatures that were 20°F lower than those on the hill. For the five-year period beginning with the winter of 1932–33, the total number of freezing nights noted at the hill station was 38; at the garden the number was 263. The frost season, the period between the first and last freezes of a winter, averaged 36 days on the hill and 157 days at the garden. During the winter of 1935–36, the frost season extended only from January 2 to January 20 on the hill, but at the garden, half a mile away, it went from October 24 to April 6.[69]

For two of the five winters, a third station was maintained by an observer at Summerhaven, 22 miles away in the Santa Catalina Mountains at an elevation of 7,600 feet. In general, minimum temperatures at Summerhaven and at the garden, 5,100 feet apart in elevation, were more closely related than minimum temperatures at the garden and on the hill. During the winter of 1933–34, Summerhaven registered 57 freezing nights; the garden had 39 freezing nights; and the hill had 6. During the same winter, the coldest temperature recorded on the hill, at Summerhaven, and at the garden was 28°F, 17°, and 16°, respectively. "These data indicate the high mountains of southern Arizona do not experience minimum temperatures very much colder than . . . those of the desert lowlands."[70]

The great differences between hill and valley, and the similarity between valley and mountain, can be attributed largely to cold air drainage. The low humidity, clear skies, and long winter nights of the desert provide optimum conditions for rapid nighttime radiative cooling of

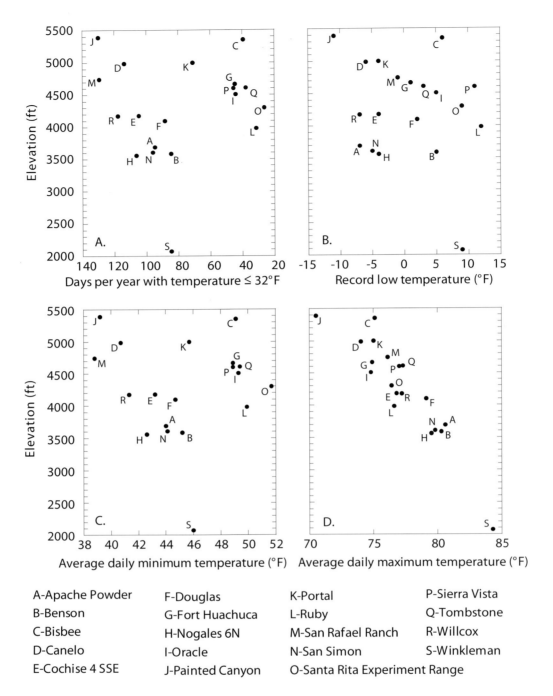

A-Apache Powder   F-Douglas   K-Portal   P-Sierra Vista
B-Benson   G-Fort Huachuca   L-Ruby   Q-Tombstone
C-Bisbee   H-Nogales 6N   M-San Rafael Ranch   R-Willcox
D-Canelo   I-Oracle   N-San Simon   S-Winkleman
E-Cochise 4 SSE   J-Painted Canyon   O-Santa Rita Experiment Range

Figure 1.3 The elevational distribution of various data of temperature for nineteen Arizona stations.

the ground.[71] The air in contact with the soil surface cools by convection and conduction,[72] and the lower air temperatures propagate upward, producing an inversion condition in the lower atmosphere in which temperature increases with height, the reverse of daytime lapse conditions.

The air next to the earth, being coldest and therefore densest, tends to drain down slopes and away from high places into lower-lying areas, where it accumulates, contributing to a well-developed inversion layer that may be several hundred feet deep in some desert valleys. In the

Desert Laboratory experiment, the hill station lay near the warmer top of such a "cold lake,"[73] the garden at the bottom.

With this explanation in mind, the scatter of figures 1.3a and 1.3c becomes easier to interpret. The stations in the lower left of each diagram are located in valleys where cold-air drainage produces a relatively high frequency of freezing temperatures (figure 1.3a), and relatively low average temperatures (figure 1.3c).[74]

The cold "floods" resulting from such air flows have been recognized for many years as contributing to the interdigitation of life zones at the upper edge of the desert and above. Where two zones make contact, the higher commonly reaches down into the lower along canyons and valleys, the paths for cold-air drainage; the lower reaches into the upper along ridges. In some cases, trees may descend along streams as much as 3,000 feet below their lowest occurrence on north slopes.[75]

Nighttime inversion layers may also account partially for the differences in vegetation between upper and lower *bajadas* (outwash slopes). On the upper slope nearest the mountain, one commonly finds in the Arizona Uplands such plants as saguaro, paloverde, and ironwood. Downslope this community grades into creosote bush and triangle-leaf bursage, which, in turn, give way on the cold bottom lands to mesquite. The transitions are usually attributed to soil,[76] which typically ranges from coarse-textured and well drained near the mountain, to fine-textured and poorly drained in the bottoms. The soil factor may well be reinforced by inversion effects.[77]

The Duration of Freezing Weather

The duration of cold need not be closely related to either intensity or frequency. As noted above, because of cold-air drainage, average 24-hour minima, usually occurring at night (figure 1.3c), do not show strong elevational relations. Average 24-hour maxima, however, occurring in daytime after inversions lift, do show strong elevational relations (figure 1.3d). Although Summerhaven and the garden station may experience approximately comparable nighttime freezing temperatures, the latter station with dawn rapidly proceeds to warm and to assume its appropriate position on the curve of temperature with respect to elevation. At the former station, freezing may persist well into the day, or even through it. The duration of freezing periods, then, may be substantially unrelated at stations with similar intensities or frequencies of cold.[78]

Since the patterns, because of cold-air drainage, are quite irregular with respect to elevation, the intensity and the frequency of freezing temperatures evidently do not control the elevational limits of life zones. The duration of freezing spells, however, may be "the most potent cold temperature factor on the high mountain slopes of southern Arizona insofar as the absence of desert plants is concerned."[79]

Some important facts about duration stem from elementary considerations about diurnal variation in winter. Daily maxima commonly occur about 1500 hours, and minima occur around 0600. Most freezing periods are limited to the hours of a single night, setting in after dark and terminating shortly after dawn, but in the event of a prolonged freeze, associated usually with an influx of Arctic air, temperatures below 32°F may endure well into the next day. The freezing period can terminate with or before the diurnal maximum, in which case its duration will be limited to around twenty-two hours, or it can continue through the warmest part of the day, in which case it is intensified by the declining temperatures of evening and night. Under these latter conditions, barring an unusual advection of warm air, the freeze will continue at least until the dawn of the following day and will result in a duration of thirty-six hours or more. A frequency distribution for the length of freezing periods thus is discontinuous. There are many of twenty-two hours and less; there are some of thirty-six hours and more. Only rarely will there be intermediate values.

Experiments with small saguaros have shown that between twenty and thirty-six hours of freezing temperatures were fatal. In their desert habitat, "the occurrence of a single day without midday thawing coupled with a cloudiness that would prevent the internal temperature of the cactus from going above that of the air, would spell the destruction of *Carnegiea*." Recognizing the existence of the quantum jump in durations, Shreve suggested that the line along which the jump occurred should coincide with the northern limit of the saguaro.[80] Turnage and Hinckley carried Shreve's generalization one step further: "The duration of freezing temperatures throughout a night, the following day and the following night . . . coincides with the northern limit of the Sonoran Desert and with the vertical limit of desert vegetation on mountain slopes and tablelands."[81]

Although values for thirty-six-hour duration are hard to come by, another datum is readily available: the number of days per year when the maximum twenty-four-hour temperature does not rise above freezing. For reasons given above, the two measures are roughly equivalent. In figure 1.4, the latter datum has been plotted for all weather stations listed in *Arizona Climate*.[82] The line separating the locations that have never recorded such a day from those that have, coincides closely with the northern boundary of the Sonoran Desert as drawn on the basis of vegetation and flora.[83] It also agrees closely with the known distribution of the saguaro.[84]

The Climatological Limits of the Sonoran Desert

Bearing in mind that the term "desert" is imprecise and that there is seldom a sharp boundary between an arid region and its semiarid borderlands, fairly definite climatological limits for the Sonoran Desert may now be stated: on the south it is bounded by the farthest advance of frost; to

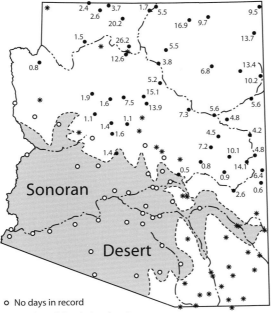

o No days in record

• Number of days below freezing per year

✻ Some days, but mean value < 0.5 days per year

Figure 1.4 Mean number of days per year with no rise above freezing temperatures.

the east and north, and on the mountain slopes within it, the limit occurs where there is no midday interruption of freezing temperatures; and to the west its extent is apparently fixed by inadequate summer rainfall. Within these broad confines the climate is varied yet homogeneous enough to produce a distinctive and diverse vegetation.

Biotic Communities

The plant life at any given place in the desert region depends upon its location with respect to axes running east-west, north-south, and vertically. Climatic gradients occur along all three: the vegetation tends to respond generally to the gradients but with strong disruptions due to edaphic and microclimatic influences. On the basis of the vegetation, and considering for a minute only the vertical axis, one can distinguish six "life zones" or biotic communities bounded by roughly horizontal surfaces that dip or rise with variation in rainfall and temperature.[85]

Some or all of the zones may be seen in ascending a single large mountain range within, or adjacent to, the Sonoran Desert. As one leaves the desert, the small-leaved plants, or microphylls, become less conspicuous, and plants armed with spines, thorns, or prickles are left behind. Grasses increase in importance and, on gentle terrain, often predominate. Mixed with the grasses are scattered leaf succulents such as agaves and yuc-

cas. Above the grassland, widely spaced live oaks introduce the woodland zone, in which perennial grasses and herbs create a rich flora. Higher still, the oaks become more closely spaced and are joined by other evergreen sclerophylls, or hardleaved plants.

Where ascent of a mountain range is interrupted by extensive level to gently rolling terrain at an elevation of roughly 5,000 feet, another biotic community asserts itself. Plains Grassland, characterized in part by tall grasses such as cane bluestem, tanglehead, sideoats grama, and plains lovegrass, reaches its western limits in our region on suitable terrain where it merges below with Semidesert Grassland and above with Oak Woodland.[86] Its ecotone with the Semidesert Grassland is subtle at best, and adding to this ambiguity is the observation that under grazing this biotic community has shifted toward Semidesert Grassland.[87] Although several of our plates show vegetation arguably referable to Plains Grassland, the only camera station that is clearly within this biotic community is on the Appleton-Whittell Research Ranch.

Blending with the upper edge of the woodland is the fourth life zone, a forest dominated by long-needled pines. The lower portions of this zone still include several species of oak in a subordinate role. With increasing elevation, the vegetation of the forest becomes simpler as oaks drop out and pine species become fewer in number. A fifth zone at yet higher elevations is signaled by the appearance of white fir and Douglas fir and the decreased importance of the pines. And, finally, the two highest and largest mountain masses in the area support at their tops forests of spruce and fir.

Such is the vertical zonation that one finds in the regional vegetation: Desert, Semidesert Grassland, Oak Woodland, Pine Forest, Douglas Fir–White Fir, and Spruce–Subalpine Fir Forest in order of increasing elevation. Only the first three zones, collectively constituting the first mile above sea level—the changing mile—will be dealt with here. Of these, the desert encompasses the largest expanse. The grassland and the woodland are confined either to the upper reaches of isolated mountain ranges within the desert or to the adjacent continuous highlands. Latitude, the size of the mountain mass, its geologic composition, and the elevation of the basal plain all play roles in determining the upper and lower elevational limits of each zone, as do various climatic, edaphic, topographic, and cultural factors.[88] The influence of living creatures on vegetational distribution may seem insignificant compared to the forces exerted by soil, temperature, and water; nevertheless, various vegetation patterns have been shaped by humans and other animals.

Chapter Two The Influence of American Indians, Spaniards, Mexicans

In assessing the extent to which humans changed the desert region, one is tempted to postulate that a "natural," static vegetation existed about a century ago, and then to estimate the extent to which Anglo-American settlement "disturbed" a previously "undisturbed" situation. One is tempted to invoke this simple model, yet it does not fit the historic and prehistoric evidence for our region.

Certainly long before cattle drives moved stock from Texas into the Sonoran Desert, human activities played an integral part in the biological balance of this area. Well before 1880, Mexican cattle roamed some parts of the grassland, and Spanish cattle were present in some areas long before then. Eons before the Spanish arrived, a diverse fauna of now-extinct grazers and browsers, akin to that seen in Africa today, affected the ecosystems of our region, producing a biological balance of yet another sort.[1]

At least four centuries ago, and possibly eight, the American Indian population of the desert was larger than the total population, European and native, that the area supported in 1880 or even as late as 1935.[2] Because the aboriginal peoples subsisted on the resources of their immediate environment, their impact on other living creatures, animal and plant, may have been substantial. For example, at Chaco Canyon, in northwestern New Mexico, the daily fuel needs of large pueblos eliminated the surrounding pinyon pine–juniper forests a millennium ago, leaving a treeless landscape that persists to the present.[3]

What we have to consider, then, is not a "natural" situation disrupted by "artificial" events following Anglo-American occupation, but rather a fluid environment shifting with the centuries and millennia under the impact of a succession of cultures, each differing somewhat from the preceding in its relation to the life around it. In the northern part of the region, the Mexican phase gave way in the 1850s to the Anglo-American society that has since remained dominant. In some respects the new culture impinged more on its surroundings than did its predecessors, in some respects less.

As we considered our photographs, we asked whether the effects of these cultures were sufficient to account for any of the vegetational changes that have taken place since 1880.

The American Indians

The influence of most American Indians can be known only from archaeological remains, contemporary observers, and later inference. Based on their societal economy, we can distinguish three groups: nonagricultural bands, band peoples, and ranchería peoples.[4] The individual cultures are shown in figure 2.1.[5]

Nonagricultural Bands

Along the coast of the Gulf of California roamed the nomadic Seri, who subsisted entirely by hunting, fishing, and gathering edible plant foods. At the time of his work among the Seri late in the nineteenth century, W. J. McGee reported that cactus fruits from the saguaro, cardón, and prickly pear constituted about 9 percent of their diet. Flour made from mesquite beans, seeds collected from sedges and grasses in the mud flats along the gulf, nuts gathered on the mountains, and small quantities of a local seagrass, in that order of importance, completed his list of their wild plant foods.[6] Today, the Seri utilize as food 25 percent of the 374 plant species that they recognize within their homeland. Although plants were an important feature of their diet, meat and fish were their staple foods.[7]

The Band Peoples

A second group, which included the Apache, practiced a little agriculture and, though nomadic, had favorite locations in which they recurrently settled. Like the Seri, they hunted and gathered wild foods. Although the Apache are important to Western history, by and large they are peripheral to an ecological discussion of the desert region with the exception of their horse and cattle raids, which assumed considerable importance. Prior to the advent of livestock and Apaches, such people as the Pai preyed upon agriculturalists, thereby affecting settlement patterns of the Southwest.

The Ranchería Peoples

The third group, which occupied by far the greatest part of the Sonoran Desert, can be called the ranchería peoples after their diffuse, more-or-less permanent villages. They lived by a mixture of farming and wild-food gathering, with the former tending to predominate but varying in impor-

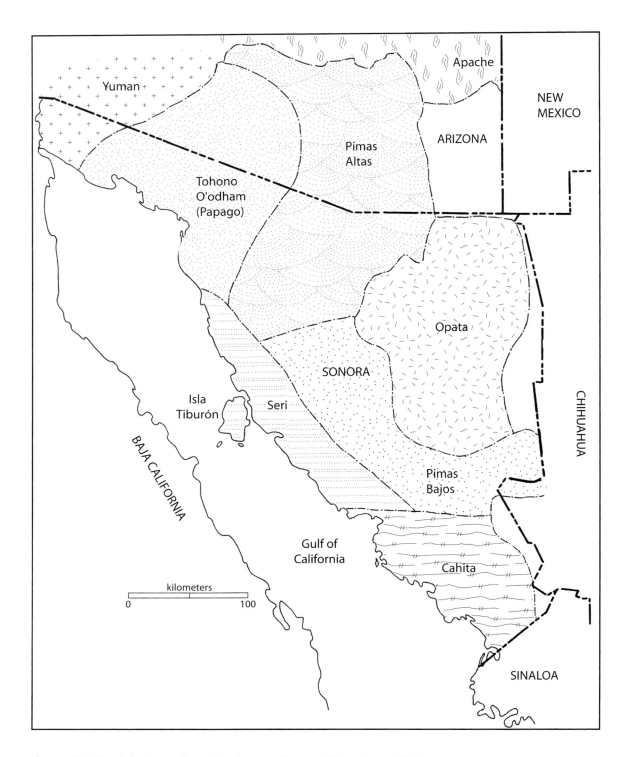

Figure 2.1 Aboriginal peoples of the Sonoran Desert. (After Sauer 1934.)

tance among the different subgroups. The Yuman, who occupied the valleys of the lower Colorado and the lower Gila, depended much more heavily on wild foods than on agriculture. To their south and east dwelt the Tohono O'odham, formerly known as the Papago Indians, and the Akimel O'odham, formerly called Upper and Lower Pima Indians. In the summertime, ranchería peoples commonly migrated from town-like settlements to valleys where surface water was available for irrigation. The Tohono O'odham shared the arid lowlands with the Yuman and also tended to be simpler in their habits than the Yaqui and Opata, village peoples who lived to the south. The Upper Pima inhabited the uplands of south-central Arizona and northern Sonora; the Lower Pima, the plains and the foothills of Sonora. Farther south, the Yaqui lived along the banks of the Yaqui River, which had perennial flow large enough to maintain extensive agricultural activities and support permanent, well-organized settlements. The Opata inhabited the uplands around the Sonora, San Miguel, and Bavispe Rivers.

The Size of the Aboriginal Population

Estimates of the native population at the time of Spanish contact vary within relatively narrow limits. C. O. Sauer places the number of Akimel O'odham, Opata, Seri, and Yaqui at about 155,000.[8] Subtracting the population that lived outside the boundaries of our region, and adding the Yuman peoples of the north, we estimate that in 1600 the desert region, excluding Baja California, sustained a population of about 150,000, comparable to E. H. Spicer's estimate of 120,000 for the same area at about the same time.[9] After the conquest, social upheaval and newly introduced diseases brought about a precipitous decline in the aboriginal population.[10] Nearly three centuries of European colonization were required to restore the former density, and by 1935 the population of northwestern Mexico was much the same as the aboriginal rural population.[11] Regardless of the overall population, vast areas probably remained unoccupied because of predatory bands.[12]

The Ecological Impact of Food Gathering

The extent to which the American Indian population influenced the ecological balance of the region in pre-Spanish times cannot be stated with any precision. Certainly American Indians impinged on the plant and animal life at many points. A partial list of the plants eaten by the Indians of the Vizcaíno region includes the roots of yucca and *Amoreuxia palmatifida*; the stems, roots, and buds of agaves; fruits or seeds from many plants, among them organpipe cactus, cardón, sinita, prickly pear, barrel cacti, mesquite, ironwood, paloverdes, jojoba, canyon grape, elephant tree, and Mexican tea.[13] Saguaro, saltbush, cottonwood, catclaw, pencil cholla, staghorn cholla, Mexican crucillo, desert cotton, and *Solanum elaeagnifolium* were common food plants. Mangrove, ironwood, and mesquite were important woods. Creosote bush, yerba de pasmo, and a number of herbs were valued for their medicinal qualities.[14]

The ecological significance of these lists is not obvious. Under some conditions, food gathering may aid in the distribution of a species, as when grass seed is harvested by flail and basket.[15] When fruits are wholly consumed and digested, however, their seeds are not dispersed. Likewise, when humans harvest roots and growing points, the results may also be adverse. McGee,

for instance, cites the rarity of barrel cacti in Seriland and states that the few plants he saw were either wounded or dwarfed. The pulp provides a ready source of water, and in the dry reaches of the Central Gulf Coast, "its dearth suggests destruction nearly to the verge of extinction by improvident generations."[16] In the Vizcaíno region, the absence of agaves from old mission sites and water holes has been ascribed to harvesting by the native peoples.[17]

In judging the many qualitative assumptions that must be made about the impact of food gathering, any quantitative information seems welcome. In his study of the Seri, McGee reports that the 300 "full eaters" of that tribe consumed 27,000 pounds of cactus fruits per year, or 216,000 individual fruits. Per person this amounts to 90 pounds or 720 individual fruits.[18] Although of passing interest, this human impact should be weighed against that of ubiquitous and often abundant wildlife species such as doves.

There is, of course, no way of knowing what proportion belonged to each of the species involved, the cardón, the saguaro, and the prickly pears; in any event the proportion would vary from year to year with the harvest. If one assumes the saguaro's characteristics for all species, there are about sixty fruits per plant[19] and an average plant density of perhaps six per acre.[20] On this basis, a "full eater" would have consumed the output of about 2 acres.

If each of an imaginary 100,000 "full eaters" matched this performance, food gathering Seri-style would have accounted for the fruit from 200,000 acres, or about 400 square miles. When one considers the range of the saguaro, this is not an impressive total, and it becomes less impressive still when divided among several species of cacti, some of which have larger density values.[21]

In terms of the individual seeds consumed it is, of course, an enormous number: this many fruits are capable theoretically of producing 1.4×10^{10} new saguaros.[22] But successful establishment is a rare event; virtually all saguaro seeds perish regardless of human activities. A more illuminating calculation is that one saguaro produces about twelve million seeds during the century that constitutes its active reproductive life. In a stable stand, one individual matures for each that dies. The end product of twelve million seeds, therefore, is one plant that grows to maturity. Calculated on this basis, the total aboriginal consumption across the entire expanse of the desert resulted in the loss of only one thousand plants per year.[23]

Moreover, in view of the uncertainties of qualitative ecology, one cannot be sure that food gathering tended, even this weakly, to interfere with saguaro establishment. Many cactus seeds remain viable after passing through an intestinal tract,[24] and the effect of human consumption might merely have been their wider dissemination.

The Ecological Impact of Cultivation

Agriculture may have been a more potent factor in pre-Spanish ecology than food gathering. Because many of the ranchería peoples practiced simple irrigation, disturbance was almost inevitable wherever water and tillable land existed. In the uplands, farming was confined to the upper reaches of canyons because most streams were dry by the time they reached the plains. Opportunities for extensive agriculture existed along the lowlands of perennial rivers such as the Yaqui, the Colorado, and the Gila. Numerous clearings must have dotted the valley floor. As rivers and crop yields shifted with time, the fields may also have shifted, so that over the course of several centuries a continuous ribbon of disturbance can be envisioned, some parts actively cultivated, some recently abandoned, some long overgrown.

The presence of water and tillable land was not the only combination making possible agricultural pursuits. In the Tucson-to-Phoenix area, evidence of dryland farming has been found at several sites. Located on gently sloping bajadas and marked by systems of thousands of minor check dams and rock piles used for agave cultivation, these agricultural endeavors undoubtedly had an impact on the landscape.[25]

An informative picture of the conditions prevailing along the San Pedro River can be recreated from the records of a trip in 1697 by the missionary Francisco Eusebio Kino. His party found about two thousand American Indians living in twelve villages scattered along the valley between Babocomari and Aravaipa Creeks.[26] In 1697, Santa Cruz, the southernmost of the villages, was inhabited by about one hundred people who practiced irrigation and tended a herd of cattle that Kino had given them on an earlier trip. About 3 miles downstream lay Quiburi, the largest of the rancherías, populated by about five hundred Indians.[27]

For some 50 miles downstream, the valley contained villages recently abandoned because of an internal war among the Sobaipuri. Below these, within a stretch of 35 miles, Kino found ten occupied rancherías. All practiced irrigation and raised corn, beans, cotton, and squash. Phrases like "good pastures" and "fertile lands" abound in the descriptions.

The history of Quiburi, which can be traced in some detail, illuminates both the mobility of the villages and the ease with which disturbance propagated along watercourses. In 1692, the Quiburi Indians lived near present-day Redington at Baicatcan.[28] Five years later, on the occasion of Kino's visit, they were also at Quiburi. In the following year, they moved to Los Reyes de Sonoidag on Sonoita Creek near present-day Patagonia.[29] By 1704, the old site at Quiburi had been reoccupied.[30] In 1762, the Sobaipuri abandoned all settlements along the San Pedro in the face of increasing Apache pressure,[31] and the valley remained largely unoccupied until Anglo-Americans arrived about a century later.[32]

A pattern of cultivation and abandonment gives plant invaders an opportunity to become established.[33] In spite of aboriginal disturbance, however, and a plentiful supply of seeds, mesquite did not become dense in the San Pedro and Sonoita Valleys until after 1880.[34]

American Indians and Fire

Various interpreters of the historical record have asserted that broad, brush-free grasslands are the product of repeated fires set by aborigines during hunting activities.[35] In the United States, in particular, Sauer and others attribute the Midwestern prairies to frequent, widespread fires of human origin.[36] In the Sonoran Desert region, as well, various investigators have asserted that the Semidesert Grasslands owe their existence to unrestricted burning by aborigines.[37] Indians commonly set hunting fires in grassland and woodland, as newspaper accounts attest;[38] extensive areas were sometimes burned.[39] Other reports describe use of fire by specific tribes in desert, woodland, or forest for hunting[40] or for clearing recently harvested fields.[41] Fires no doubt originated from other sources, as well, such as abandoned campfires, mescal roasting pits, and especially lightning.[42]

Two additional pieces of evidence point to the importance of fire in Semidesert Grassland. First, many plants and animals indigenous to the grassland zone are well suited to survive fire through post-burn sprouting and germination enhancement.[43] Second, abundant fire-scars in the tree-ring record show that woodland and forest—and, by inference, the grassland downslope—have long coexisted with fire in our region.[44]

The importance of fire in vegetation zones above the desert does not necessarily apply to the desert proper. Except in small, moist areas following exceptionally wet years, vegetation of the desert is typically too sparse to carry fires for great distances.

The Ecological Role of American Indians

The indigenous cultures of the Sonoran Desert as they existed about 1600 were the product of many thousands of years of slow evolution.[45] As

American Indian societies evolved, they undoubtedly made greater demands on the environment, at least during those times when their populations were waxing. We can surmise that the effects of food gathering were probably localized around villages. Farming, an important source of disturbance, was confined principally to the valleys where there was surface water.[46] There is scanty evidence about use of fire for hunting; the safest assumption is that American Indians added little to the natural incidence of burning.

The Spaniards

The Age of Exploration

By 1521, Hernán Cortés had destroyed the Aztec empire, and the Spaniards, using Mexico City as a base, raced out in all directions across the new land. Within ten years Nuño de Guzmán founded Culiacán, 700 miles away from the fresh ruins of Tenochtitlán. His slavers, raiding among the Yaqui, made the initial penetrations of the Sonoran Desert.

By 1536, Alvar Nuñez Cabeza de Vaca had journeyed across Texas and down the Río Sonora, where he was hospitably received by the Opata. In 1537, Fray Marcos de Niza retraced part of Cabeza de Vaca's route and saw, he said, the cities of Cíbola. Two years later Francisco Vázquez de Coronado—trailing sheep, horses, and cattle in his wake—followed the Friar, descended the San Pedro, crossed the highlands of east central Arizona, and found to his disgust not the fabulous cities but instead the mud pueblos of New Mexico. One of his lieutenants, Melchior Díaz, crossed the Sonoran Desert from the Río Sonora to the Colorado River and partially explored the upper reaches of Baja California before falling victim to his own lance. Hernando de Alarcón, cooperating with Coronado by sea, reached the Colorado and proceeded up it, possibly past the mouth of the Gila.[47]

Spectacular though they were, the conquistadors impinged very little on the life of the desert, and the same can be said of the explorers who followed them at intervals during the remainder of the sixteenth century. Not, in fact, until the end of the century—when missionaries reached the Yaqui River without fanfare—did the Spanish make their presence felt.

The Mission

The Sonoran Desert peoples fell under Jesuit influence in 1617 when Andrés Pérez de Ribas led an expedition to the Yaqui River.[48] Thereafter missionary activity spread rapidly across the region.[49] The Jesuit impact, which was profound out of proportion to the small number of Europeans involved, brought about a revolution in the cultural and economic life of the native population, considerably modifying their ecological role. The unique institution through which it operated was the mission. These self-sufficient units not only spread Christianity but also played a key part in the imperial policy of Spain by serving as a tool of acculturation and assimilation.[50] In theory, missions existed only on the frontier, and their lifetime was only ten years. During that time a mission was expected to concentrate American Indians in settlements nearby, to Christianize them, and to teach them artisan skills, farming, and animal husbandry. Having done this, the mission was then supposed to move forward with the advancing frontier. The new converts, sufficiently Christianized to depend on a secular priest, were in theory ready to assume the role of yeoman farmers and to assimilate themselves into the culture of the recently arrived Spanish settlers.

By 1650 the mission system extended up the Yaqui Valley to the headwaters of the Sonora, San Miguel, and Bavispe Rivers—nearly to the present International Boundary. For the next four decades, few new establishments came into existence. Then, beginning about 1687, Kino sent a second wave of missionaries across the homeland of the Akimel O'odham. These missions marked the northernmost expansion of New Spain in the inland west. After Kino's death in 1711, sporadic attempts to convert the original inhabitants of the San Pedro, Gila, and lower Colorado Valleys met with no permanent success. In fact, the Jesuits, engaged in an unequal struggle with the settlers, the Apaches, and, finally, the crown itself, were hard put merely to maintain what they had won.

The Colonization of Sonora

During the early missionary period, white settlers were absent. By 1600, a few ranches existed on the coastal plain south of the Yaqui and, in the mountains, a few mines. The main stream of colonization came a generation later in 1637, when Pedro de Perea received viceregal permission to found a colony on the San Miguel River at Tuape, several hundred miles in advance of the other Spanish settlements. This colony eventually became a nucleus for rapid secular expansion.[51] By the time Kino undertook the reduction of Pimería Alta, gold and silver diggings dotted the uplands throughout the province, soon followed by ranches in the grasslands and farms in the valleys. As the native population dwindled, European settlers encroached on the Indian villages and in many places jostled the missions themselves. Increasing friction between the two cultures brought a growing demand for secularization of the mission centers and for distribution of their lands. The Jesuits resisted this movement for two reasons. On a static frontier, there were no new areas in which to proselytize. Moreover, they knew that acculturation of the native tribes required longer than the allotted ten years and that to leave their charges essentially meant abandoning them.[52]

The Quarrels of the Eighteenth Century

The period after 1711 became increasingly a time of conflict for Sonora, involving a many-sided struggle among missionaries, colonists, and American Indians in which, by and large, the civil administration took the part of the settlers. The most obvious manifestation of the conflict was a general revolt by American Indians starting in 1740, when a rebellion spread northward out of the Yaqui and Mayo villages. In 1751 the Upper Pimas rebelled.[53] The Seri, implacable enemies of all Europeans after the abduction and enslavement of Seri women in mid-century, harassed the settlements from havens on Tiburón Island, along the coast of the Gulf of California, and in the Cerro Prieto. Presidio after presidio came into existence as the Spanish king attempted to control the unrest and to combat yet another enemy, the Apache.

The Apaches

It is impossible to follow the movements of these seminomadic bands with any degree of certainty. They seem to have moved southward at about the time of the Pueblo Revolt in New Mexico in 1680. By 1692, they were raiding into New Spain and threatening the Sobaipuri of the San Pedro River Valley.[54] As the presence of the Spanish colony expanded, so did the tempo of the Apache raids on missions, towns, mines, and the growing herds of livestock. Horses and cattle were driven to strongholds where troops rarely dared penetrate. The ill-armed, ill-fed, ill-trained garrisons of New Spain were virtually powerless against these hit-and-run tactics.[55]

Eighteenth-century accounts are replete with references to the destruction and depopulation attendant upon the Apache raids. J. Nentvig wrote that "Sonora is on the verge of destruction, for we have seen . . . scarcely ten places inhabited by Spaniards, while there are more than eighty ranches and farms destroyed by the enemies."[56] I. Pfefferkorn stated, "No one can engage in agriculture in outlying areas without risking his life." He added that stock-raising had been "almost completely ruined."[57]

Other contemporary accounts present a somewhat different picture. In a matter-of-fact survey of his flock in 1763 or thereabouts, the bishop of Durango makes it clear that there were considerably more than "ten places inhabited by Spaniards" in Sonora. Indeed, of the fifty-two he lists, five had a population in excess of one thousand.[58] By his account, the total population of the province was at least seventeen thousand, which suggests continuous growth rather than near desolation. Thus, although the Apache-Seri warfare undoubtedly resulted in bloodshed and destruction, it does not seem to have interfered with the fundamental processes of colonial development.[59]

The Reforms of the Enlightenment

By 1759, a variety of internal dissensions and external disasters combined to reduce Spain to a second-rate power. Over this decline there had presided a succession of kings ranging from merely incompetent to literally imbecilic. When Charles III came to the throne in 1759, Spain finally had a ruler capable of addressing the whole complex of domestic and colonial ills. He dispatched José de Gálvez to the New World on a tour of inspection.[60] This visit resulted in the creation of the Interior Provinces, a sprawling jurisdiction partially independent of the viceroy and governed initially from Arizpe by its own commandant-general. Under the direction of Gálvez, able administrators supervised the rejuvenation of the northern frontier. A policy that coupled force with feeding brought the Apache under control, and for the next fifty years Sonora enjoyed a measure of peace.

Much of the good Charles did was undone in 1767 when, to ameliorate political turmoil in Spain, he expelled the Jesuits from all his domains. This act destroyed the equilibrium between American Indians and European colonists. The Franciscan order was appointed to replace the Jesuits, and, despite the efforts of the Franciscans to restore the system to its old vigor, many mission centers became secularized. With further encroachment by European settlers, native tribes were rapidly assimilated into a semi-Europeanized people that outnumbered both indigenes and colonists.[61]

The Ecological Impact of the Spaniards

During their two centuries in the Sonoran Desert region, the Spaniards came to exercise an ecological importance commensurate with their political dominance. First, they altered the relation of American Indians to the environment. Not only did the number of natives decline following the European invasion, but their geographical distribution was transformed as well. After they had been concentrated in mission centers, native peoples necessarily relied more heavily than before on cultivated foods (although game and wild plants were still an important part of their diet). If there were one-third as many natives in 1800 as in 1600, and if they relied only half as much on wild food, their impact on the wild life would have been, on the average, only one-sixth as great as it had been in 1600 when the Europeans arrived. In this respect, the Spanish presence must have lessened the direct impact of people on the environment, at least locally.[62]

Although disturbance in the immediate vicinity of missions probably remained high, food-gathering beyond the mission centers must have dropped to inappreciable levels. This suggests that, after 1600, most of the region should have experienced a slow recovery from whatever pressures the native population had once imposed. Unfortunately, any tendency toward overall recovery was doubtless swamped by the Spaniards, who, through mining and livestock raising, played a new and profound role in the ecology of the region.

SPANISH MINES AND ECOLOGY. Mining constitutes an important way in which Spanish activity impinged on the ecology of the desert region. On the alluvial plains the effect may have been slight; in the uplands, where the principal ore deposits lay, the impact may have been substantial. One observer, for example, concluded that the scrub grassland around the Chihuahuan mining town of Parral used to be oak woodland and that the transition from one to the other occurred during colonial times.[63] Making charcoal, timbering shafts, smelting ore, and keeping inhabitants warm made such heavy demands on local fuel resources that oak was wiped out. At the same time, overgrazing by pack animals reduced the grass cover on the public lands and induced erosion. Observers in another area to the south noted the following in about 1604: "At the time of the discovery of the mines the hills and canyons around Zacatecas were covered with extensive stands of trees, all of which have been cut down for use in the smelters. Only a few wild yucca remain. Wood is very dear in this town, for it is brought in carts from a distance of eight to ten leagues."[64] There is little room for doubt that mining, where it was carried on extensively, exercised an important effect on the plant life and by 1800 had been a strong factor in shaping parts of the landscape. Some of our photographs, taken perhaps a century later, were selected to show conditions near mines and will address the impact of mining in southern Arizona.

SPANISH CATTLE RAISING. The most far-reaching action of the Spaniards was the introduction of domestic grazing animals—cattle, sheep, and goats. In the first half century after the conquest, the American Indian population on the Central Plateau fell to half its former level while the number of sheep grew to about 2,000,000 and of cattle to about 200,000. During the next fifty years, when the human population again dropped by about half, the number of sheep and cattle increased four- and five-fold.[65] Similar figures for the livestock population of the desert northwest are lacking, but it is known that by 1610 there were ranches near the southern fringe of the Sonoran Desert between the Sinaloa and Fuerte Rivers.[66] By 1678, ranches existed at Huépac and in the valleys of Bacanuchi and Cedros.[67] By 1694, according to one estimate, there were more than 100,000 head of stock on ranges from Janos on the east to the headwaters of the San Pedro and Bavispe Rivers on the west.[68] This value, which presumably included all forms of livestock, is probably an exaggeration.[69]

Cattle spread via the mission chain, as well, because each new establishment received a herd from the older missions. Kino's home herd at Dolores provided 1,400 head for Baja California and San Xavier del Bac in 1701, and 700 more in 1702. In 1703, yet another 3,500 were available for distribution.[70]

Except for very arid tracts along the lower Colorado River Valley and the coast of the Gulf of California, by 1700 cattle had probably found their way to the vicinity of well-watered sites in many parts of mainland Mexico. By the mid-eighteenth century, private herds of 4,000 or 5,000 head were commonplace.[71] Because large numbers of cattle are known to have existed despite Apache raids, it seems likely that during this period the direct effect of the Apache on cattle numbers may have been only slight.[72] The Apache preferred horse meat, for one thing.[73] For another, their bands were too small to consume in excess of the replacement capacity of the herds. Moreover, to move swiftly and to avoid detection, the Apache took only small numbers of cattle in each raid. In the immense province of Nueva Vizcaya, for example, the Apache and Comanche took 68,000 cattle, horses, and mules between 1771 and 1776,[74] an annual loss of only 1 or 2 percent.

The Mexicans

In 1810 the Mexican movement for independence formally began, ending in 1821 with the appearance of a new nation. During the few decades when the Republic of Mexico ruled the Sonoran Desert, the ecological trends inaugurated by Spain continued for the most part. In northwestern Mexico, the years until 1830 were consumed in a quarrel over the question of state boundaries. In 1823 Sonora established its own government at Ures. Two years later, a state comprising both Sinaloa and Sonora came into existence, then, for the next five years, Ures, Arizpe, Culiacán, Fuerte, and Alamos quarreled over which should be the capital. In 1831 Sonora was again constituted as a separate entity.[75]

During this troublesome time, Sonora was plagued by other misfortunes. The Opata rebelled in 1820, the Yaqui in 1825, in 1832, and at intervals thereafter until the twentieth century. Federal edicts expelling the Spaniards from the Republic and secularizing the remaining missions gave rise to further dislocations.[76] Amid the prevalent civil and political disorder, the policies initiated by Gálvez had, by 1800, virtually put a stop to Apache raiding, and the truce apparently continued through the otherwise troubled period when Mexico was asserting independence from Spain.[77] From 1810 until 1831, when the raids began again,[78] the Sonoran cattle industry migrated northward into Arizona.[79] Insofar as this study is concerned, that expansion is the principal cultural fact bearing on the ecology of the region during the Mexican period.[80]

The Northward Expansion of Stock-raising

Other than the missions at Guebabi, Tumacácori, and San Xavier and the Presidio at Tubac, no important, permanent Spanish settlements were founded north of the present International Boundary until the nineteenth century. Then, with the lessening tempo of the Apache raids, large-scale ranching began. In 1806–7, Spain granted title to the Indians for large tracts of land around Tumacácori and Calabasas on which to raise their livestock.[81] In the years between 1821 and 1843, ten grants were made to Spanish and Mexican ranchers in southern Arizona.[82] But even before the last of these grants was made, the period of peace with the Apache came to an end.

By 1831, Apache activity brought to a halt the attempts to develop a grant at Tres Alamos on the San Pedro River below modern Benson. In 1833 and again in 1836, the Herreras family was driven from its lands on Sonoita Creek.[83] None of the grants later upheld by the U.S. courts was made after 1831, and one can infer that after that time conditions for settlement again became too adverse.[84] By 1846, all of the estates were abandoned. By the time of the earliest American accounts—on the heels of the Mexican War—the area north of the present boundary was uninhabited except for a few ranches along the Santa Cruz River that saw sporadic occupation as conditions permitted.

Because of the area's early abandonment, cattle-raising in the 1820s and 1830s did not attain the proportions of the 1880s, which is when arroyo cutting and shrub invasion began. Early American accounts describe the impressive remains of the industry, suggesting a minimum level of activity that must have taken place. Philip St. George Cooke, of the Mormon Battalion, described San Bernardino in 1846 as follows:

This old ranch was abandoned, I suppose, on account of Indian depredations. The owner, Señor Elias, of Arizpe, is said to have been proprietor of above two hundred miles square extending to the Gila, and eighty thousand cattle. Several rooms of the adobe houses are still nearly habitable. They were very extensive, and the quadrangle of about one hundred and fifty yards still has two regular bastions in good preservation. In front and joining was an enclosure equally large, but it is now in ruins.[85]

In 1851, John Russell Bartlett, lost along the Mexican Boundary he was supposed to be surveying, stumbled onto the old hacienda of San José de Sonoita, "A cluster of deserted adobe buildings . . . which seemed to have been abandoned many years before, as much of its adobe walls was washed away."[86] Upon returning to the San Pedro River, he wandered into the headquarters of the old Babocomari ranch:

This hacienda, as I afterwards learned, was one of the largest cattle establishments in the State of Sonora. The cattle roamed along the entire length of the valley; and at the time it was abandoned, there were not less than forty thousand head of them, besides a large number of horses and mules. The same cause which led to the abandonment of so many other ranches, haciendas, and villages, in the State, had been the ruin of this. The Apaches encroached upon them, drove off their animals and murdered the herdsmen; when the owners, to save the rest, drove them further into the interior, and left the place.

At Calabasas he inspected the "ruins of a large rancho"[87] and at San Lazaro "a large deserted hacienda"[88] where three years earlier H. M. T. Powell had noted "vast piles of bones of cattle. . . . The corral which was burnt up, could have easily held a 1000 head of cattle."[89] Whether Tubac, Tumacácori, Guébabi, and other locations described in these accounts were deserted also depending evidently on the current extent of Apache raiding.[90]

Not only the ruins of the haciendas but also the size of the remnants of the old herds help to establish the nature of the Mexican undertakings: "Many of the cattle . . . remained and spread themselves over the hills and valleys near

[Babocomari]; from these, numerous herds have sprung, which now range along the entire length of the San Pedro and its tributaries."[91] In 1846, Cooke, encamped at a watering place a few miles east of Agua Prieta Creek, wrote that "the wild cattle are very numerous. Three were killed today on the road and several others by officers. Around this spring is a perfect cattle yard in appearance, and, I suppose, I myself have seen fifty. . . . It is thought that as many as five thousand cattle water at this spring."[92]

All the way along the deserted San Pedro Valley, the battalion dined on beef. At the junction of the river with Babocomari Creek, they fought their first engagement of the Mexican War—with a herd of wild bulls: "One of them knocked down and run over Sergeant Albert Smith, bruising him severely. . . . One of them tossed Amos Cox of Company D into the air, and knocked down a span of mules, goring one of them till his entrails hung out, which soon died. . . . Lieutenant Stoneman was accidentally wounded in the thumb."[93]

The frequency with which the journals mention them between 1846 and 1854 leaves no room for doubt about the abundance of the wild herds.[94] Yet, despite the many early observations of large numbers of cattle, the numbers were probably exaggerated and ultimately gave rise to the myth of the great herds.[95] Moreover, evidence suggests that any large herds were gone by 1854, scarcely twenty years after they were formed. Records kept that year by cowboys passing through southeastern Arizona on large Texas trail drives reveal that no signs of wild cattle were seen.[96]

In the absence of the centrifugal windmill, which did not appear before the 1870s, livestock would have been clustered around natural water sources.[97] Perhaps the passersby extrapolated the numbers in these locally dense herds over the vast adjacent areas. One recent estimate of cattle numbers in Mexican Arizona suggests the number never exceeded 20,000 to 30,000 animals.[98] By perhaps doubling that number to account for the animals within the adjacent border area of present-day Mexico, we arrive at a rough maximum of 40,000 to 60,000 animals, far below the livestock population there in the 1880s.

Ecologically, then, cattle-raising was extensive enough to result in a thorough dissemination of mesquite seeds; insufficient to reduce the grasses to the impoverished condition observed by 1890; and sufficient to constitute the severest cultural stress yet imposed on the region. Certainly the pressure of 60,000 head of livestock on the vegetation dwarfed any food-gathering activities of the earlier American Indian population, numbering some 10,000–15,000. And finally, this initial period of cattle grazing was not synchronized with arroyo cutting and shrub invasion, which were delayed until half a century later.

That cattle produced major changes in vegetation during this early period is unlikely. Nothing in the early American accounts indicates that the country looked any different than it had at the time of early Spanish occupation.[99] Reports from both periods indicate that grass was plentiful, that the landscape was open, and that mesquite, although widespread, had nothing like its present density. The streams at both times were marshy, open, and largely unchanneled. The image cannot be resolved more closely.

The Transition to the American Period

By 1846, the native cultures, except for that of the Apache, had seen their best day; the Spanish had come and gone; the Mexicans remained, but in the northern part of the desert their influence had waned. To the east, the Manifest Destiny of the United States was carrying that nation vigorously westward. In 1846 the old and the new cultures met, and by 1848 everything north of the Gila River belonged to the United States. In 1853–54, Antonio López de Santa Anna, giving his final performance before the Mexicans, sold their land between the Gila and the present International Boundary. Thereafter the desert region knew two political masters, and its ecology was divided along an artificial line into two distinct and, in many ways, divergent sets of cultural forces.[100]

Chapter Three The Influence of Anglo-Americans

The earliest American accounts of the Sonoran Desert region date from shortly after the Mexican War for Independence when the mountain men — the early fur hunters — trapped beaver along the Gila River and its tributaries.[1] As the western migration swelled, and well before title to the lands south of the Gila passed to the United States, travel across the region became commonplace. For the decade prior to the Gadsden Purchase, a score of journals exist that describe the country, most of them written by troops during the Mexican War or by forty-niners on their way to the gold fields of California.[2] Taken as a whole, the early accounts present a rich and detailed picture of the natural conditions that prevailed in the 1840s and 1850s.

The landscape that they describe lasted for another generation, and the cultural factors that had helped to shape it changed very little in the 1850s. Some settlement took place immediately after the Gadsden Purchase; a few military posts came into existence,[3] a few mines opened in the hills. By and large, however, the years until the Civil War constituted for the region's ecology merely an extension of the Mexican period. From the early journals a picture of the country emerges that in many respects contrasts sharply with present conditions. In particular, the appearance of the valleys and the grasslands has changed.[4]

Conditions Before the Civil War

The Valleys

Before the Civil War, the streams of our region wound sluggishly for much of their course through grass-choked valleys dotted with ciénegas, pools, and small coppices of riparian trees. In spite of the efforts of trappers, beaver dams were still numerous, and as late as 1882 a settler on the San Pedro could report: "Our ditch was just above a beaver dam and if the water was low and we tried to irrigate at night the beaver would stop up our ditch so that the water would run into their dam."[5]

Of the San Pedro, one observer in 1846 noted a fact equally improbable from the perspective of the twentieth century: "Fish are abundant in this pretty stream. Salmon trout are caught by the men in great numbers; I have seen them eighteen inches long."[6]

Although they flowed more regularly than they do today, the rivers nevertheless did go dry at times or along particular stretches:

> The Sanpedro river as they Call it — is a stream one foot deep six feet wide & runs a mile & half an hour & in ten minutes fishing we Could Catch as many fish as we Could use & about every 5 miles is a beaver dam this is a great Country for them — & we have went to the river & watered & it was running fine & a half mile below the bed of the river would be as dry as the road — it sinks & rises again.[7]

Where the present reporter took quantities of fine trout in March and April 1858 not a drop of water was to be seen [in September 1858].[8]

River channels were less deeply cut in the pre–Civil War period than today. At many places, rivers could not be seen from afar.[9] Near Tucson, at a place where its channel is now ten feet deep and seventy-five feet wide, the Santa Cruz in 1849 ran in an unentrenched, but well-contained stream with multiple channels: We encamped in a grassy bottom, much covered with saline efflorescence. The river has divided to a mere brook, the grassy banks of which are not more than 2 yards apart.[10]

The degree of entrenchment varied. A distinct channel trench existed in places along the San Pedro and probably along the other streams:

> The valley of the San Pedro River [near the mouth of Dragoon Wash] . . . was anything but luxuriant [in 1851]. It consists of a loam, which if irrigated might be productive, but as the banks are not less than eight or ten feet high, irrigation is impracticable. . . . The grass of the vicinity is miserably thin and poor, growing merely in tufts beneath the mezquit bushes which constitute the only shrubbery, and in some places attain a height of ten or twelve feet. . . . In order to cross the river, it was necessary to level the banks on both sides, and let the wagon down by hand.[11]

A few miles farther north, A. F. Bandelier, writing just before the severe erosion of the 1890s, found "a cut with abrupt sides . . . 10 to 15 feet deep, and about 25 wide."[12]

About the presence of brush, there is similarly no simple generalization that will hold. It is apparent from many of the written accounts that parts of the valleys were more open than at present. Nevertheless, brush and even bosques are frequently mentioned.

> The valley of this river [the San Pedro near its junction with the Gila] is quite wide, and is covered with a dense growth of mezquite . . . cotton wood, and willow, through which it is hard to move without being unhorsed. The whole appearance gave great promise, but a near approach exhibited the San Pedro, an insignificant stream a few yards wide, and only a foot deep.[13]

> The bottom of the San Pedro is one mile broad [at the same site in 1846], and of the character of those on the Gila above, dusty dry soil, grown in places with cottonwoods and willow, in others with grass and again mesquite chaparral, other places bare.[14]

> The land on each side of the Pedro river bottom [in 1846] is a dense thicket of bramble bush, mostly muskeet, with which millions of acres are covered.[15]

A description published in 1858 in the documents pertaining to the Leach Wagon Road makes it clear that the contrast between mesquite forest and open valley was sometimes sharp:

> A forest of heavy mesquite timber about one mile in width extends from the [San Pedro]. . . . Running nearly due north a road lies opened in March last through the Mezquite forest . . . for a distance of about three miles. . . . Leaving the forest it enters upon a tract of the bottom lands of the San Pedro . . . from one fourth of a mile to one mile in width. . . . The entire body of these lands were covered with a dense growth of sacaton grass.[16]

A second account by the builders of the wagon road describes the same part of the river:

> The San Pedro, at the first point reached in the present road, has a width of about twelve (12) feet, and depth of twelve (12) inches, flowing between clay banks ten or twelve feet deep, but below it widens out, and from beaver dams and other obstructions overflows a large extent of bottom land, forming marshes, densely timbered with cottonwood and ash, thus forcing the road over and around the sides of the impinging spurs.[17]

The foregoing appears to be a reasonable explanation, but four years earlier and 2 miles upstream, the riparian zone was somewhat different:

> At the Tres Alamos [crossing] the stream is about fifteen inches deep and twelve feet wide, and flows with a rapid current over a light, sandy bed, about fifteen feet below its banks, which are nearly vertical. The water here is turbid, and not a stick of timber is seen to mark the meandering of its bed.[18]

Here a treeless condition is equated with the presence of a channel trench. But this account, in turn, must be read in light of another, five years earlier, of the same crossing:

> This river, the San Pedro, is extremely boggy & has to be crossed by making a brush bridge. . . . I cannot agree with Colonel Cook, who calls this a beautiful little river, although where he crossed it, some 10 or 15 miles above, it may have presented more amiable qualities. Here it is lined with a poor growth of swamp willow & other brush, so that it cannot be seen till you come within a few feet of it; & then the bank is perpendicular, not affording an easy access to its waters, which, though not very clear, is good. The banks & bed are extremely boggy, & it is the worst place for cattle & horses we have yet been, being obliged to watch them very close.[19]

For the abundance of marshes along the valleys, there is extensive documentation, as seen in this description of the Santa Cruz Valley as it appeared in about 1875:

> The wild turkeys hatched in places where the marshes were, during those days, and they were protected by the grass that grew waist high. There were marshes along the Aqua Fria [a ranch on the Santa Cruz River], right there at Calabazes, and up at Buena Vista. . . . These marshes have all dried up.
>
> This bottom down here just back of and south of A mountain [near Tucson] was marshy. We used to have a ranch in there when we first came here from Calabazas. There was a thirty-seven acre lake, just south of A mountain, in that bottom. This lake was made by Solomon Warner. There was marshy ground over there, and he put in a dyke and planted willows along the dyke to protect it.[20]

From the court records of an early suit over water rights, one can establish the existence of a large ciénega extending along the San Pedro from about modern Benson to old Tres Alamos.[21] The boggy—but treeless and trenched—location already discussed lay at the foot of it. In 1846, one traveler complained about the heavy going in the upper San Pedro Valley southeast of the Huachuca Mountains, and, in 1851, another mentions a "boggy plain" in the same area.[22]

Indirect evidence for the degree of marshiness appears in the high incidence of malaria, a disease almost unknown in that area today. In 1868, so much of this disease existed at Camp Grant, located at the junction of Aravaipa Creek and the San Pedro River, that the army set up a temporary convalescence camp 28 miles away.[23] Similar conditions prevailed along Babocomari Creek, Sonoita Creek, and the Santa Cruz River.[24]

Today the streams in the Arizona part of the region—the San Pedro, the Santa Cruz, and their tributaries—maintain mostly intermittent flow regimes with few perennial reaches. Except in their upper reaches or where rock formations force water to the surface, the beds much of the time are either dry, sandy wastes that support little, if any, vegetation or are narrow forests of cottonwoods and willows. From 5 to 30 feet above the channel and set apart from it by abrupt vertical banks, one typically finds a bosque dominated by mesquite. During the summer rainy season, flash floods render the streams impassable. At such times the channel is filled bank to bank with a raging, muddy torrent that may carve new incisions into the flood plain, sharpen the edges of the old, or deposit sediments in the channel to raise its bed.

These accounts suggest that there was considerable complexity in riparian zones in the 1800s. The valleys were wetter and more open than today, ground-water levels were higher, and channels were not as deep. But the precise conditions varied from place to place and from time to time. As the tributary washes dumped greater or lesser amounts of debris, depending upon where heavy summer rains may have struck, the rivers had to transport varying loads of sediment at different points along their course. Channeling and filling, aggradation and degradation—all may have been going on simultaneously, in various stages of development along various parts of the stream.[25]

Vegetation necessarily reflected this dynamic situation. Along the parts of the river where there was no trench, the water table was high, the bottoms marshy, and the soil waterlogged and too poorly aerated during at least part of the year to support woody plants. Where an arroyo existed, in contrast, the bottom of the trench fixed the top of the water table. Soil on top of the bank was sufficiently well drained to support mesquite and other plants whose roots require more oxygen.[26] As mesquite invaded a flood plain that had been drained by a temporary trench, elsewhere a former bosque was being replaced by marshes where the plain was aggrading. The old accounts present a picture that is neither homogeneous nor static. By postulating a dynamic situation, one can reconcile the variety of conditions that evidently existed.[27]

The Grasslands

After the river valleys, the grasslands of the desert region appear to have changed most. The overall tendency has been toward shrubbiness and a less open, less expansive landscape. On the bajada of the Sierra San José, which in places now supports a dense cover of spiny shrubs, one early group of emigrants went without breakfast because there was no wood for cooking.[28] Another journal relates that "the men picked up every fragment of wood or brush [they] passed."[29] In the Sulphur Springs Valley, a member of a railroad survey party described the Willcox Playa as "bounded by smooth and grassy plains."[30] At Croton Springs on the northwest side of the Playa, other travelers reported that wood was "distant,"[31] and "fire wood [had to be] dug from the ground, the roots of the Mesquite being the sole dependence."[32] All of these areas are characterized today by woody vegetation.

Yet mesquite clearly was present on upland sites in 1858. The situation was very far from being as black and white as interpretations have sometimes made it.[33]

The situation along San Simon Creek may have been typical of large expanses of the Semidesert Grassland:

> This plain presents the same features as the [Willcox] playa just described, as far as regards vegetation, without being absolutely bare, as that was; yet its growth is of that thorny, worthless, desert character. Fouquieria, larrea, yucca, palmetto, and agave, are the only growths on the slope of the plain, down to half a mile from the river, where the mesquite tree begins to appear, and the willow is found collected round some of the water holes in the bed of the stream.[34]

It would be an exaggeration to think of the grassland as being uniformly like the Midwestern prairies: open, rolling, and treeless. Parts were—and still are—but the chances are that most parts were not. Although the past century has seen a striking increase in the number of shrubs and small trees, the species involved were already present (except for exotics like Russian thistle). The invasion was not an encroachment by new plants into new areas.[35]

Newspaper accounts and journals from the 1840s to the 1880s establish that large wildfires were frequent in southeastern Arizona grasslands.[36] Numerous studies have shown that, before the late nineteenth century, fires burned in nearby forests and, by inference, across adjacent grasslands at frequencies greater than once every ten years.[37] Lightning was the main cause of wildfires then, as today, and fire frequency was controlled by the dynamics of fuel availability and climate.[38] Some have argued that the grassland fires were started almost exclusively by the newly arrived Europeans,[39] but recent studies place greater emphasis on lightning and American Indians. The

Apache, for example, burned off extensive areas as a side effect of their raiding and warfare.[40]

The American Civil War and After

The outbreak of the Civil War brought to a temporary halt the American development of the Gadsden Purchase, such as it had been. Faced by what they thought to be imminent Confederate invasion, the Union garrisons at Forts Buchanan and Breckenridge burned their posts and fled to New Mexico, leaving the inhabitants to arrange for their own protection. The Santa Rita Mining Company closed its doors, and the territory's first newspaper ceased publishing. Tucson, a center of secessionism, stayed within its walls while the Apache ruled the countryside, if indeed their reign had been interrupted at all.[41]

The arrival of the California Column in 1862 under General James H. Carleton restored Union authority,[42] and in 1863 President Lincoln signed the bill organizing Arizona as a territory. The first set of territorial officials arrived in December. The years following Appomattox saw settlement again proceed. In 1867, homesteaders moved into the San Pedro Valley, bringing an end to its abandonment.[43]

Through the 1870s, as the army had increasing success against the Apache, farming spread along the arable valleys while mines multiplied in the mountains.[44] In 1880, the transcontinental railroad arrived in Tucson. The following year, a line connecting Benson, Arizona, with Guaymas, Sonora, by way of the San Pedro River, Babocomari Creek, Sonoita Creek, and Nogales was completed.[45]

The 1880s witnessed the spectacular rise of Tombstone to its silver pinnacle as the Territory's first city. Its fall was equally spectacular when water flooded the mines and the price of silver plummeted. The army found the final solution to the Apache problem; the "ancient scourge of Sonora" thereafter stayed on the reservation. By 1890, Arizona Territory had a population of 60,000.[46] With increased human population and safe access to open range, a new industry—cattle ranching—developed in the country south of the Gila River.

The Growth of Livestock Production

By and large, the story of the 1880s is the story of ranching and its expansion out of Texas and Sonora onto the lush ranges of the Semidesert Grassland.[47] Although there were only about 38,000 cattle in all of Arizona Territory in 1870,[48] the following decade laid the foundation for a truly massive industry. Stimulated by government contracts for supplying beef to military posts and Indian reservations, large pioneering establishments like the Ciénega, Vail, and Sierra Bonita Ranches came into existence. Trail drives out of Texas and Sonora began, and any excess above consumption went to stock the ranges. In 1872, two ranchers alone brought in four herds totaling 15,500 head.[49] In 1880, a Texas herd of 2,500 grazed in Mule Pass east of the San Pedro; 3,600 were on the Babocomari Ranch; and 3,000 sheep grazed along the San Pedro near Tombstone. The census figures for that year show that the San Pedro Valley supported perhaps 8,000 head of cattle and 10,000–12,000 head of sheep, more than the entire territory ten years before. Of the 35,000 cattle in the territory, about 20,000 grazed south of the Gila River.[50]

The 1880s witnessed a rapid expansion of the cattle industry throughout the West. The new railroads made it possible to raise beef for markets to the east and west. The Apache had ceased to harass the ranchers, and a debilitating depression in the 1870s vanished. But the biggest factor, perhaps, was the ease with which the cattle industry attracted investment capital. A great ballyhoo campaign waged by railroad prospectuses, livestock journals, and territorial legislatures trumpeted to an eager public that the West held easy riches and that grass was gold.

> A good sized steer when it is fit for the butcher market will bring from $45 to $60. The same animal at its birth was worth but $5.00. He has run on the plains and cropped the grass from the public domain for four or five years, and now, with scarcely any expense to his owner, is worth forty dollars more than when he started on his pilgrimage. . . . With an investment of but $5,000 in the start, in four years the stockraiser has made from $40,000 to $45,000. . . . That is all there is of the problem and that is why our cattlemen grow rich.[51]

Having found in the 1870s that feeding corn to livestock was more profitable than marketing the corn, farmers in the Midwest started feedlots for fattening both western and local cattle. However, once the western boom began, the feeders found greater profit in sending their livestock west than in buying animals to fatten in their feed lots. From 1882 to 1884 as many head traveled west as east.[52] As prices and profits soared, more eastern and English capital rushed to participate. In the scramble to stock the ranges and participate in the easy money, the normal flow of transport reversed itself, and instead of the West feeding the East, the East fed the West.

In Arizona by 1883–84, "every running stream and permanent spring were settled upon, ranch houses built, and adjacent ranges stocked."[53] By the middle of the decade, the governor reported that:

> The number of cattle assessed this year has been 435,000 and 50 per cent not assessed 217,500, making a total of 652,500 in the Territory at the present time (October 20, 1885). This number is being rapidly increased, and within another year it is expected that ranges with living springs and streams will be fully stocked.[54]

About the same time, a reaction set in with the appearance of protective associations, formed by established ranchers to protect the ranges against the flood of new cattle. They sought to adjudicate grazing rights by voluntary agreement and to restrict further immigration by means of quarantine laws aimed at keeping out "Texas fever."[55]

In 1885, the editors of the *Southwestern Stockman* wondered if the ranges were not overstocked but concluded that the heavy recent increases were natural, and that the day of overproduction was "remote."[56] The ranchers along the San Pedro Valley seemed less certain. The Tres Alamos Association passed a resolution in 1885 stating that the ranges were "already stocked to their full capacity" and demanding that the influx be controlled.[57] A year later the Tombstone Stock Grower's Association spelled out the details:

> Whereas, We, the members of this Association, perceive with deep concern, that a crisis is fast approaching . . . demanding immediate action; and
>
> Whereas, For several years past, there have been such large importations of cattle into Cochise county, from Mexico, and from our own States and Territories,
>
> whereby these herds and their increase have so stocked our ranges, to the extreme limit of their capacity [that they] . . . leave us no surplus grass to tide us over . . . and,
>
> Whereas, Sundry well disposed but misinformed persons, within our midst, have been, and are now, from time to time, circulating extravagant and highly colored reports concerning the resources of our county . . . therefore be it . . .
>
> Resolved, If, notwithstanding our remonstrance and protest, any person or persons persist in driving cattle into our section, without first legitimately securing sufficient grass and water for their herds, we will deny them all range courtesies.[58]

But the time had passed when putting a stop to either courtesy or immigration would help. In anticipation of a better market, association members allowed their own herds to pile up.[59] The census of 1890 showed more than a million range

cattle for Arizona.[60] The *Southwestern Stockman* conceded in 1891 at last that "the malady of overcrowding is with us in an aggravated form," and it reported that disaster had been averted that summer only by the "phenomenal" late rains.[61]

For 1891, the official assessment roll showed 720,940 cattle, and the governor wrote that there were "closer to 1,500,000."[62] The summer rains of 1891 were well below normal. In the arid foresummer of 1892, the livestock began to die. The summer rains of 1892 again were scanty, and by the late spring of 1893 the losses were "staggering."[63] "Dead cattle lay everywhere. You could actually throw a rock from one carcass to another."[64] The governor estimated the mortality at 50 percent and possibly 75 percent of the herds; the assessment rolls for Pima and Cochise Counties, at least, bear out his assertion.[65]

The overgrazing and drought leading to the disaster of 1893 had an enormous impact on the ecology of the region. Even in arid or semiarid ecosystems, prolonged drought is injurious to many plants. Certain perennial rangeland grasses are susceptible to drought, and their populations fluctuate rather rapidly in response to changes in soil moisture.[66] Prolonged heavy grazing has many deleterious effects, including loss of biodiversity and disruption of nutrient cycling.[67] Thousands of square miles of grassland, denuded of their cover, lay bared to the elements. Grazing unquestionably weakened the old plant communities, leaving them open to invasion; it unquestionably upset the balance between infiltration and runoff in favor of the latter.

How widespread was the grazing impact? The prolonged heavy grazing of the rangeland would have been concentrated around water sources and, except for brief periods of abundant rain, would not have been spread evenly across the landscape. Thus, if viewed from above, widely separated patches of ground with low plant cover surrounding water sources would be seen imbedded in a matrix of heavier plant cover. Even these remote grasslands would be grazed heavily during periods of abundant rainfall and might support little vegetation for protracted periods, especially if heavy grazing were followed by drought. But grasslands are resilient and should eventually recover from periods of drought or heavy grazing once the rains return. On a large scale, the effects of climate on geomorphic phenomena will usually overwhelm grazing effects.[68] Perhaps overgrazing cannot be blamed solely or even principally for the events that followed.

The Onset of Arroyo Cutting

Floods in the desert region can be documented well back into the Spanish period,[69] and, indeed, geomorphic studies show that they have been part of our natural scene for the past 8,000 years.[70] In some respects, the arid and semiarid climate leads to flooding, because sparse, intermittent rainfall prevents development of a dense ground cover. Without such a cover to impede runoff and promote infiltration, rainfall intensities, particularly in summer, occasionally exceed the capacity of the ground to absorb water. Runoff results, reaching flood proportions in the event of unusually heavy or intense storms.

Before the mid-1880s, heavy flooding occurred often enough that bridges were among the Americans' earliest construction projects.[71] Even so, high water seems to have spread out in a shallow layer over the flood plains, causing some inconvenience but not much damage. Beginning about the mid-1880s, the frequency and severity of the floods increased.

In 1886, "the water in the San Pedro River was . . . higher than it was ever known to be. Between Contention and Benson there was four feet of water on the side of the [railroad] tracks."[72] At its junction with the Gila, "an avalanche of water swept down . . . like a wave, 6' high."[73] On Sonoita Creek, in the same summer "water flowed down the valley some places 10 feet deep. At Calabasas the Santa Cruz overflowed its banks and swept a part of the valley."[74] In Nogales "the new adobe building of Mr. Samuel Brannon happened to be right in the course of a young river [and] . . . was totally destroyed, as was also his fish pond."[75] A Tucson newspaper reported that "some of the valley fields dived under water, and have not reappeared yet, they will probably be up in time to be assessed next year."[76]

During the following year, 1887, the San Pedro again had "higher water than . . . ever . . . known before."[77] "For nearly the entire length of the river from Benson down to the Gila the crops with the exception of hay . . . [were] destroyed."[78] The flood carried away a dam at Charleston.[79] At Nogales, "the streets were a perfect sea of water and bridges and dams were no more than straws."[80] The Santa Cruz River at Tucson was "more than a mile wide and deep enough to float a mammoth steam boat."[81]

There were floods in 1888 and again in 1889.[82] In 1890, the flooding reached its climax when, during a two-week period at Tucson, the Santa Cruz carved a channel into its formerly shallow floodplain:

[August 5] The flood yesterday washed a deep cut across the hospital road, so that the road now is not only impassable but extremely dangerous for teams or travel as the embankment of the cut is perpendicular and the water below deep, and pedestrians might easily endanger their lives.

[August 6] It is thought that the washout in the Santa Cruz, opposite this city, will reach Stevens Avenue this morning. Boss Levine says that the Santa Cruz was higher last night than at any time during the last twenty-five years, and he ought to know as he has lived on its banks during that time.

[August 7] The channel or cut being made by the overflow of the Santa Cruz river, is now one mile and a half long, by from one to two hundred yards wide—in other words—it extends from the smelter to about two hundred yards this side of Judge Satterwhite's place.

[August 8] More than fifty acres of land which has formerly been under cultivation in the Santa Cruz bottom, has been rendered worthless by being washed out so as to form an arroyo.

[August 9] The single channel which was being washed out through the fields of the Santa Cruz by the floods resulted in considerable damage but this danger has been greatly increased from the fact that the wash or channel has forked at the head, and there are now several channels being cut by the flood, all of which run into the main channel. If the flood keeps up a few days longer there will be hundreds of acres of land lost to agriculture. As these new channels or washes are spreading out over the valley, they will cut through and greatly damage the irrigating canals.

[August 13] The "raging" Santa Cruz continues to wash out a channel and the head of it is now opposite town. It may reach Silver Lake before the rainy season is over.[83]

On Rillito Creek, a tributary of the Santa Cruz, a flood during the same period "so cleaned out and deepened the channel that a third more water could be carried."[84]

On the San Pedro River, the chronology cannot be traced so precisely, but during the same two weeks, much the same thing seems to have occurred:

Of the country down the San Pedro, from Tres Alamos to the Gila [Captain Van Alstine] . . . says, "all of it is gone, destroyed, torn up, 'vamoosed' down with high water." He never saw such a destruction in all his life. . . . The San Pedro never was as high as it was this time, and will not probably be for the next ten years. The losses sustained by the people will reach into the thousands.[85]

At Dudleyville, near the mouth of the San Pedro, the river "caved within 15' of Cook's [store]."[86] Upstream at Mammoth, flooding cut a channel 30 feet deep, exposing archaeological relics.[87] The continuous present channels of these two rivers can be definitely dated from August 1890.

For the other streams of the desert region, the evidence is scantier and less reliable. During the same month, a newspaper account states that San Simon Creek in eastern Arizona "was over a mile wide and running strong."[88] There is, however, no mention of channeling. Entrenchment may have begun as early as 1883; it certainly had commenced by 1905.[89] For the Sonoyta River, 200 miles away on the Mexican Boundary, there is similarly no firsthand information.[90] On the authority of an old inhabitant, C. Lumholtz states that there used to be a series of ciénegas near the town, but that the flood of August 6, 1891, ripped a channel through them. "The swamps dried up in three years. . . . Where there had been before only a llano, a forest of mezquite trees sprang up."[91] The channeling had certainly taken place by 1893 when the Boundary Commission noted that:

> Sonoyta was formerly quite a flourishing little agricultural village, but heavy rains caused the river bed to sink so deep below the level of the surrounding lands that irrigation was attended with many difficulties, and . . . the village fell into decay, family after family moving away, until now scarce a half dozen Mexican families remain.[92]

Causes of Arroyo Cutting

Determining causality is one of the ultimate goals for scientists, but discovering what caused initiation of arroyos is difficult because the major variables cannot be separated.[93] Although we may never know exactly why arroyos formed along the San Pedro and Santa Cruz Rivers, some of the details give us insight into the processes of long-term change in the Sonoran Desert region.

During the first half of the twentieth century, it was commonly asserted that overgrazing was the cause of arroyo cutting,[94] an assertion based partly on circumstantial evidence, partly on observation, and partly on assumption. The circumstantial evidence stems from what appeared to be a close timing of large-scale grazing and the cutting of channels. Although cattle had been introduced to the region in the 1700s, grazing on a very large scale started in southeastern Arizona in the 1880s. Flooding began shortly thereafter, and extensive channel cutting, at least along the two major streams, was readily apparent and making irrigation difficult by 1890.

Range ecologists have observed and measured the effect that plant cover has on increasing the amount of water infiltrating the ground. Removal of the plant cover by any mechanism—including grazing—increases runoff and surface erosion.[95] Trampling decreases infiltration rates, also increasing runoff. In addition, well-worn ruts— wagon roads, game and cattle trails, footpaths, drainage ditches—tend to enlarge rapidly into gullies and then into arroyos.[96] Livestock grazing thus promote erosion in three ways: by removing plant cover, by compacting soil, and by wearing trails in floodplains.

Several studies argue against livestock grazing as the sole agent creating arroyos. The first, developed by Kirk Bryan,[97] holds that the arroyo cutting of the late nineteenth century was merely the most recent erosion cycle of many in the Southwest since the late Pleistocene.[98] Arroyo cutting has alternated with periods of alluviation, during which the channels filled with sediments, sometimes to a higher level than before the arroyo cutting. The second argument holds that climatic variations leading to more intense rainfall and larger floods occurred at the same time as arroyos formed.[99] According to this argument, arroyos would have formed even if grazing animals were not on the landscape, although the channels might have been smaller. Another group holds that intrinsic geomorphic processes create arroyos independent of either climate or land use.[100] Finally, and coupled with the climatic argument, arroyos are thought to have been formed during floods that were of a different origin and a much larger size than previous floods.[101]

Bryan, the first to note problems with the grazing argument, wrote:

> Arroyos similar to and even larger than the recent arroyos were cut in past time. As these ancient episodes of erosion antedate the introduction of grazing animals they must be independent of that cause. . . . It seems reasonable to believe that the present arroyo is essentially climatic in origin. The introduction of grazing animals handled by optimistic owners may have reduced the already impoverished vegetation, and precipitated the event. Overgrazing thus becomes merely the trigger pull which timed the arroyo cutting in the thirty years following 1880.[102]

He reiterated his conviction that grazing was a secondary factor, calling on drought as the primary mechanism for arroyos. This idea stemmed from the observation that ground-water levels had dropped historically, leading to die-off of the riparian vegetation that held floodplains together. Overgrazing helped determine when cutting actually began.[103] J. T. Hack and L. B. Leopold took positions similar to Bryan's, which was that climatic change had lowered the land's resistance, and heavy grazing was the germ or infection that touched off an epidemic of erosion.[104]

Scientists contemporaneous with Bryan and who actually observed arroyo formation were sharply divided in their opinions.[105] A little later, researchers elaborated on these themes to develop more sophisticated positions. E. Antevs also attributed past cycles of cutting to a drier climate but attributed the present cycle to overgrazing. He asserted that:

Clearly, if left alone, the native vegetation could have weathered the droughts during which the arroyo erosion set in during the 1880s. The impoverishment of the plant cover which permitted the channeling must have been caused by new and foreign detrimental agencies, and the new factors during the 1870s and 1880s were large herds of cattle and sheep and numerous settlers. Therefore, the reduction of the vegetal cover which allowed arroyo-cutting to begin during the 1880s in the Southwest was caused by livestock and man.[106]

Others regarded past channeling as random and local rather than systematic and regional. Discontinuous channeling is normal and proceeds in slow waves up the course of a stream. An exceptionally big storm of high intensity may unite the segments, producing a continuous trench. C. W. Thornthwaite, C. F. S. Sharpe, and E. F. Dosch maintained that, if the vegetation is impoverished, "rains of . . . even moderate intensity can initiate a period of accelerated erosion."[107] In their view, "overgrazing by livestock introduced by the white man has reduced the resistance of the land to erosion and . . . intense storm precipitation has been the germ or infection."[108]

Working at several places in the Rocky Mountain West, Leopold refined the chronology for cut and fill and established almost beyond question its regional extent and cyclic nature. Recognizing a deficiency in Bryan's hypothesis—a failure to specify the exact climatic factor that brought about the recent shift to cutting—he made two suggestions. The first relates to the discontinuous trench segments that existed in 1880s (and evidently had existed for some time):

The discontinuous channels and the lack of heavy alluviation during the period A.D. 1400–1860 possibly indicate an instability of the fill. The climate was not quite humid enough to cause further alluviation, nor was it sufficiently arid to cause degradation. On such a stage, post-settlement grazing could play a quick-acting and decisive role.[109]

He noted secondly that the weather records for New Mexico between 1850 and the 1870s showed a significant decrease in the number of small rains compared to the number of large rains, although there was no change in the average annual precipitation over the same period. This shift, he argued, had an adverse effect on the plant cover.[110]

Although Leopold argued that climatic varia-

tion inaugurated the current cycle of arroyo cutting, he recognized that land use may either reinforce or counteract the effect of climate with the statement that "recognition of the current climatic variation does not require an abandonment of measures for improvement of land use."[111] In contrast, S. Judson maintained: "An enlightened land policy will, of course, be important in preserving any gains made by nature. But it is extremely doubtful that even the strictest control of grazing, combined with 'upstream engineering,' will bring alluviation of the arroyos unless it is accompanied by sufficiently effective precipitation."[112]

The climatic argument receives considerable support from studies of prehistoric arroyo formation. Both the Santa Cruz and San Pedro Rivers, as well as many other ephemeral and perennial rivers in the region, developed arroyos as early as 8,000 years ago, with well-preserved sequences of cutting and filling common in the last 4,000 years of river-bank stratigraphy.[113] From a geologic-time perspective, arroyos are reasonably synchronous regionally and represent a geomorphic record of climatic change.[114]

Regional differences did not enter into the debate until much later. H. E. Gregory remarked that: "For the Navajo country . . . human factors exert a strong influence but are not entirely responsible for the disastrous erosion of recent years. The region has not been deforested; the present cover of vegetation affects the runoff but slightly, and parts of the region not utilized for grazing present the same detailed topographic features as the areas annually overrun by Indian herds."[115] The same point was made by H. V. Peterson, who stated that he had observed trenching in the Fort Bayard Military Reservation, where grazing had been either excluded or rigidly controlled.[116]

R. U. Cooke and R. W. Reeves wrote the first comprehensive treatise on arroyo cutting and concluded that a complex set of factors contributed to arroyo cutting.[117] Because of the lack of a close temporal relationship between overgrazing and channel cutting, they reasoned that valley floor entrenchment was more likely the result of new roads and other features that concentrated runoff. Certainly, one entrenched reach of the Santa Cruz River south of Tucson has exactly this history. A segment of the old Tucson-Nogales Road, with wheel ruts deepened by runoff from the nearby Sierrita Mountains, served as a conduit for storm waters entering the main channel. Ultimately, the new channel along the road alignment captured the main stream, and the old course was abandoned.[118]

The particular climatic circumstances able to initiate channeling have also been a fertile source of controversy, even among those who hold climate responsible or who recognize distinct periods during which cutting occurred. There is, first, disagreement as to whether secular trends or short-term variations are involved. Among those who favor trends, the consensus among some of the earliest observers as well as among some recent ones[119] is that increasing aridity brought on the erosion. Some of these observers neglect the fact that it takes water to erode channels, and water typically is in short supply during droughts. Among the short-term variations suggested are periods of particularly intense and heavy precipitation and seasonal shifts in the relative amounts of summer and winter precipitation.[120] Those concepts will be evaluated later in this book.

The Regional Extent of Arroyo Cutting

The synchroneity of arroyo cutting has been long debated. Bryan reviewed the dates at which the current cycle of erosion began in Arizona, New Mexico, Utah, and Colorado; he concluded that, though channeling began at slightly different times in different parts of the area, the period from 1860 to 1900 encompassed nearly all of the cases.[121] Others have revised his dates of incision and potentially narrowed the period, but not by much.[122] The consensus is that initiation of arroyos occurred mainly between 1870 and 1920 (southerly drainages) and 1880s and 1940 (northerly drainages). Although some have made a great deal out of the variation in the exact dates of downcutting, no individual storms have affected all drainages simultaneously throughout the Southwest, and therefore the range in dates of downcutting would reflect the variability of storms in time and space.

Dates can be reported with greater precision on individual drainages. W. P. Cottam and G. Stewart demonstrated that erosion started at Mountain Meadows, Utah, during one protracted period in 1884.[123] H. E. Gregory and R. C. Moore reported that arroyos began cutting elsewhere in southern Utah in the 1880s.[124] R. H. Webb and others showed that the Kanab Creek channeling commenced in 1882 and ended in 1909.[125] Similarly, Webb found that arroyos on the Escalante River in southern Utah formed between 1909 and 1932, or much later than in adjacent drainages.[126] In New Mexico, Bryan established dates of incision for the Rio Puerco and Rio Salado in the late 1880s.[127] C. O. Sauer and D. Brand concluded that in northern Sonora, a particularly critical area because of its early colonization, accelerated erosion began in the period from 1881 to 1891.[128]

Part of the problem of synchroneity stems from definition. Many rivers, particularly in southern Arizona, had at least limited reaches that were incised before the period known as arroyo cutting occurred. These channels have been termed "discontinuous ephemeral streams,"[129] although discontinuous arroyos occurred on perennial streams as well. On most rivers, entrenchment proceeded as a coalescing of formerly discontinuous arroyos. For some arroyos, such as Kanab Creek, there was no doubt when the arroyo had fully coalesced. For other channels, the dates are more ambiguous.

Although the dates of incision for many arroyos remain unknown, we now know that most arroyos in the southwestern United States—from the Altar Valley, Sonora, to Kanab Creek, Utah, and from the Sonoyta River, Sonora, to Rio Puerco, New Mexico—began downcutting in the twenty-year period from 1875 to 1895. All started downcutting as a result of some flood that was markedly larger than those experienced previously. Allowing that channel downcutting seldom occurred at the same time—even in adjacent drainage basins—the coarse synchroneity (in a geologic sense) that was observed raises interesting questions about regional mechanisms that led to the large floods.

Many observers, noting the close relation between the dates of settlement and arroyo cutting, assume a causal relationship.[130] But cattle-raising at Sonoyta dates from 1695, and arroyos became incised much later, in 1891. "Settlement" in the Altar Valley of Sonora dates from the early 1700s, and channel incision began about 1890. "Settlement" of the San Pedro Valley by Caucasians can be traced to the 1760s when the Spanish founded the presidio of Santa Cruz, or to the 1820s and 1830s when Mexicans undertook large-scale cattle-raising, or to 1867 when Anglo-American settlers arrived. Accelerated erosion in the San Pedro River dates from 1890.[131]

Assuming that settlement caused arroyo cutting raises several problems; not least is the possibility that the two might merely be associated in time. Moreover, given the uneven pattern of settlement, it does not seem likely that an area of 297,000 square miles could have been so equally affected as to result in widespread erosion in one twenty-year period, particularly when the two preceding centuries of colonization and grazing in the Mexican watersheds of the region did not see channel erosion. That overgrazing accompanying "settlement" may have influenced arroyo cutting is not in question. Rather, the historical evidence indicates that the human influence may have been a mere auxiliary to the broader influence of climatic variability.

The impact of cattle is easily perceived and appreciated; in the semiarid Southwest, one can see it every day in the contrast between the grass on a fenced right-of-way and the adjacent range. But there is no similar easement through time along which one can observe the cumulative consequence of small changes in climatic factors. Grazing impacts are readily perceived, but climatic effects are more subtle. The lack of what might appear to the casual observer to be immediate and clear impacts should not be used to discount climate in favor of grazing.

Fuelwood Cutting

Until recently, little was known about the extent and magnitude of fuelwood cutting in southeastern Arizona in the late 1800s. Recent work closely documents this practice, which previously was only assumed to be widespread. During the 1870s and 1880s in southeastern Arizona, there were about twenty-seven major mining centers that required cordwood at almost every stage in the process. Tombstone, one of the most important centers, had more than fifty different mines and 150 stamp mills. From 1879 to 1886, Tombstone stamp mills consumed an estimated 47,260 cords of fuelwood.[132] Contemporary anecdotes also point to intensive wood cutting; mines in the vicinity of Bisbee, for instance, used so much cordwood that the surrounding hills were reportedly denuded of trees.[133] Cordwood was also required for heating, running steam engines of every sort, firing kilns, and constructing buildings and fences. The most intensively harvested species were pine, juniper, mesquite, and oak. Riparian species—particularly cottonwood—were largely avoided because they did not burn as hot as the other available fuelwood. Between 1890 and 1940, the amount of wood removed from the Tombstone area "woodshed" probably exceeded the total cordage now growing there.[134]

If the Tombstone area was typical, it is not difficult to imagine that, by the time of most of our earliest photographs (late 1800s), the oak woodland was already depleted. Indeed, as has been noted, old photographs taken in the vicinity of mines typically show this sort of denudation.[135] Because several of our camera stations are located near old mines, we are able to examine this possibility. Past fuelwood harvesting is also considered a major force in fashioning riparian woodlands, a habitat recorded in several of our photographic sequences.[136]

Culture, Nature, and Change

This chapter and the previous one have presented the basic historical facts about human occupancy of the desert region during the past three or four centuries. In addition, we have speculated about the ecological role of humans during that time. The historical evidence illustrates a continuum of cultural influence for as long as people have inhabited the region. Well before the time of Anglo-American settlement, disturbance by humans attained a considerable magnitude, but there is also little question that change accelerated after about 1865.

Having established the cultural and natural setting of recent vegetation changes, we now examine its extent and causes. To discuss why and where changes have taken place, we must first establish what those changes are, and the next three chapters are therefore devoted to triplicated photographs of the oak woodland, grassland, and the desert, the life zones lying within our changing mile region.

Chapter Four The Oak Woodland

The life zone lying above grassland and below pine forest has been referred to as *encinal*, after the Spanish word for oak.[1] The zone is divided into two subdivisions, Upper Encinal and Lower Encinal. Dominated by trees, mainly small oaks, the Upper Encinal is dense and contains a high proportion of shrubby species. At its upper edge in our region (at about 6,500 feet above sea level), the oaks (Emory oak, Arizona white oak, silverleaf oak) are closely spaced on steep slopes, and between them are crowded other woody species (Arizona madrone, Wright's silktassel, Mexican pinyon, and pointleaf manzanita). Above this boundary, Upper Encinal merges with the pine forest through an intermediate belt of vegetation sometimes called Pine-Oak Woodland[2] where scattered tall pines (Apache pine, Chihuahua pine, Arizona pine) occur among the oaks. In Mexico, the Pine-Oak Woodland becomes broader, attaining dominance on some Mexican mountains (e.g., the Ajos and Oposuras).[3] With decreasing elevation, Upper Encinal gradually gives way to Lower Encinal.

The Lower Encinal, found at moderate elevations (3,650–5,000 feet) on relatively gentle terrain, is often referred to as "oak woodland" or simply "woodland." In this orchard-like community, widely spaced oaks (Mexican blue oak, Emory oak, Arizona white oak, gray oak) are interspersed with other trees, among them Arizona rosewood, one-seed juniper, and mesquite. The matrix of grasses includes numerous gramas, as well as many species of three-awn, *Muhlenbergia*, and *Andropogon*. Several plants with succulent leaves or a succulent caudex occur with the oaks: beargrass, sotol, and species of agave and yucca. Small, woody members of the community include skunkbush, mimosa, and fairyduster. The broad sweep of zones encompassed by F. Shreve's Upper and Lower Encinal and J. T. Marshall's Pine-Oak Woodland is now widely referred to as Madrean Evergreen Woodland.[4] The transition between the two Encinals marks the approximate upper limit of this study. We consider "Oak Woodland" to mean the same as "Lower Encinal."

Toward the lower elevational limit of the encinal zone, the oaks retreat to relatively moist, cool sites on north-facing slopes and in canyons. Lower still, below 4,000 feet, oaks are confined to canyon bottoms that are separated by ridges of grassland. Ultimately, the Lower Encinal merges completely with the grassland or, in cases where grassland is missing, with the upper edge of the desert.

Although Oak Woodland extends eastward into New Mexico[5] and southward into the Sierra Madre of Sonora and Chihuahua,[6] the plates in this chapter are taken from an area of about 2,000 square miles, centered roughly on the Santa Rita Mountains of south-central Arizona (figure 4.1). Annual rainfall in this area is 18–20 inches, with much of it (65–70 percent) falling in summer. Because of the relatively mesic climate, Oak Woodland occurs at a lower elevation in this area than it does elsewhere in southern Arizona. At comparable elevations along the San Pedro River, for example, the predominant community is Semidesert Grassland.[7]

Four species of oak appear in the plates. The most common is Emory oak (known also as *bellota*, or blackjack), a tree or shrub of valley and upland habitats. Mexican blue oak, a foothill species, is restricted to Lower Encinal. Arizona white oak, a tree of more mesophytic habitats than Emory or Mexican blue oak, is restricted to cool north slopes and valleys. Arizona white oak ranges freely into the higher elevations. Toumey oak, a shrub or a small tree of limited distribution, appears only in plates 9 and 28. Some woody species in Oak Woodland, including Emory oak and Mexican blue oak, can be found well within the desert zone, typically along streams.[8]

It was sometimes difficult to decide whether to assign plates to Oak Woodland or to Semidesert Grassland. The transition between the two can be subtle, and the pronounced interdigitation of life zones also complicates matters. Moreover, the oldest photograph of a plate may portray woodland, whereas the new matched view may lack oaks altogether. In spite of these complications, Oak Woodland is easily recognized and relatively homogeneous wherever it occurs. As the pictures that follow will show, the secular trends within it are for the most part well defined.[9]

Threading through Upper and Lower Encinal are thin lines of gallery forest or canyon forest, a distinctive vegetation restricted to watercourses. Its species composition changes little from encinal to grassland to desert. Goodding willow, velvet ash, cottonwood, walnut, Arizona sycamore, Texas mulberry, soapberry, and netleaf hackberry, the principal trees, may be associated with such shrubs as burrobrush, seep willow, and rabbitbrush, and by canyon grape and poison ivy, both lianas. These plants span a greater range in elevation than most of the species on the adjacent uplands because their canyon bottom habitat is relatively homogeneous in temperature and moisture at all elevations.

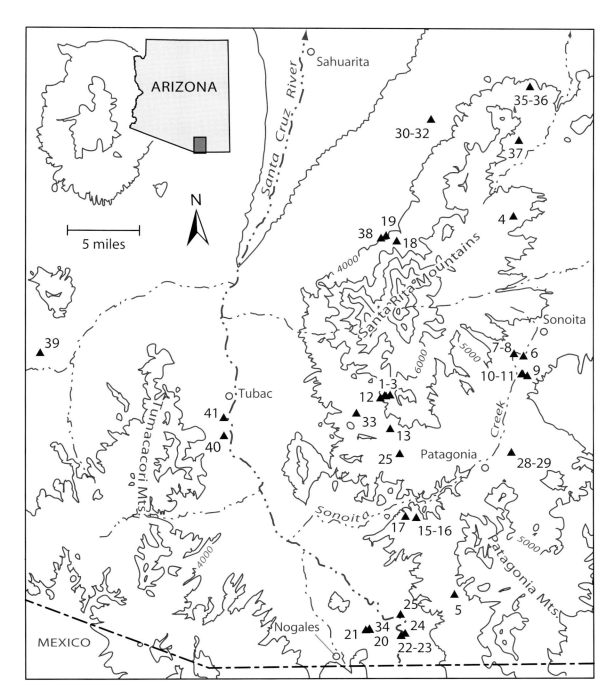

Figure 4.1 Location of photographic stations in the Oak Woodland and the grasslands around the Santa Rita Mountains. (See figure 6.3 for Oak Woodland stations in Mexico [stations 14 and 26].)

1a

Plate 1a (1891) The view, looking southwest toward El Plomo Mine on Alto Ridge on the southwestern side of the Santa Rita Mountains, is also the site of the next two sets of plates. Local legend attributes the discovery of El Plomo to early Spanish missionaries, but similar stories have sprung up about most of the mines in this region. The mine was worked intermittently for at least fifteen years before this photograph was taken.[10] The tree on the right is Emory oak. The others are also oaks, either Emory or Mexican blue. A stack of fuelwood is seen just left of center at midground. The elevation is 5,500 feet.

Plate 1b (1962) In the intervening years between the photographs, two mesquites have invaded the south-facing slope just left of center, but not the opposing north-facing slope. Two other mesquites, seen on the skyline, are rooted on the south-facing slope beyond the ridge top. At this elevation, mesquite is near the upper limit of its range[11] and therefore avoids colder, north-facing habitats. An alligator juniper occurs in the midground at the right near the ravine. The rest of the trees are oak. The fence marks the boundary between the Coronado National Forest (foreground) and the Salero Ranch (background).

Plate 1c (1995) The increase in trees is almost entirely the result of oak establishment. The Emory oak on the right side of the 1891 view is still visible, although a mesquite has grown in front of it. No new mesquites have appeared on the disturbed north-facing slope, although two oak saplings have become established there. The small mesquites on the south-facing slope are about the same size as before, probably having been kept in check by severe freezes. A sotol has appeared, a replacement for the one found nearby in 1962. A barrel cactus is in the foreground.

2a

Plate 2a (1891) This view is north-northeast toward an oak-dotted saddle about one-quarter of a mile west of the diggings that appear in plate 1. All of the El Plomo views, including six not reproduced, show oak to be more abundant in 1962 than in 1891; the size of the woodpile in the foreground suggests why. The stack is probably Emory oak, which fetched premium prices at the mines.[12] Mexican blue oak and Arizona white oak were regarded as inferior. Mesquite, another good firewood, was probably not abundant nearby or worth the effort of hauling in by burro train over steep mountain trails. Note the stump, probably of oak, to the right of the tent. The elevation is 5,500 feet.

Plate 2b (1962) The oaks in and about the camp continued to decline after 1891. The few remaining are restricted to the rocky slope at the right side of the photograph, except for one on the site of the cabin. In this view and on south-facing slopes in the area, about three-quarters of the oaks are Emory, the rest Mexican blue. A third species, Arizona white oak, becomes important on north-facing slopes. The remaining trees in the photograph are mesquite, and although none appear in the earlier view, they now clearly outnumber the oaks. Throughout the woodland and the grassland, level saddles where cattle rest frequently become excellent establishment sites for mesquite, as shown in the midground. The large number of mesquites on these east and south exposures is a contrast to the nearby north-facing slope, where it is absent (plate 1).

Plate 2c (1995) The mesquites have not gained size in thirty-three years. Just right of center foreground, one has died. All have the pollarded appearance typical of frost-injured plants. The "cabin-site" oak has increased in size, as have those on the far hill.

3a

Plate 3a (1891) This northerly view at El Plomo Mine shows a work crew walking along the trail that leads from the shaft in plate 2 over the hill to the vicinity of plate 1. All of the trees appear to be oaks, many severely pruned. The scattered small clumps are probably beargrass, a likely source of the thatch on the small structures. The larger leaf succulents are sotol. The elevation is 5,500 feet.

Plate 3b (1962) The trees are mesquite, Emory oak, and Mexican blue oak, none clearly dominant and all either new or more abundant in 1962 than in 1891. The principal grass is sideoats grama. Among the smaller plants present are desert cotton, western coral bean, mala mujer, wait-a-minute, amole, and yucca. Two or three of the dead trees in the photograph appear to be remains of individuals in the earlier view.

Plate 3c (1995) Status of the woody plants has changed little in thirty-three years, although one oak on the skyline just left of center has died. The remaining oaks (trees with dense foliage) are slightly larger, and the mesquites (trees with open crowns) are smaller or unchanged. An alligator juniper has replaced the oak at center.

3b

3c

4a

Plate 4a (1925) In this view, along the old road between Tucson and Sonoita looking southwest toward the north spur of the Santa Rita Mountains, the dominant trees are oak. A few Emory oak, one-seed juniper, and mountain mahogany also are present. In this locality the woodland is largely confined to ravines and north-facing slopes, a fact most evident, perhaps, in the background of the newer photograph. The elevation is 5,100 feet.

Plate 4b (1962) Although the two photographs span only thirty-seven years, they register some critical changes. Juniper has increased while oak has declined. Many dead plants can be seen in the photograph; of 134 oak trees below the dotted line, 40 percent are dead. This is an extremely high mortality rate for oaks. The highest death rate has occurred on the upper one-third of the hillside, where 60 percent of oaks are dead. Most junipers have sprung up beneath the canopies of mature oaks, a fact that may point to dispersal by birds. Although mesquite has increased in the ravine at lower right and on some of the warmer slopes, it is of minor importance. Skunkbush and beargrass occur on the slopes, yerba de pasmo and burroweed on the ridges.

Plate 4c (1987) In 1994, a count of living trees on the foreground ridge yielded 152 Mexican blue oaks, 6 Emory oaks, and 97 junipers. In addition, there were 33 dead oaks, of which 22 were next to living junipers. The area covered by the 1994 census was slightly larger than that of the 1962 census, with large overlap. Oak mortality is about 17 percent, a reversal (or at least a slowing) of the trend in 1962. Mesquite has continued to increase.

5a

Plate 5a (1887) In the Patagonia Mountains east of the Santa Cruz River, about 5 miles north of the International Boundary, this view is due east up Cañada de la Paloma. Part of Guajalote Peak appears at the far right. A simple savanna vegetation of oaks and grasses prevails. Trees are confined to ravines and north-facing slopes. Emory oak dominates the bottomlands and shares the uplands with Mexican blue oak. The elevation is 4,750 feet.

Plate 5b (1962) A much more complex plant assemblage exists. Ocotillo has become dense on the south-facing slope at the left. Mesquite has invaded extensively along the ridges and to a lesser extent on north- and south-facing slopes. Winter temperatures at this elevation are not low enough to restrict mesquite to the warmer slopes. Oaks may be fewer than in 1887. The tussocks on the nearest hill are clumps of grass, fairyduster, and velvetpod mimosa. Heavy grazing has produced the terraces.

Plate 5c (1994) The foreground oak in the 1887 photograph has died; its remains are out of view. On the midground ridge at left, the two trees present in 1887 dwindled to one in 1962 and zero in 1994. Otherwise, the number of oaks is probably little changed. Ocotillo has continued to increase on steep slopes at left midground. Cattle trails along the slope contours are not as conspicuous as before, perhaps indicating reduced grazing pressure. Several plants of sotol are new since 1962. Throughout most of the view, mesquites have grown in size but have not increased in number. New mesquite plants appear at the extreme right and left on the foreground ridge and on north facing slopes of Cañada de la Paloma. A single graythorn carries over from the 1962 view. Other species seen in the vicinity are western coral bean, Arizona cottontop, and Texas bluestem.

5b

5c

6a

Plate 6a (ca. 1895) This view is north-northeast toward the north spur of the Santa Rita Mountains from a station three miles southwest of present-day Sonoita. In the midground, Sonoita Creek winds through a trestle on the Fairbank-Guaymas railroad, completed in 1881 by Santa Fe and later sold to Southern Pacific. The predecessor of this trestle was destroyed by a flood in 1892.[13] Erosion along the creek is in its infancy. Woody vegetation on the hillside in the foreground is sparse. The trees are clearly oaks, probably Arizona white oaks. Although the shrubs cannot be positively identified, some at lower right may be cliff rose. The elevation is 4,750 feet.

Plate 6b (1962) The railroad track is being dismantled. The trestle timbers appear taller, an indication of continued channeling. The near hillside has a denser cover than before, including skunkbush (1), alligator juniper (2), Arizona white oak (3), yerba de pasmo (4), beargrass (5), *Yucca* (*baccata* or *arizonica*) (6), a species of *Brickellia* (7), *Rhus choriophylla* (8), and mesquite (9). The dominant grass is sideoats grama. Shrubbier areas are designated by letters. In area A, cliff rose predominates, with some mountain mahogany and *Ceanothus greggii*. Rabbitbrush occupies area B adjacent to the channel and, out of view, grows even more densely in the coarse alluvium of the eroded flood plain. Soapberry occupies the cliff above B, with Goodding willow in the streambed nearby. In the rolling country beyond the track, there has been less change. The oak population—Emory oak rather than the Arizona white oak of the hillside—is about the same as before. Some mesquite has become established in the flatter areas and along the ravines. One large cottonwood (10) is visible in the left midground beside the railroad track. The hills in the background appear virtually unchanged.

Plate 6c (1994) Removal of the railroad trestle and track appear to have shifted this scene back to a presettlement condition except for the six or seven houses that now dot the landscape. Mesquite has expanded into the formerly grassy areas. The alligator junipers identified at midground are larger, as is the Arizona white oak at extreme left. Except for the increase of mesquite in the canyons and juniper on the uplands, the Oak Woodland on the hills beyond the valley of Sonoita Creek has changed little. At center foreground, Engelmann prickly pear is new to the scene.

6b

6c

7a

Plate 7a (ca. 1895) On Sonoita Creek, this view is upstream (southeast) from a point one-quarter of a mile below the trestle in plate 6. The unweathered vertical banks suggest recent downcutting. A few young cottonwoods within a small fenced area at Cottonwood Spring represent the only trees visible along the stream. The recent erosion has toppled two of them. Heavy use by livestock is evident. The elevation is 4,600 feet.

Plate 7b (1962) Erosion has claimed the original camera station. This view, taken from a point perhaps 30 feet away from the old location, shows the same rocks on the hillside. A mesquite bosque covers the flood plain. Seep willow, which fringes the creek on both sides, is new on the scene. Other species in the new riparian community along the valley floor include walnut, netleaf hackberry, skunkbush, canyon grape, Goodding willow, Texas mulberry, and poison ivy. The cottonwood trees (right of center) probably persist from the previous view. The vegetation of the north-facing slope in the right background is largely Arizona white oak, with possibly a few Emory oak, some cliff rose, and, along the crown of the hill, some mesquite and a solitary one-seed juniper.

Plate 7c (1998) Wholly changed and completely overgrown with mesquites, the previous camera station, itself shifted from the original, is moved back again to capture a rock and peak seen in the original photograph. In the foreground, sacaton and mesquite grow on what was the valley floor before the downcutting seen in plate 7a. The deepened channel of Sonoita Creek stands several yards below the old valley floor and supports a rich growth of riparian plants. The tall tree at left midground is netleaf hackberry. The peak seen in plate 7a is just visible through the right side of the tree's crown.

7b

7c

8a

Plate 8a (ca. 1895) Near Sonoita Creek, a mile east of the preceding station, this view is west-southwest across the ruins of Fort Crittenden, left and center midground, toward the south end of the Santa Rita Mountains. Fort Crittenden, founded in 1867 near the site of Old Fort Buchanan, was closed in 1873,[14] and the corral belongs to a ranching operation that postdates the military complex. A pure, or nearly pure, stand of Emory oak meanders across the grassland along a runnel. Mesquites may be present, but, if so, they cannot be identified. "For miles and miles the country is covered with grass and free of all sage, greasewood, or other worthless shrubs. Beautiful thrifty . . . oaks give much . . . beauty and charm."[15] The elevation is 4,700 feet.

Plate 8b (1962) The density of the adult Emory oaks has remained about constant, and many individual trees remain. However, the young oaks in the lower right-hand corner of the old photograph have not survived. Around the confines of the old corral, an appreciable mesquite invasion has taken place, a fact of some interest in connection with the overgrazing hypothesis. Downstream from the oaks, cottonwood trees (1), none of them evident in the old view, dominate the scene, with a line of mesquites, also recently established, at their base. New mesquites dot the pasture beyond the trees, and in the grassy foreground an incipient invasion is underway; arrows indicate the young mesquite plants.

Plate 8c (1994) The mesquite increase that began before 1962 has continued to the present. The small mesquites have grown and are joined by many one- or two-year-old plants not visible in the photograph. Deaths of mesquite are far outnumbered by establishments. Dominant grasses are sideoats grama, hairy grama, and alkali sacaton. Scattered among the grasses in the foreground is Wright buckwheat. Other plants found near the camera station are one-seed juniper, burroweed, wait-a-minute, annual goldeneye, telegraph plant, Engelmann prickly pear, and cane cholla. The two cactus species are frequent beneath the crowns of the oaks. The cottonwood trees and oaks along the arroyo are much as before. The arroyo bank at right appears to have healed since 1962 and looks much as it did in the 1890s. Just right of the old corral, the arroyo is incised 10 feet. Mesquite in the background has increased markedly along the tops of the mesas but not on the steeper slopes.

8b

8c

9a

Plate 9a (ca. 1895) This photograph was taken from a hill on the east side of Monkey Canyon about four miles southwest of present-day Sonoita, looking west across Monkey Lake toward the massif of the Santa Rita Mountains. Mt. Baldy and Mt. Hopkins are at left, Monkey Spring is just out of view to the right, and Sonoita Creek is out of sight behind the midground ridge. Oaks dominate the foreground; cottonwoods, leafless for the winter and still fairly young, grow around the lake. A few oaks can be seen on the valley floor near the spring. The elevation is 4,750 feet.

Plate 9b (1962) There has been a dramatic increase in the complexity of the vegetation. Most of the oaks in the foreground have died and have been replaced by one-seed juniper (1); nevertheless, three oak species are still represented: Emory (2); Mexican blue (3); and Toumey (4). Other shrubs are pointleaf manzanita (5) and alligator juniper (6). The smaller plants include fairyduster, wait-a-minute, velvet-pod mimosa, sideoats grama, slender grama, and a species of three-awn. In the rich valley flora are cottonwood (7), Arizona white oak (8), Goodding willow (9), red willow (10), velvet ash (11), Arizona sycamore (12), netleaf hackberry (13), coffeeberry (14), *Rhus choriophylla* (15), desert broom, and a great many mesquites, all unlabeled. Mesquite, juniper, and desert broom are present along the road descending the ridge in midground; juniper, agave, and ocotillo are elsewhere on the ridge.

Plate 9c (1994) This photograph is slightly too far left for an exact match. Branches of Toumey oak block some of the foreground from view and hide several new plants of Palmer's agave. The small alligator juniper present in 1962 (right foreground) was dead by 1980. Several new Emory oaks have sprouted in the vicinity of the camera station. The most prominent grasses in the photograph are bull grass and plains lovegrass. Numerous plants of heavily browsed Arizona carlowrightia are scattered among the grasses. A new cluster of ocotillos is visible at left foreground. Many dead oaks, all lacking axe marks and often ringed by junipers, are still found downslope from the camera station. Woody plants have increased over the valley floor, both on the dry flat at left midground and along the riparian strip of the former Monkey Lake. More and larger one-seed juniper and velvet mesquite are visible on the low ridge running across the center of the view.

9b

9c

10a

Plate 10a (1889) From the ridge in the midground of the preceding plate, this view faces west-northwest across Sonoita Creek and up Adobe Canyon toward the Santa Rita Mountains. Most of the trees are probably oak. Three plants of a yucca can be seen near the tripod legs. Note the open, shrub-free appearance of the Sonoita Valley. The elevation is 4,600 feet.

Plate 10b (1962) A natural gas pipeline, constructed in the late 1940s, accounts for the faint tracks cutting diagonally across the foreground. In the foreground, dead oaks have been engulfed by living clumps of one-seed juniper, a sight common throughout the lower woodland in this area (compare plate 4b). This photograph, which shows only those foreground slopes with a south-facing component, may convey a false impression of the region's juniper density. The species grows abundantly here on north-facing slopes but seldom reaches ridge lines or crosses over them to the south faces. In the valley, only a few cottonwoods, velvet ash, and oaks have managed to compete with the tide of mesquite. The bosque is densest in the lowest part of the valley, progressively thinning toward higher ground. At about 4,600 feet, the elevation at the base of the heavily dissected hillslopes, mesquite becomes relatively uncommon. Above 4,600 feet, oaks, velvet ash, and Arizona sycamore dominate the canyon, and oaks dominate the uplands. Roughly the same demarcation can be detected in the old picture; there, however, it represents the transition between treeless and wooded areas. Presumably the terrain above the ancient terraces across the valley is relatively unchanged in its plant cover; the lower terrain is much changed.

Plate 10c (1994) Enough time has elapsed since pipeline construction for a good cover of grass and fairyduster to have developed. The population of Palmer's agave, represented by a single plant in 1962, has increased several fold. The group of one-seed junipers visible just over the ridgetop in the foreground has more and larger plants, and the dead oaks, still standing in 1962, have fallen to the ground. Young oaks and point-leaf manzanitas now grow among the junipers. Ocotillo has increased noticeably on the ridge facing the camera at right center. The cultivated land on the valley floor was abandoned sometime after 1962 and now supports many small mesquites. Although not discernible in the photograph, the mesquites in the dense forest on the valley floor at right show abundant signs of frost damage, presumably from the catastrophic freeze of 1978.[16] Several houses are now scattered among the low hills across the valley. Woody plant growth on the distant hills has increased slightly since 1962.

10b

10c

11a

Plate 11a (1889) George Roskruge's surveying party is standing above Monkey Spring. The camera site is perhaps 350 feet northeast of the preceding plate. The view is slightly east of north, looking up the creek toward the spring, which is located at the foot of the hillock on which the group is standing. The stream banks, the valley floor, and the south-facing slopes beyond are almost devoid of large plants. The elevation is 4,550 feet.

Plate 11b (1962) The camera station is almost precisely the same as for the old photograph, but the camera has been elevated to avoid a shrub. The principal foreground plants are *Baccharis neglecta*, a relatively rare species; Goodding willow; one-seed juniper; *Rhus choriophylla*; Bermuda grass, an exotic; and watercress, also exotic. In the midground, desert broom, canyon grape, and *Baccharis neglecta* are present. Mesquite, Arizona white oak, and one-seed juniper hide the skyline, but, in contrast to the valley floor and the stream banks, the surrounding hills look much the same. If, as seems likely, the spring has been continuously fenced off to avoid pollution, few of the changes immediately around it can be attributed to cattle.

Plate 11c (1994) The foreground plants have continued to close in upon the camera station. Clearly visible in the foreground is velvet mesquite. Other plants in the view are one-seed juniper, Goodding willow, and Fremont cottonwood.

11b

11c

12a

Plate 12a (1891)   Ten miles northwest of Patagonia, Arizona, this view is east-northeast across Alto Gulch and up the western escarpment of the Santa Rita Mountains. The camera stations for plates 1–3 are situated about half a mile away on the top of the ridge. The trees in the photograph, some of which still survive, are Emory oak and Mexican blue oak. Almost no shrubs are present. The elevation is 4,650 feet.

Plate 12b (1962)   The large tree in the right foreground is a Mexican blue oak, as are most of the oaks on the opposite slope. At the summit, Emory oaks are scattered among many mesquites. A careful inspection shows that oaks present in 1962 have counterparts in the old photograph. No new ones have become established, and mortality among the old trees has been severe. Arrows mark four of the more prominent dead oaks. The highest survival is on the rocky outcrop in the right midground and at the higher, cooler elevations near the summit. The substantial thicket of ocotillo above the outcrop appears to be new. The small, dark shrub so abundant on the slopes is turpentine bush, also a recent invader.

Plate 12c (1995)   The oak population has remained stable since 1962. Furthermore, all visible oaks were established by 1891. Mesquite biomass has increased during the last one-third of a century, although numbers of plants have remained about the same.

2b

12c

13a

Plate 13a (1909) The Salero Mine, in the southwestern foothills of the Santa Rita Mountains, is one of the oldest in Arizona, having been worked sporadically since 1828.[17] Note what appear to be ricks of firewood in the background to the left of the mill and hoisting works. Many of the trees between the camera and the buildings carry over into 1962. The camera faces north-northeast; Salero Mountain is at the left. On the upper reaches of the mountain, Mexican blue oak predominates. The elevation is 4,500 feet.

Plate 13b (1962) The buildings show less change than the plant cover. Of the trees in front of the line, all are mesquite (1) except the Emory oaks (2), the Mexican blue oaks (3), and one gray thorn (4). There has been little or no recruitment of oak; in fact, all visible individuals can be seen in the old photograph, as well. Ocotillo (5) has become established on the foothills. The two trees labeled 1 are mesquites that have persisted from the 1909 view. Although Mexican blue oak has decreased in density on the dry south- and southwest-facing slopes of the mountainside, mesquite has not appreciably increased.

Plate 13c (1994) The buildings are more hidden than before, one indication that mesquites have increased in stature and number over much of the view. The Emory oak at right foreground has grown, blocking more of the scene.

13b

13c

14a

Plate 14a (1931) This classic development of the Oak Woodland is about three miles south of Nogales, Sonora. The camera faces east-northeast. Emory oak predominates; a few Mexican blue oaks may also be present. Arrows point to two young trees, an Emory oak and a mesquite, that are prominent in the later photograph. The elevation is 4,300 feet.

Plate 14b (1962) The evergreen oaks and the partially defoliated mesquites are easily distinguished in this December view. The two trees marked by arrows in the earlier photograph now stand just beyond the railroad track at center midground. The attrition among oaks in just thirty-one years is remarkable, especially when contrasted with the simultaneous increase in mesquites. How much of the decrease may be due to woodcutting is difficult to determine, although the fact that mesquites (also a favorite firewood) have increased argues against active woodcutting and indirectly supports the view that the oaks were killed by drought. The dense grass cover apparent in plate 14a has disappeared. Beyond the fence, heavy grazing may be responsible. The absence of grass on the semiprotected right-of-way in the fore- and midground again suggests death from drought. Note the advance of gullying on the other side of the railroad tracks.

Plate 14c (1995) The town of Nogales, Sonora, has advanced toward this site, and a commercial building now occupies the camera position. From a new vantage point several yards to the right and closer to the railroad tracks, the scene is dominated by human impact and artifacts. Beyond the railroad tracks, evidence of woodcutting is seen: the mesquite at right midground in the 1962 photograph has been replaced by root sprouts; the oak that stood behind it in the old view is completely gone; and the oak beyond the fence has been heavily coppiced by woodcutters. Despite removal of some trees, the general impression is one of increased woody biomass across the hill slopes, with fewer but larger oaks than before and with many mesquites of varying sizes. Houses have been built to within a few hundred feet of this site and will undoubtedly dominate the scene in another third of a century.

14b

14c

15a

Plate 15a (1890) This view toward the west from a station 7 miles southwest of Patagonia, Arizona, shows what George Roskruge, the photographer, calls the Hill of San Cayetano. The Grosvenor Hills are at right; the San Cayetano Mountains are at left. At this time the area evidently lay on the lower edge of the Oak Woodland. The trees are widely spaced and confined to ravines and north-facing slopes. A few junipers may be scattered among the oaks. The small, round tussocks are probably mostly sotol and some beargrass. The elevation is 4,200 feet.

Plate 15b (1962) Not a single living oak can be seen in the photograph, although some relict Mexican blue oaks grow in sheltered spots nearby. Death occurred recently enough for an impressive number of dead fallen oaks to remain. None of them bears axe or fire marks, and the isolation of the area rules out any overt interference by humans. Mesquite, the new dominant, grows most densely along ravines and on north-facing slopes, as did oak in the earlier view. Another recent invader is ocotillo (right midground and scattered over the hill). Also present are desert broom, wait-a-minute, beargrass, Santa-Rita cactus, kidneywood, gray thorn, netleaf hackberry, a few one-seed junipers, and a species of *Yucca*. The hummocks are composed mainly of fairyduster and mimosa. Sotol and beargrass have markedly declined but are still common.

Plate 15c (1994) The former Oak Woodland of the valley floor continues its evolution toward a mesquite thicket. On drier exposures, such as the open grassland of the Hill of San Cayetano, the mesquite is less dense but advancing nonetheless. On the low hills below the camera station, increased biomass probably reflects enlargement of individual plants rather than an expanding population; on the slopes of the hill, both processes have been active. Ocotillo has increased in prominence; fairyduster and wait-a-minute are still prominent low shrubs. With the exception of two one-seed junipers, all the trees are mesquite. As in 1962, oaks still grow in sheltered draws nearby. Beargrass and sotol are still relatively sparse.

15b

15c

16a

Plate 16a (1890) This view of Sanford Butte is toward the northeast and from the same camera station as for the preceding plate. At the foot of the butte, Sonoita Creek is visible. As in plate 15a, oaks line the ravines and grow on the north-facing slope, and sotol appears on the hillsides. In contrast to its upper course (plates 7 and 10), Sonoita Creek here has heavily vegetated banks, dominated probably by cottonwood. The elevation is 4,200 feet.

Plate 16b (1962) All of the midground oaks are dead. On the rocky slope in the right foreground, sotol has generally declined in importance, having been supplanted by beargrass. A few mesquites have become established. Ocotillo, the most noticeable invader on these predominantly south-facing slopes, appears densest where sotol has disappeared. The Sonoita Valley looks about the same. Cottonwoods line the creek; the mesquite-studded spur indicated by arrows may be a little less open than in 1890.

Plate 16c (1994) In the last third of a century, woody plants have increased in size as well as number. There appear to be no new additions among the twenty or so velvet mesquites on the nearby hills facing the camera station. On the slope at right background, mesquite has apparently increased in size as well as number. Ocotillo is more prominent, an apparent result of size increase. At midground, sotol appears little changed since 1962. Beargrass plants in the foreground persist from 1962. No new oaks have become established to replace those present a century ago. Directly behind the camera, a north-facing slope now dominated by mesquite, kidneywood, sotol, wait-a-minute, and banana yucca once supported an open Oak Woodland, as shown by standing and fallen oak carcasses.

16b

16c

17a

Plate 17a (1890) This photograph was taken at the head of Guevavi Canyon looking southwesterly toward the Santa Cruz River valley and the Pajarito Mountains beyond. The site is about 1 mile from perennial flow of Sonoita Creek and about 10 miles from Calabasas, former *visita*[18] of the Jesuit fathers and short-lived frontier town. Sheep, goats, horses, and cattle have probably grazed here for more than a century. The barren, rocky soil surface is typical for that era and is probably caused by lack of rain and heavy grazing.[19] The sparse woodland of oaks on the rounded hill at midground is similar to that seen in plate 15a. The trees are closely trimmed from below by browsing animals, suggesting heavy use by domestic livestock. The elevation is 4,260 feet.

Plate 17b (1962) That cattle use this elevated saddle is readily apparent. Velvet mesquite now occupies the site densely and blocks the view. The large, partially buried rock at right (arrow) can be seen in the previous photograph.

Plate 17c (1996) Because the dense mesquite thicket blocks the original view, the camera has been moved forward about 20 yards to a position near where the man in plate 17a is standing. The shorter growth beneath the interlocking mesquite crowns is primarily wait-a-minute. Examination of the hill at midground in 1996 revealed the remains of the oaks. As with the many other dead oaks in the area, no axe marks are evident. The ground cover in the foreground, sparse through the combined effects of heavy grazing and recent drought, is composed of such grasses as spidergrass, blue grama, and sprucetop grama. Utility poles and a house roof (visible above the ridgeline, left background) are signs of housing development, a new force for change in the region.

17b

17c

18a

Plate 18a (1911) On the northwest side of the Santa Rita Mountains, 10 miles southeast of Continental, the view is south-southeast up a woodland slope in the Santa Rita Experimental Range. Florida Canyon is at far left. Along the ravine in the foreground, oaks dominate with an understory of young mesquites that are seasonally defoliated. Young mesquites have also gained a hold on the lower part of the slope in the center foreground. Oaks dominate the hillsides; the compact, dark shrubs are sotol, probably interspersed with a few clumps of beargrass. The dense pine forest of the upper mountain region can be glimpsed in the distance at the far left. The elevation is 4,100 feet.

Plate 18b (1962) The young mesquites are now adults. Their summer foliage contributes to the cluttered appearance of the photograph. A substantial population of ocotillo is present, all of it evidently recent. On the upper half of the hill at right, sotol and beargrass show a marked increase. The Mexican blue oaks along the wash in the foreground look much the same; intermingled with them are desert hackberry, Arizona sycamore, and low, spiny velvet-pod mimosa. On the upper part of the hillsides, the oak population remains unchanged, but downward the mortality increases, and many dead oaks litter the mesquite woodland at the bottom of the slope in center midground. A number of mesquites are established higher in the hills, and some are, in fact, to be found almost at the summit at the right.

Plate 18c (1994) The Oak Woodland seems about as dense as in 1962. Many individual trees, including the three Mexican blue oaks in the foreground, are the same. On the slope at midground, coverage of wait-a-minute has increased since 1962, and sotol, made prominent by the white flowering stalks on many plants, seems more numerous. Juniper, which was not apparent in the 1962 view, is now represented by at least six individuals on the midground slope.

18b

18c

19a

Plate 19a (1911) The camera station for this view, which looks southeast toward the ridge in plate 18, is about 200 feet lower and a mile west-northwest of the preceding location. Over that distance, Oak Woodland gives way to grassland. In fact, three life zones appear in this photograph. From bottom to top: grassland, already densely covered by adult mesquites on the flats, but still open on the slopes; the encinal, ranging from the open Oak Woodland below, reaching down into the grassland on north-facing slopes and along ravines, to the dense Upper Encinal above; and the pine forest, visible along the ridgecrest of the highest peak. The large size of the nearby mesquites suggests that they became established one or two decades earlier. The elevation is 3,950 feet.

Plate 19b (1962) The old camera station cannot be recovered precisely, but the new one is within 50 yards. A dense mesquite thicket has replaced grassland in the foreground. Common associates include desert hackberry, catclaw, burroweed, and several species of prickly pear. Mesquite and ocotillo, both recent, dominate the low, formerly grassy hill at left center and have engulfed a few blue paloverdes that carry over from the old view. Behind the hill and to the left, mesquite, sotol, and beargrass have made some inroads on the only grassland slope that remains. Even there, a tide of mesquite has inundated the lower valleys in Faber Canyon and has filtered up through the lower edge of the Oak Woodland.

Plate 19c (1994) The foreground looks strikingly like that in 1911, the result of a reduction in mesquite cover and an increase in grass cover, both artificially induced. Trees were apparently removed during construction of a drainage ditch near the camera station. The dominant grass at this well-watered site is Lehmann lovegrass, an African native that was introduced on the Santa Rita Experimental Range in 1937. This species has spread aggressively through much of the immediate area and elsewhere in southern Arizona. On the distant slopes, there is as much or even more woody vegetation as in 1962. About the same number of oaks are present in the fingers of woodland extending downslope into the former grassy patches. Other woody species such as one-seed juniper and mesquite have increased among the oaks.

20a

Plate 20a (1887) From a saddle on Proto Ridge, just east of Nogales, this view is northwest across Proto Canyon toward Mt. Benedict, left center, and the Cerro Colorado Mountains, far left. Young mesquites are clearly recognizable around the monument. Oak dominates the runnels in the midground, but the intervening slopes are open and probably grassy. Many of the shrubby-looking plants are probably *Yucca arizonica*. The elevation is 4,000 feet.

Plate 20b (1962) The new camera station is about 30 feet to the west to avoid a dense stand of mesquite. The remains of the monument still exist out of view. Fewer oaks (Emory and Mexican blue) grow along the drainage channels on the other side of the Nogales-Patagonia highway. Mesquite has replaced them and dots many upland sites as well. Ocotillo dominates the south-facing hill at far right and also grows densely on the steep southern exposures below Mt. Benedict.

Plate 20c (1994) In the foreground and on the distant hills, mesquites have grown larger and are joined by many young plants. Several of the oaks along the highway persist. Yucca has increased on the distant slopes. Ocotillo appears about as before. Eight houses have been built near the base and to the right of Mt. Benedict; they appear in plate 21c, as well.

20b

20c

21a

Plate 21a (1887) The view is similar to plate 20, but the camera station is about one-quarter mile farther west. Oak, the dominant, is restricted here to relatively cool, moist sites in ravines and on favorably oriented slopes. The smaller shrubs are probably yucca. The elevation is 4,000 feet.

Plate 21b (1962) Emory and Mexican blue oaks are fewer, although one hardy survivor (an Emory) still stands beside the highway at left center. Mesquite dominates the ravines and shares the slopes with yucca.

Plate 21c (1994) Industrial and residential buildings across the foreground now mark the outskirts of Nogales, Arizona. At least eleven houses, all new since 1962, can be seen near Mt. Benedict. Mesquite appears more numerous, and persistent plants are larger than before. Stands of oak on north-facing slopes appear little changed except that mesquite is more prominent. Yucca at center midground is little changed except the plants are larger.

21b

21c

22a

Plate 22a (1890) This is the first of two photographs taken by George Roskruge from the same hill on the Maria Santissima del Carmen land grant just north of the Mexican Border in the Santa Cruz Valley. The view is to the southwest. Oaks are sparse and restricted to the moister north slope, while the south slope is pure grassland, devoid of shrubs and trees. The open-crowned trees on the summit of the distant hill are mesquite. The other trees right of the summit appear to be oak. The elevation is 3,950 feet.

Plate 22b (1962) The remains of a mesquite persist. In general, the old division of oak on the north slopes and grass on the south has given way to mesquite on the north and light mesquite and ocotillo (none of which is visible here) on the south. The principal grasses are tanglehead and sideoats grama; also present are wild buckwheat, fairyduster, and a species of ragweed.

Plate 22c (1994) The mesquite at midground has died, and its remains are visible on the ground at the base of the monument. Another dead mesquite lies just out of view to the left. The mesquites on the distant hill have increased in size and are joined by a few new plants. Remains of the oaks noted in plate 22a are gone, possibly having been harvested for firewood or fence posts. A few sprouting stumps indicate that mesquite has been cut, too. The number of fairyduster, the dark plant in the foreground, is the same as in 1962. Other plants seen nearby include ocotillo, rainbow cactus, Palmer agave, western ragweed, and sideoats grama.

22b

22c

23a

Plate 23a (1890) The view is north-northwest down the Santa Cruz Valley toward the San Cayetano Mountains at center. The Santa Cruz River enters the photograph at lower right, bends sharply out of sight, and re-enters at right center to follow a course marked by a line of cottonwoods. Several *Agave palmeri* plants are in the foreground, and many more grow along the ridgecrest at midground. The tree at lower left is an oak, as is the one midway along the bare slope beyond it and slightly to the right. The terraces across the valley appear to be covered with almost pure grassland. The elevation is 3,900 feet.

Plate 23b (1962) The Yerba Buena Ranch occupies the valley bottom, which has been extensively disturbed by cultivation. In the foreground are mesquite, prickly pear, grama grass, tobosa, fairyduster, three-awn, and *Amoreuxia palmatifida*. On the hills at left, mesquite occurs with some sotol. Cottonwoods still dominate the stream banks; intermingled with them are yewleaf willow and Goodding willow. The arrow marks a relatively undisturbed section where mesquite, catclaw, and gray thorn predominate. The terraces on the far side of the valley are shrubbier. The river channel is broader and the bend appearing in the right midground sweeps farther left.

Plate 23c (1994) The Engelmann prickly pear in the middle foreground has died, and another now grows at the far left along the fence line. *Agave palmeri* is no longer found at this site. Mesquite has increased in size and number in many places. The golf course fairway at center midground is now dotted by small trees, and the grassy expanse is interrupted by two houses. Fewer cottonwoods line the river channel. Foreground mesquites make evaluation of channel changes difficult.

23b

23c

24a

Plate 24a (1890) From midway up the east terrace of the Santa Cruz River, the view is west toward the hill from which the two preceding sets of photographs were taken. In the foreground, cottonwoods flourish in the channel of the Santa Cruz. Oaks and probably mesquites occur on the far side of the valley. On the hillside, oaks are confined to moister microhabitats. The trees on the skyline at the right appear in the lower left-hand corner of plate 23a. The elevation is 3,800 feet.

Plate 24b (1962) The oaks have disappeared. Mesquite has become established on the slopes across the river and in front of the camera. The Santa Cruz is wider but not yet entrenched in this reach. The branch in the foreground belongs to a dead mesquite. Its death and that of several others on a knoll just beyond the distant hill (plate 22b) are noteworthy because we seldom saw dead mesquites in our photographs.

Plate 24c (1996) The foreground mesquite has fallen to the ground. The riparian trees are larger and more numerous and include velvet ash (right foreground), Goodding willow, and Fremont cottonwood. Mesquite has increased noticeably all over the hill. Many of these plants are so small they must have become established since 1962. A solitary single-seed juniper grows on the far hill just out of view to the right. Other plants seen here are fairyduster, catclaw, desert honeysuckle, gray thorn, and cane cholla.

24b

24c

25a

Plate 25a (1887) This scene shows a small tributary valley of the Santa Cruz River east of Nogales. The view looks northeast from the river's first terrace toward Guajalote Peak, right, in the Patagonia Mountains. From individuals still surviving, the trees can be identified with reasonable accuracy. The wispier ones along the wash, center and right, are adult mesquites. The darker trees on the slopes and in the bottoms are mainly oaks. Palmilla is also visible. The elevation is 3,800 feet.

Plate 25b (1962) Mesquite has increased notably in the uplands as well as in the washes. A single mesquite on the ridge persists from 1887. The oaks have all but disappeared; a white arrow marks the only remaining oak, an Emory. Juniper has increased in density and is increasingly widespread. (The juniper on the far left [black arrow] is visible in plate 25a as a small plant.) Also seen here are palmilla and gray thorn.

Plate 25c (1994) All of the foreground mesquites are larger, and, in most instances, have sprouted from the base. Many new small mesquites have become established near and below the camera station, confirming the impression of increased numbers on the far hills. The many one-seed junipers present in 1962 have persisted and grown. Few if any new individuals have appeared during the intervening thirty-two years. Thus, the population density appears to have stabilized since 1962. The one Emory oak is still present. Other plants noted here are palmilla, *Yucca arizonica*, fairyduster, gray thorn, Engelmann prickly pear, and cane cholla.

25b

25c

26a

Plate 26a (1931) In the lowest reaches of the woodland, Emory oak drops out and Mexican blue oak, the only oak species represented, is confined to small patches on north-facing slopes. This photograph, taken along the Pan-American Highway about 16 miles south of Nogales, shows one of the lower occurrences of Mexican blue oak in 1931. The grasslands in the valley already support a substantial mesquite population. The elevation is 3,650 feet.

Plate 26b (1962) The oaks have disappeared, and mesquites have migrated upslope into the former savanna. Even where cultivation has not destroyed it, the grassland has deteriorated. The proximity of a small brick factory raises the question of woodcutting, as do axe marks on some of the oak stumps. For two reasons, it seems likely that the oaks were chopped down after they died. First, one would expect to find stump sprouts if living trees had been cut, but no sprouts were seen. Second, it is difficult to imagine a woodcutter climbing the slopes to cut and then cure an inferior firewood when living mesquites, equally easy to cure and an excellent fuel, were abundant on the valley floor.[20]

Plate 26c (1996) Regardless of the cause of death, the oaks have not regained a foothold in this former ecotone between Oak Woodland and grassland. The mesquites have continued to increase in number and are especially dense on the north-facing slope formerly occupied by Oak Woodland. In 1962, the season's crop of grasses on the slopes has been fully utilized by cattle. Woody plants such as wait-a-minute and velvet mesquite are the dominants now. Desert honeysuckle is common at the foot of the hill. The field in front of the camera, plowed in 1962, now supports a dense growth of feather fingergrass, three awn, amaranthus, and alkali sacaton. The tree in front of the building, a canyon hackberry, persists from 1962. The tree branches extending into the picture at far left belong to a large Goodding willow.

Changes in the Oak Woodland

Before 1962

The twenty-six sets of plates devoted to the Oak Woodland span a period of 31–75 years and reveal some general trends:

1. At most stations below 4,700 feet, oaks died faster than they became established, and at lower elevations the failure to repopulate was complete.

2. The mortality was most severe at the lower edge of the woodland, where mortality approached 100 percent and the boundary separating woodland from grassland migrated upward.

3. The oak decline post-dated our early photographs.

4. Because of shrub encroachment, the woodland became less open at all elevations.

5. The encroachment, mainly by mesquite, evidently antedated the decline of the oaks.

Out of twenty-six Oak Woodland plates, one showed increase in oak biomass, three showed no increase, and eighteen showed attrition (plates 2, 4–6, 9–10, 12–17, and 21–26). (In four photograph pairs [plates 7, 11, 19, and 20], changes in the oaks could not be evaluated.) In Barrel Canyon, 54 of 134 oaks were dead in 1962 (plate 4). The species involved were Arizona oak and Emory oak, which typically live 200–250 years or longer.[21] In the Southwest, mortality of very-long-lived (>100 years) woody plants should be less than 0.2 percent per year.[22] If all 54 oaks died in the thirty-seven year interval between the original photograph (1925) and its match (1962), mortality in that population was 1.1 percent per year, more than five times the expected value.

Many of our woodland sites showed considerable disturbance and were, therefore, more likely to reflect human impact than long-term climatic trends.[23] Even when we omit the ten plates that depict mines, railroads, cultivated fields, and watering holes (plates 1, 2, 3, 10, 11, 12, 13, 14, 23, and 26), a large proportion of the remainder (twelve of sixteen) showed moderate to severe attrition in the oak population (plates 4, 9, 13, 15, 16, 17, 18, 20, 21, 22, 24, and 25).

Although the decline in oaks appeared limited to lower elevations, other striking changes occurred throughout the Oak Woodland. By the early 1960s, woody plants such as mesquite and juniper often exceeded oaks in number and volume. Remarkably, all of the Oak Woodland stations registered an increase in mesquite.[24] Oneseed juniper occurred at four stations (plates 9, 10, 11, and 25) and increased at all four. Ocotillo occurred at seven stations in the Oak Woodland; it increased at six stations (plates 5, 13, 15, 16, 17, and 18) and remained stable at one (plate 9).

After 1962

During the decades after 1962, the decline in oaks stopped. Of the twenty-six camera stations in the Oak Woodland, seventeen showed no change in oak biomass, five showed an increase, and none declined. (Results for four stations were not tallied because oaks were not clearly shown on the pictures.) Whatever caused the loss of oaks during the early part of the century had mostly ceased operating during the last third. Most of the lost oaks were not replaced by new ones during this period; our post-1962 replicates showed increased density of Emory oak in a single view, taken at El Plomo Mine at 5,500 feet.

The Upward Retreat of the Woodland

Between the late 1800s and 1962, the oaks disappeared completely or nearly so at seven lightly disturbed sites: those shown in plates 15, 16, and 17 at 4,200 feet; in plates 22 and 23 at 3,950 feet; in plate 25 at 3,800 feet; and in plate 26 at 3,650 feet.[25] At the site of plate 25 (3,800 feet), also only lightly disturbed, only one oak remained. (This plant was still alive but unhealthy in 1996.)[26] Although these locations lay at the lower edge of the Oak Woodland in the 1880s, this was no longer true by 1962. In some cases, the nearest oaks could be found only several miles away and several hundred feet higher. Other photographs, as noted above, showed a steep decline in oak numbers below 4,700 feet. From these facts, we concluded that the lower boundary of Oak Woodland had shifted upward.

In examining our original findings,[27] Bahre discerned no overall change in the lower limit of Oak Woodland in southeastern Arizona.[28] The discrepancy between his conclusions and ours can be explained largely by differences in selection of photographs. Of Bahre's nine matched views from the woodland, eight were located above 4,700 feet. We identified oak woodlands as low as 3,650 feet; in fact, nineteen of our twenty-six woodland views were made at or below 4,700 feet. By examining populations at the lower, arid edge of the woodland, we documented its "retreat," whereas Bahre, working in a more mesic, central part of the woodland zone, saw little change. In addition, Bahre based some of his conclusions on aerial photography, a platform that is not always well suited for identification of species. Without field checking at the time of overflight, it might be difficult to recognize a situation in which individual junipers, say, replaced individual oaks in the same position. (See, for example, plate 10.)

Writing in the early 1960s, Hastings and Turner were not yet able to assess the severity of the 1950s drought. This drought, among the most severe on record for our area,[29] would have been extremely stressful for plants that were already growing at their xeric limit. Such plants, growing as stringers on the north-facing slopes of low hills, died in large numbers. Where soil and topography ameliorated droughty conditions, a few trees survived, but the woodland was gone. As we visualize it, the retreat of the Oak Woodland did not involve upslope movement by a continuous fringe of trees but, rather, high mortality in the lowermost, scattered populations. Throughout geologic time, more massive woodland retreats most likely began the same way.

Several life-history traits suggest that oaks will reoccupy these low-elevation sites only slowly, if at all. The seeds require animals, mainly birds, for transportation and burial. Acorn dispersers such as the acorn woodpecker and the Mexican jay are woodland inhabitants and seldom visit areas that lack trees. Acorns lose viability quickly[30] and, in the case of Emory oak, at least, require above-average rainfall for germination and seedling survival.[31] Even if acorns are reintroduced to low-elevation sites, climatic variability ensures that most will perish without becoming established.

Species interactions may delay or deter oak regeneration, too. The current abundance of mature mesquite trees in the lower woodland might prevent the establishment of oaks either because mesquite occupies the best regeneration sites or because it preempts water and nutrients. Furthermore, because mesquite encroachment into the Oak Woodland began several decades before the loss of oaks, resource competition between the two might have played some role in the oaks' decline.

Enrichment of the Woodland

At all elevations, even on undisturbed sites, the modern woodland has become infused by a group of invasive species. Mesquite, the most notable of these, was newly established at all twenty-six Oak Woodland stations by 1962. The upward trend diminished only slightly during the next third of a century, when mesquite increased at twenty-two stations while showing no change at four.[32] Other increasers to 1962 were: beargrass (plates 16, 18, and 19); cottonwood (plates 7 and 8); ocotillo (plates 5, 13, 15, 16, 17, and 18); one-seed juniper (plates 4, 9, 10, and 11); seep willow (plate 7); sotol (plates 18 and 19); and turpentine bush (plate 12). Most of these plants continued to increase during the next third of a century. After 1962, changes in other species identifiable in the Oak Woodland photographs are: Palmer agave increased in two, no change in one; sotol decreased in two, increased in two; ocotillo increased in six, no change in one; one-seed juniper increased at all four stations at which it was found. Our data set suggests that the proliferation of long-lived woody invaders of the Oak Woodland has hardly slowed or stopped since the early 1960s.

Chapter Five The Grassland

Our camera stations within grasslands represent two different assemblages: Semidesert Grassland and Plains Grassland.[1] Semidesert Grassland occupies an elevation range between 3,000 and 4,000 feet but may go higher or lower depending upon microclimate and soils. It is best developed on bajadas and broad areas of gentle relief and lies just below evergreen woodland, chaparral, or Plains Grassland. Where mountain masses rise abruptly from the desert floor, it may be completely absent. Perennial grasses such as black grama, slender grama, spruce-top grama, bush muhly, three-awns, Arizona cottontop, and slim tridens grow with a mixture of shrubs, short trees, and leaf succulents.

Annual average precipitation ranges from about 10 to 16 inches, with slightly more in summer than winter. Summers are hot, and winters are relatively warm. The frost-free period is about 265 days long. Most of the grasses are dormant in winter and spring; they resume growth with the onset of summer monsoons. A few species are active in March and April if rainfall and temperature are favorable.[2]

Plains Grassland occurs at higher elevations where temperatures, especially in winter, are lower. Thus, growth of the dominant plants is confined almost entirely to the summer rainy season. Within our region, Plains Grassland is found on level plains and low hills at elevations as low as 4,500 feet; it may extend upward to elevations as high as 7,000 feet on south-facing slopes where it contacts the evergreen woodland above.

In its maximum development, Plains Grassland may occur as uninterrupted shrub-free prairie or as a luxuriant matrix of grass studded by semiwoody plants like yucca, beargrass, and sotol. Woody species, including fairyduster, Mexican crucillo, all thorn, and several mimosas and acacias, are present in some areas, perhaps in response to soil factors that are at present poorly understood. Grasses commonly encountered include blue grama, side-oats grama, plains love-grass, vine mesquite grass, and wolftail. There is considerable overlap in grass species between Plains and the Semidesert Grassland. The black-tailed prairie dog, once common in Plains Grassland, has been virtually extirpated in our region. Two large prairie dog colonies still occur in our area between Cananea, Sonora, and the south end of the Huachuca Mountains.

The grass flora of these grasslands is exceedingly rich, particularly at higher elevations. The flora of "grasslands" has been described for several areas in our region.[3] More than eighty species were reported for a single site in Plains Grassland.[4] Recent local floras from southeastern Arizona indicate that "sky island" mountain ranges can support a hundred or more grass species.[5] Some variation in species dominance has been noted with elevation: black grama dominating many lowland sites, for instance, and blue grama, the higher elevations.[6]

We have selected photographs from one camera station within the Plains Grassland to depict changes there (plate 27). Plains Grassland is poorly represented in our region. Examples include grasslands around Sonoita, Arizona, and within the San Raphael Valley.

For this discussion, the Semidesert Grassland has been divided into two geographical areas. Fourteen plates deal with the belt centered around the Santa Rita Mountains within the Santa Cruz River drainage (plates 28–41; see figure 4.1). The final eighteen grassland photograph sets portray the grassland of the San Pedro Valley (plates 42–59; figure 5.1). The latter area has typically been invaded by Chihuahuan Desert plants, often Chihuahuan whitethorn; the western area has been invaded by Sonoran Desert species, frequently mesquite. The grass species composition of these two areas is not necessarily different.

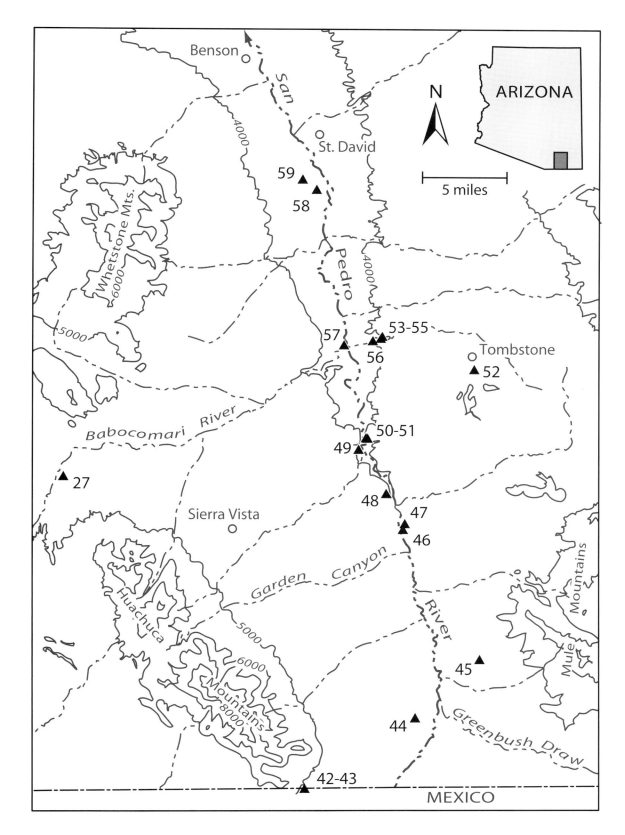

Figure 5.1 Location of photographic stations in the San Pedro Valley grasslands.

27a

Plate 27a (1974) The view is west-southwest across O'Donnell Canyon from a ridge near the north boundary of the Appleton-Whittell Research Ranch. At the time of this photograph, the ranch had not been grazed by livestock for seven years. The dominant grasses are cane bluestem and plains lovegrass, characteristic species of Plains Grassland, with an understory of blue grama, a species found also in the Semidesert Grassland. Several clumps of beargrass and a dead agave are visible in the foreground. The valley floor at midground is dominated by sacaton. Mesquites are the common, widely scattered woody plants on the midground ridge and distant mesa. The Santa Rita Mountains appear in the background. The elevation is 4,650 feet.

Plate 27b (1975) A lightning-started fire, an unusual event for the month of May, swept across the camera station six months after the first photograph was taken. This photograph, taken only a few days after the fire, shows that the top of the midground ridge was not burned, nor was the mesa beyond O'Donnell Canyon. On the sacaton flat, flames missed only the incised channel and a few small areas.

Plate 27c (1995) Twenty years after the fire, the foreground slope is much the same as in 1974 except for the appearance of snakeweed and the addition of low shrubs, partly hidden by the grass. These shrubs include yerba de pasmo and wait-a-minute. Mescal and desert spoon have also increased substantially. Mescal is apparently able to reoccupy burned sites within a few years. Beargrass, rarely eliminated by fire, has changed little in the period spanned by the photographs. Mesquites on the midground ridge and the distant mesa are larger and more numerous than in 1974.

27b

27c

28a

Plate 28a (1895) The view is south-southeast across the valley of Red Rock Canyon east of Patagonia. Level areas and slopes with southerly aspect support grassland; the north slopes support Oak Woodland. On the rounded hills across the midground, there is a dense chaparral that seems to be controlled not by slope and aspect but by edaphic factors, since its occurrence is confined to island-like extrusions of shallow, red soil. In the valley, the lowest terrace is probably covered by sacaton and the flood plain by burrobrush. In the channel, scattered deciduous trees, leafless for the winter, include desert willow, mesquite, and velvet ash. A few large mesquites interrupt the grass cover of the upper terraces. The elevation is 4,450 feet.

Plate 28b (1965) The most dramatic changes have occurred in the old patches of grassland, where mesquite now prevails. The patch of chaparral is apparently unchanged; it is dominated by Toumey oak and pointleaf manzanita and contains no mesquite. Much of the valley bottom is similarly free of mesquite; where the plant does occur here it falls within sharply delineated patches. On the slopes opposite the camera, the Oak Woodland has been altered by the addition of many mesquites. The total number of oaks appears little changed, however.

Plate 28c (1996) The foreground mesquites and those on the slope below the camera station have all increased in size and partially block the view of the valley below. Burroweed has greatly increased in prominence since 1965, partly replacing the former dominant, sacaton. The crowns of the mesquites on the terrace at right midground appear to form a continuous cover with almost no open spaces. Across the valley, the open slopes of a century ago have sustained a steady increase in mesquite biomass since 1965. Populations of oak and manzanita have remained relatively stable through time.

28b

28c

29a

Plate 29a (1895) This view is directly east from the station for the preceding plate. Here, because of the direction of the camera, it is possible to observe more clearly the sharp habitat preference of the oaks for ravines and north-facing slopes, and of the grassland for drier sites. What appear to be large mesquites grow along the base of the hill opposite the camera. The arrows mark two prominent plants that persist to the present; the black arrow denotes a Mexican blue oak, and the white arrow, an Arizona white oak. The elevation is 4,450 feet.

Plate 29b (1964) The presence of large, dead mesquites on the hill in 1964 supports the assumption that this plant was already well established in 1895. Aside from the notable increase in mesquite, two other changes, mostly out of the view, have taken place. A high proportion of the Mexican blue oaks standing in sheltered habitats along the ravine below the photo station are dead. Also, small one-seed junipers grow beneath the oak skeletons, an invasion that is evident elsewhere in the panorama. (Arrows mark a few of the many junipers visible.) As in the preceding photo pair, the creek bed appears narrower and has more pronounced meanders in 1964 than in 1895.

Plate 29c (1996) Mesquites that were too low to be in view have grown and now block part of the scene. The plant just right of center in 1964 has died. Off camera, several new mesquites have become established. For mesquite, as for most long-lived, woody plants, mortality is generally low. The co-occurrence of dead plants and young recruits suggests that the population may be approaching equilibrium. The slopes across the valley are more densely covered by junipers and mesquite than before, although comparison is made difficult because of the different seasons and sun angles of the two photographs. The channel of Redrock Creek is wider, its course is altered, and several large trees growing near its bank have disappeared. The dense patch of chaparral midway up the slope at far right appears little changed since 1895.

29b

29c

30a

Plate 30a (ca. 1899) This view faces west across the old mining camp of Helvetia, then in its heyday, from the northwestern foothills of the Santa Rita Mountains. Huerfano Butte stands alone at left center; across the Santa Cruz Valley, but not visible in the old photograph, are the Sierrita Mountains. The ground cover in the foreground appears to be largely amole, the dried inflorescences of which are more prominent than the plants themselves. Interspersed with them are larger clumps of beargrass, a few yuccas and, along the lower edge at the right, mortonia. The dark trees at the left are Emory oak. A single ocotillo appears in the right midground. The elevation is 4,400 feet.

Plate 30b (1960) Both mesquite and ocotillo have registered significant increases, the former along ravines and in the level grassland, the latter on south-facing hillsides and in the midground at the right. Most of the shrubs on the light-colored slope between the camera and the building with the pyramidal roof are mortonia. This plant, too, seems to have undergone an increase. The Emory oaks have grown but otherwise are about the same. The foreground is still dominated by amole and beargrass, with sotol and *Rhus choriophylla* also present. Mexican crucillo, gray thorn, catclaw, and a few isolated blue paloverdes grow on the hillsides.

Plate 30c (1994) Only ruined adobe walls remain where the buildings once stood, and the site has been colonized by mesquite and snakeweed. New signs of human habitation have appeared across the Santa Cruz Valley, where mine tailings and the town of Green Valley are faintly visible. Throughout the view, a number of plants persist, including individuals of Emory oak, ocotillo, mortonia, beargrass, and yucca. Shindagger and beargrass still dominate the foreground.

30b

30c

31a

Plate 31a (ca. 1899) The skyline has been retouched in this view looking southeast across Helvetia toward the Santa Rita Mountains. The camera station for plate 29 is in the left midground. Helvetia lies just below the border of the Oak Woodland; oaks dominate the slopes of the Santa Ritas and, judging from relics, reach down into the valley along the two prominent ravines. A few isolated individuals grow at favorable locations in the grassland. As was the case with El Plomo Mine (plates 1–3), it is difficult to generalize about the tree population because the settlement and smelter (right), must have required large amounts of firewood. The elevation is 4,300 feet.

Plate 31b (1960) The most pronounced change has been the obliteration of the grassland community by mesquite on the tablelands at the center and in the midground at the right. In the foreground, Arizona white oak, mesquite, catclaw, gray thorn, desert hackberry, Mexican crucillo, *Sageretia wrightii*, and *Ceanothus greggii* contribute to increased shrubbiness along the wash. The hill in the midground at the left, a limestone intrusion, supports amole, beargrass, and yucca as well as calciphiles such as Arizona rosewood, mortonia, and *Ceanothus greggii*. The Emory oaks scattered across the bajada seem to have changed little; many of the same individuals are present.

Plate 31c (1994) Mesquite and snakeweed have colonized the highly disturbed foreground. Low shrubs on the ridge at midground have increased in density. Elsewhere in the view, the density of trees and shrubs seems about the same as in 1960, as does the species composition of the plant community.

31b

31c

32a

Plate 32a (ca. 1899) In this view, almost due north across Helvetia, the group of three tents at the right appear in plates 30 and 31. Of greatest interest at this early date is the presence in the fore- and midground of half-grown and adult mesquite trees. Hillslopes in back of town at the right of the photograph appear open. Arrows indicate eight large shrubs: a mesquite (extreme right), two Mexican crucillos (left), and Arizona rosewood (all others). The elevation is 4,300 feet.

Plate 32b (1962) The large increase in mesquite along the channel is striking. Prickly pear, burroweed, and yucca are the most prominent associates. Amole forms the general matrix for plant life on the hill and is responsible for the grainy, gray appearance of the slopes. In its midst are several localized colonies of ocotillo, mortonia, and sotol. Ocotillo has probably increased, although the lack of detail in the old photograph makes comparison difficult. At the bottom of the hill, a colony of mortonia occupies a triangular area that was much more open in the old photograph. This spot appears to have changed more than any other part of the slope. A single one-seed juniper, possibly present in the old photograph, appears to the left of the triangle.

Plate 32c (1994) The shrubs marked in the 1899 photo are still present in 1994, demonstrating that Mexican crucillo, mesquite, and Arizona rosewood are long lived. As in 1962, the most abundant plant on the hillslope is shindagger; also present are mortonia and ocotillo. The density of mesquite seems unchanged. Many of the same plants appear in both the 1962 and the 1994 views. Few perennial plants have colonized the mined area at left midground.

32b

32c

33a

Plate 33a (ca. 1891) This view of the Santa Rita Mountains is looking east-southeast across a sloping valley dissected by many ravines. In the midground is Hacienda Santa Rita, an early mining operation that was abandoned in 1861 because of warring Apache. At the time of this photo, about thirty years later, the camp had been reoccupied. Even at this early date, mesquite is conspicuous in the upland vegetation. Many mesquites appear at right as a dark wedge above the buildings. The white arrow points at another mesquite. Black arrows mark what are presumed to be oaks. The tree with white branches in the left foreground is a velvet ash. Branches in the right foreground belong to a netleaf hackberry. The elevation is 4,150 feet.

Plate 33b (1962) The collection of dwellings has all but disappeared among the many mesquites that now dominate the scene. What was formerly grassland sparsely studded with woody plants is now a woodland sparsely covered with grass.

Plate 33c (1994) To avoid woody vegetation, the camera has been placed forward a few feet. Mesquite is even more prominent than in 1962. A number have died and fallen. The white object on Mount Hopkins is the Smithsonian Institution's multiple mirror telescope, completed in 1979.

33b

33c

34a

Plate 34a (1887) From Proto Ridge near Nogales, this view is east across the Santa Cruz Valley toward Guajalote Peak, right center, in the Patagonia Mountains. No grass can be seen on the rocky slope in front of the camera. The slopes are bare except for an open patch of ocotillo on the nearest hill, which faces south. Several head of cattle graze on the barren slope below. The elevation is 4,100 feet.

Plate 34b (1962) The camera station has been shifted 20 feet north to get an unobstructed view around the ocotillo thicket inhabiting the earlier camera station. The density of ocotillo has markedly increased on the south-facing hill in midground. Mesquite has invaded the ridges, north-facing slopes, and ravines. South-facing slopes are largely free of mesquite. Because the camera was aimed too high for this photograph, the soil surface near the camera is not shown. Crowns of two mesquites, part of the north slope population, appear at the lower edge of the picture. The smaller shrub is fairyduster. Sprucetop grama and sideoats grama are the principal grasses.

Plate 34c (1994) The camera has been returned to the original site to avoid ocotillos at the 1962 camera station. Several foreground rocks are the same as in 1887. On the north-facing slope below the camera, mesquites form a dense, almost continuous, canopy. The south-facing slopes are still relatively free of mesquites; nonetheless, the number has tripled since 1962. The patch of ocotillos has expanded slightly since 1962 and appears more dense.

34b

34c

35a

Plate 35a (1915) From a hill west of Highway 83 between Tucson and Sonoita, the view is southeast across Davidson Canyon toward the highest peak in the Empire Mountains. In the foreground, the dominant shrub is whitethorn, with mesquite, ocotillo, an agave, a yucca, and two species of *Opuntia* also recognizable. In the midground, yucca and Mexican crucillo predominate on the uplands with some mesquite along ravines. Beyond Davidson Canyon, the western bajada of the Empires is generally open and has scattered mesquite, whitethorn, and yucca. The elevation is 4,100 feet.

Plate 35b (1962) The fore- and midground vegetation is less open than before, but has the same constituents. In fact, many individual Mexican crucillo bushes are identical in the two photographs. Sotol, turpentine bush, and ocotillo can also be seen in the midground; tobosa, snakeweed, and sideoats grama occupy the spaces between shrubs. A few one-seed junipers are scattered across the bajada amidst abundant mesquite, catclaw, whitethorn, and prickly pear. The trees near the mountain summit (5,500 feet) are mostly Mexican blue oak, many of them dead, with an admixture of mesquite. The proportion of oak to mesquite decreases downward until the last oak (see arrow), a dying Mexican blue oak, is reached at about 4,500 feet. From there to the base of the hill at 4,200 feet, all trees are mesquite. Clumps of beargrass dominate hillside A.

Plate 35c (1994) Except perhaps for the upper reaches of the mountain slopes, woody plants, especially mesquite, have increased throughout the scene. Catclaw, whitethorn, and prickly pear are still important members of the vegetation. Plants of Mexican crucillo have increased in size but not in number since 1915. Recent impacts include the houses near the base of the mountain and the new road that provides access to them.

36a

Plate 36a (1915) Looking west-southwest toward Mt. Fagan, left center, in the Santa Rita Mountains, the camera station is near Davidson Canyon. Prickly pear and mesquite trees, many of them mature, can be identified in the wash. The slopes appear to be open Semidesert Grassland with a few mesquites, some ocotillo, and many Mexican crucillo. The elevation is 4,050 feet.

Plate 36b (1962) Both the wash and the uplands are brushier. Along the drainage, many mesquite trees are common to both photographs. The abundant and varied flora includes catclaw, whitethorn, desert hackberry, netleaf hackberry, gray thorn, desert honeysuckle, burroweed, snakeweed, and Wright lippia. On the uplands, Mexican crucillo enjoys about the same density as before. Many carry over from the earlier photograph. The number of ocotillos has markedly increased, especially on rocky, south-facing slopes. Mesquite has also proliferated. The most important upland grasses are tobosa, sprucetop grama, and sideoats grama. Other common hillside plants are prickly pear, fairyduster, desert holly, and velvet-pod mimosa.

Plate 36c (1994) The formerly unvegetated watercourse is now occupied, largely by desert broom and sacaton. Their density suggests that no heavy flooding has occurred in recent years. The nearly continuous plant cover of the valley floor is less open than it was in 1962. On the slope, biomass of woody plants such as mesquite and wait-a-minute has increased. Mexican crucillo, the dark plants in the earliest photograph, have not changed in number, although individual plants have grown. Engelmann prickly pear is much more numerous on the slope than before.

36b

36c

37a

Plate 37a (1909) The Helena Mine, located about 1 mile west of the road between Sonoita and Tucson on a tributary of Davidson Canyon, is seen in this view looking toward the north-northwest. The mine was established in 1894[7] and is still being worked. Note the ricks of firewood stacked near the mill. The spring in the canyon bottom supports cottonwood trees. The slopes are open, show little grass cover, and support a scattering of mesquites, one-seed juniper, and, in one minor canyon at upper left, oaks. The oaks at this site, on the flanks of a mountain composed mainly of rhyolite, are at their lower elevation and are found here only in relatively protected moist north-facing ravines and slopes.[8] The elevation is 4,600 feet.

Plate 37b (1973) Little is left of the abandoned mine except scattered construction debris and the mine shaft opening, which is out of view to the right. Woody plants, including mesquite and ocotillo, are much more abundant than before, and grass cover is greater. Close study will show that many shrubs have persisted, including the juniper seen part way up the slope at right and the mesquite above it on the horizon. The oaks are gone, as are the cottonwoods, whose remains lie on the ground in the canyon bottom. Shindaggers give the slope at right its dappled appearance. The top of a banana yucca is visible at the bottom just left of center.

Plate 37c (1994) Only twenty-one years later, some noteworthy changes have taken place. Coverage of woody plants has increased and prickly pear cactus has greatly proliferated in the view and even more so on many nearby slopes. Two oak carcasses lie on the ground in the minor canyon at upper left. Neither has axe marks. The persistence of many large mesquites and the presence of oak carcasses suggests that the mine did not rely upon local wood.

37b

37c

38a

Plate 38a (1911) From a ridge about half a mile southwest of the camera station in plate 19, the view is southwest toward Elephant Head on the westernmost escarpment of the Santa Rita Mountains. The long outwash of the Santa Rita bajada drains left to right toward the Santa Cruz River, about 8 miles away. When David Griffiths called attention to mesquite encroachment in 1904, he was referring to this area (see introduction). Young mesquites and young ocotillos are apparent in the foreground. Juvenile and half-grown mesquites have already become established in the grassland as well; by and large, however, the bajada still appears open and grassy. What may be oaks can be seen protruding from above the banks of the wash in the midground at the left (arrows). The elevation is 3,900 feet.

Plate 38b (1962) The new camera station is a few feet away from the old. A rock (arrow 1) and the mesquite behind it can be recognized in both photographs. The foreground ocotillo and mesquite have reached maturity. Kidneywood, catclaw, fairyduster, sotol, hopbush, velvet-pod mimosa, *Opuntia chlorotica*, several species of grama grass, and three-awn also grow on the hillside. On the bajada, the proliferation of mesquite is obvious. The fairly open area marked by arrow 2 now contains hopbush, desert broom, and sotol. Arrow 3 indicates a dense and recent growth of ocotillo. The oaks highlighted in the previous plate cannot be seen.

Plate 38c (1996) In the foreground and on the distant bajada, mesquites are larger but not more numerous. Perhaps by 1962 mesquite had reached its maximum density for this habitat type. The area marked by arrow 2 in plate 38b is a dense growth of hopbush with many small plants among the more scattered old ones. A visit to the wash with the three oaks highlighted in plate 38a revealed that the tree farthest to the right is a Mexican blue oak and is still alive, but with many dead branches; the remains of the other two oaks are on the ground, dead. Engelmann prickly pear is present in increased numbers at the site. Desert spoon is more abundant now than before at the camera station. Other plants noted in the camera vicinity are hopbush, Palmer agave, fairyduster, ocotillo, catclaw, Wright buckwheat, western coral bean, rainbow cactus, and pincushion cactus.

38b

38c

39a

Plate 39a (1892) This view is north-northeast toward the Cerro Colorado Mine, worked by the Arizona Mining Company before the Civil War. J. Ross Browne wrote in 1864, "The headquarters lie on a rise of ground, about a mile distant from the foot of the Cerro Colorado, and present at the first view the appearance of a Mexican village built around the nucleus of a fort. . . . The works are well protected by a tower in one corner of the square, commanding the plaza and various buildings and storehouses, as also the shafts of the mine which open along the ledge for a distance of several hundred yards. . . . At the time of our visit it was silent and desolate—a picture of utter abandonment. The adobe houses were fast falling into ruin; the engines were no longer at work."[9] Despite the mine, there are many woody plants and a number of large, presumably old, mesquites. The site lies near the lower edge of the grassland. The elevation is 3,650 feet.

Plate 39b (1962) The camera station is not in exactly the same place. The rock formation in the right foreground of the old photograph is farther left and more distant from the camera. Although the grass on the foreground slope is still dense, the grassland in general is shrubbier. Among the plants that have recently increased are ocotillo, catclaw, and mesquite.

Plate 39c (1996) The camera is in the same place as in 1962. Erosion of the adobe structures at the Cerro Colorado Mine has left only remnants, barely visible in the background. Overall, there has been an increase in woody plant biomass. The ocotillo to the right of center is gone. This species has increased in number, especially across the midground. Several catclaw plants are conspicuous across the foreground. Besides mesquite, other plants present are snakeweed and fairyduster.

39b

39c

40a

Plate 40a (1891) Looking southeast across Tumacacori Mission, this camera station is on the west terrace of the Santa Cruz River. The San Cayetano Mountains are in the background. This was the site of a Pima village when first visited by Eusebio Kino in 1691. A mission visita was established here six years later. This was an important ecclesiastic and agricultural center for more than a century before the structure seen here was built (1800 to 1822); it was abandoned in 1840 as the result of Apache pressures.[10] By the time traveler J. Ross Browne visited Tumacacori in 1864, a half century had passed since it had been the hub of extensive cattle operations on the Tumacacori and Calabasas land grant. Browne noted: "The remains of the acequias show that the surrounding valley-lands must have been at one time in a high state of cultivation. Broken fences, ruined-out buildings, bake-houses, corrals, etc. afford ample evidence that the old Jesuits were not deficient in industry." Disturbance during the Spanish and Mexican periods, in other words, reached a fairly high level at this location. It is doubtful that Anglo-Americans exceeded or even reached that level. Browne observed that "[a] luxuriant growth of cotton-wood, mesquit, and shrubbery of various kinds" grew along the Santa Cruz River. The large cottonwoods in this photograph are probably the same individuals. They almost certainly grow in an entrenched channel that formed well before Browne's visit. Mesquite has colonized the abandoned fields. From the open condition of the terrace slopes, it is apparent that Semidesert Grassland prevailed on the uplands, even at this low elevation of 3,350 feet.

Plate 40b (1962) The mission, now a National Monument, appears to be in a better state of preservation than before. Apart from recent cultivation and loss of most of the large cottonwood trees along the Santa Cruz River, the valley floor looks about the same. The terrace slopes, however, are now dominated by whitethorn and jumping cholla, and the open, grassy appearance is a thing of the past. The modest vegetative changes apparent in 1891 foreshadowed the more extensive ones evident in 1962. A decline in the water table probably caused the loss of cottonwood trees.[11]

Plate 40c (1994) Mesquites have grown and partly blocked the view of disturbed ground immediately below the camera station. Overall, there is more woody plant growth throughout this panorama of former grassland. Virtually the entire sweep of country from the Santa Cruz River to the base of the background mountains is being developed. One new house is visible across the river, and new roads crisscross the slopes. Riparian trees are much more abundant than in 1962, the result of two unusual circumstances. First, in 1972, the sewage treatment plant serving Nogales, Sonora, and Nogales, Arizona, began to release effluent into the Santa Cruz, assuring a constant flow of surface water in a reach where the water table had dropped 20–30 feet.[12] Second, most of the trees in this view became established after the flood of October 8, 1977, which destroyed much of the earlier riparian community. Vigorous growth of these trees is assured by year-round flow of sewage effluent.[13]

40b

40c

41a

Plate 41a (1890) This view, taken little more than a mile north of plate 39, is east across the Santa Cruz River toward (from right to left) the San Cayetano Mountains, the Grosvenor Hills, and the Santa Rita Mountains. White-thorn grows in the foreground to the left of the rock monument. It is a shrub of the uplands here and, by inference, across the valley as well. At the base of the foreground hill, there are mesquites and other shrubs, probably including catclaw and gray thorn. This reach of the Santa Cruz River, lined by widely spaced cottonwoods and mesquite, probably carries surface water only intermittently. What appear to be cultivated fields are marked by areas where vegetation has been cleared from the valley floor. The dark band of vegetation on the far side of the valley is a mesquite bosque. The low hills across the valley appear dark as though covered by shrubs. The elevation is 3,250 feet.

Plate 41b (1964) The most conspicuous changes have to do with buildings and cultivated land. Carmen, Arizona, and its cotton fields are new, but little change has taken place in the native plant cover. The mesquite bosque across the valley is no more prominent than in 1890, and the many large trees present attest to its great age. Any changes that may have occurred on the hills across the valley are not easily discerned, but the small area of foreground hillside that can be seen clearly shows an increase in shrubs, mainly whitethorn. No cottonwoods appear along the entrenched channel.

Plate 41c (1994) Although the foreground has been destroyed by construction of Interstate 19, much useful information remains. Many of the cultivated fields have been abandoned and are now occupied by a dense mesquite forest, showing the resilience of such lowland habitats. The eroded, mainly dry channel of thirty years ago now carries a perennial stream of sewage effluent and, as a consequence, supports a dense growth of cottonwoods and willows. The riparian trees mostly became established after the flood of 1977, which widened this reach of the river channel and produced a broad swath of open sandy soil, ideal habitat for cottonwood and willow establishment.

41b

41c

42a

Plate 42a (1891)   On the International Boundary at a point just west of the San Pedro River, the camera faces southeast into Mexico. A minor tributary flows toward the San Pedro River. The Sierra San José lies beyond. Describing this view more than three decades earlier, W. H. Emory wrote: "At this point, approaching from the east, the traveler comes within a mile of the river before any indications of a stream are apparent. Its bed is marked by trees and bushes, but it is some sixty or one hundred feet below the prairie, and the descent is made by a succession of terraces. Though affording no very great quantity of water, this river is backed up into a series of large pools by beaver-dams, and is full of fishes. West of the river there are no steep banks or terraces, the prairie presenting a gentle ascent."[14] The slopes of the terrace along the east side appear, like the valley, to be grassy and open. Palmillas inhabit the basin in front of the camera. The other shrubs may be Mexican tea or, in the case of the larger ones, mesquite. The elevation is 4,350 feet.

Plate 42b (1962)   Chihuahuan whitethorn instead of grass dominates the sides of the distant terrace beyond the valley. The deeply scoured river channel is marked by a line of trees, denser than before, which in addition to cottonwood may include Goodding and desert willow, velvet ash, and walnut. A dense growth of mesquite and sacaton blankets the flood plain. In the small basin are palmillas, several of which persist from the earlier view, as well as mesquite, desert willow, rabbitbrush, and sacaton. The curved spit of high ground half encircling the basin supports Mexican tea, wait-a-minute, threadleaf groundsel, mesquite, fairyduster, blue grama, hairy grama, *Aristida glauca*, and a species of blue stem.

Plate 42c (1994)   Velvet mesquite has continued to increase and is the dominant woody plant at the camera station and beyond. The conspicuous palmilla in the previous view is now gone. Many of the mesquites have developed multiple stems where branch tips were killed by frost. The dominant grass covering the foreground hill is Lehmann lovegrass, a recent arrival in this vicinity.[15] Along the valley of the San Pedro, the dominant trees are cottonwoods, and the understory is sacaton.

42b

42c

43a

Plate 43a (1891) The camera station is 50 feet east of the preceding station and looks west up the bajada of the Huachuca Mountains along the International Boundary. The wash at the left can be seen at the right in plate 41. Palmillas grow in the wash. The grove to the left of and behind the surveying party is probably desert willows or soapberry. Emory, thirty-seven years before the date of this photograph, described the bajada as "composed of hard gravelly soil, and supporting a close sward of grama grass, giving a peculiarly smooth shorn look to the general face of the country."[16] The scarcity of grass in this view probably reflects the combination of overgrazing and drought that decimated cattle herds all over the region in the 1890s.[17] The elevation is 4,350 feet.

Plate 43b (1962) This photograph was taken on the Mexican side of the fence and looks toward Monument 98, which has replaced the old rock cairn. The foreground, almost unchanged, supports wait-a-minute, several small mesquites, and some threadleaf groundsel, a plant that is seldom grazed and is toxic to cattle. Although mesquite, desert willow, and desert broom choke the wash, woody plants have increased on the uplands only slightly. The grasslands of the upper bajada here and those of the rolling country around Sonoita are the only major ones in southeastern Arizona that remain free from brush and substantially unchanged. In the background, Oak Woodland grows at the foot of Montezuma Peak and extends down into the bajada along drainages.

Plate 43c (1995) The encroachment of woody species, mainly mesquite, has continued to the present. Lehmann lovegrass is the most abundant grass on both sides of the fence. Out of view, about 100 yards to the north, is one of several houses in a new housing development that abuts the International Boundary.

43b

43c

44a

Plate 44a (1891) The view is southwest from the camera station on the Green Ranch, located on the east side of the San Pedro River about 2.5 miles north of the Mexican Boundary. At far right are the Huachuca Mountains. The low hills in the midground support a number of shrubs, including one-seed juniper and little-leaf sumac. In the valley at midground, small palmillas are scattered amidst grasses, probably a mixture of tobosa and sacaton. The rocky soil here lacks any tall grass cover, although low tufts of grass occur among the rocks. To the right of the hills and beyond the house is a light-colored field of grass (sacaton?), a common feature of the valleys of this region during this period. In it are a few scattered trees, possibly cottonwoods, growing near the San Pedro River, which flows year round through this reach.[18] The elevation is 4,250 feet.

Plate 44b (1962) Two species account for most of the changes: mesquite is the dominant woody plant of the valley at midground, and beyond that, Chihuahuan whitethorn has moved onto the upland and onto hillsides, imparting a characteristically dark cast. On the foreground hill are curly mesquite, black grama, and sprucetop grama, species that were probably present seventy-one years earlier. In the valley appear mesquite, palmilla (with the conspicuous white inflorescences), rabbitbrush, tobosa grass, and sacaton. In the uplands between the valley and the hills are little-leaf sumac and one-seed juniper, some persisting from 1891, and Chihuahuan whitethorn. Two large cottonwoods, seen slightly behind the hill at far right, mark the course of the San Pedro River, which has undergone considerable channel erosion since the previous photograph[19] and now flows only intermittently.[20]

Plate 44c (1998) Mesquite has taken over the camera station, necessitating pruning of one plant to permit a clearer view of the scene. The mesquites are more numerous than before. Many show the multistemmed growth habit typical of frost-damaged plants. Sacaton is still the dominant large grass in the swale at midground; beyond it, at center, is a patch of four-wing saltbush. Erosion has removed a large amount of soil from the base of the hill below the camera, and grasses are still scarce here. The downcut and widened channel of the San Pedro River has reached near-equilibrium[21] and now supports a dense continuous forest of cottonwood and willow trees. This site is just within the San Pedro River Riparian National Conservation Area, established in 1988 and administered by the Bureau of Land Management. Several houses have been built recently on private land off camera to the left.

44b

44c

45a

Plate 45a (1891) The north end of the Huachuca Mountains as seen from across the upper San Pedro Valley. The view is west down the broad valley of Spring Creek. The vegetation is a mosaic of shrubs, probably including little-leaf sumac, Mexican crucillo, fairyduster, and Chihuahuan whitethorn, and more open areas where a sparse covering of grasses is interrupted by blue yucca, palmilla, and sotol. The elevation is 4,400 feet.

Plate 45b (1962) After seventy-one years the patches of grassland are gone and the shrubby sites are even shrubbier. The camera station is covered by Chihuahuan whitethorn; the small valley in the foreground supports mesquite, desert willow, little-leaf sumac, and sacaton. The three relatively large trees growing along the channel at center midground are Arizona walnuts.

Plate 45c (1994) The camera station supports more grass than at either of the earlier times, and the foreground shrubs of little-leaf sumac and Chihuahuan whitethorn are smaller than in 1962, suggesting that a fire has burned across the site. In the runnels of the uplands is whitethorn, and on adjacent intervening areas are found such typical Chihuahuan Desert species as Chihuahuan whitethorn, mariola, tarbush, zinnia, coldenia, and little-leaf sumac as well as ocotillo, desert spoon, and Palmer's agave. The dominant grass of the valley is sacaton; the shrubs include sotol, burroweed, and velvet mesquite; the trees are Arizona walnut and desert willow. This site is currently protected from development by its status as Arizona State Trust Land; the nearest housing development is on private land perhaps 2 miles to the east.

45b

45c

46a

Plate 46a (1891) From a point on the east side of the San Pedro River about 1.5 miles southeast of Lewis Springs, the view is north toward the Lewis Hills. Bronco Hill appears left of center. The photograph shows a classic but heavily grazed grassland community. The only hint of the nearby San Pedro River is two large trees at far right. This approach to the river seems little changed since 1846 when Captain Phillip St. George Cooke, of the Mormon Battalion, wrote, "[We] saw a valley indeed but no other appearance of a stream than a few ash trees in the midst. . . . On we pushed, and finally, when twenty paces off, saw a fine bold stream."[22] Palmilla, the large dominant, is highly susceptible to fire injury because of the dead flammable material along its stem. Although not necessarily fatal, fires induce a different form, with multiple sprouts being sent up from the uninjured subterranean stem. Many of the yuccas have shaggy stems and are not clumped, indicating that they have not been exposed to fire. Several in the view are tall enough to be several decades old, suggesting that part of the area has not been burned in recent years. The elevation is 4,050 feet.

Plate 46b (1960) We moved the camera station to within 100 yards of the old one, which cannot be located precisely. The railroad cut and the dirt road to Lewis Springs have disturbed the surface so badly that it is almost unrecognizable, but, in spite of the disturbance, three major encroachments can be distinguished: beyond the road, Chihuahuan whitethorn has replaced the grassland; a mesquite bosque dominates the lower-lying area between the acacia and the river; and cottonwoods line the channels of the San Pedro River and Government Draw, which enters from the right and joins the San Pedro just to the left of the tall knob in the Lewis Hills. In the foreground at the right is Chihuahuan whitethorn; at the left, the dried inflorescence of a palmilla. In and along the cut, sacaton, Mexican tea, gray thorn, mesquite, and Chihuahuan whitethorn are apparent.

Plate 46c (1998) Judging from the positions of the fence posts and joints along the rails relative to the background peaks, the camera position is the same as in 1960. Although the cottonwood trees of the riparian strip are at least as abundant as before, those with light-colored leaves were killed by a fire about five months earlier in May 1998. The palmillas in the railroad right-of-way are new since 1960. The one next to the railroad tracks at left is about 7 feet tall. The utility poles and power lines have been removed.

46b

46c

47a

Plate 47a (1891) The photograph was taken about 0.3 mile north of the preceding one, with view to the north-northeast. The camera station is on the border between the yucca grassland community of plate 45 and the low-lying plain surrounding the San Pedro River, which is about 0.5 mile away. Tree-lined Government Draw joins the San Pedro River, as in the previous view. A palmilla is in the foreground, and three cottonwoods can be seen in the midground. Other midground plants might include burrobrush and small mesquites. The elevation is 4,050 feet.

Plate 47b (1962) Only mesquite branches can be seen from the old station. The new one, perched on a dike 50 feet away, looks out across the top of the bosque that now blankets the plain. Among the treetops at left is a relict palmilla. On the floor of the bosque are sacaton, burroweed, desert holly, cane cholla, Mexican tea, and palmilla. Burrobrush occupies some areas closer to the river. The cottonwoods, partly leafless in December, have greatly proliferated since 1891.

Plate 47c (1998) The same May 1998 fire noted in the previous plate burned across this valley bottom. The fire-killed trees appear white in this October view. These deaths shift the riparian community toward its condition in 1891. The San Pedro River is a perennial stream along this reach. The camera is moved to the right a few feet to avoid velvet mesquite branches. Many of these mesquite plants show evidence of frost damage, presumably from the severe freeze of December 1978. In the spaces among the woody plants grow sacaton, burroweed, four-wing saltbush, and Mormon tea.

47b

47c

48a

Plate 48a (1891) This camera station is 1 mile north of the preceding plate's station on the Escapule Ranch. In this easterly view, the San Pedro River, visible between the second line of shrubs and the railroad tracks, follows what appears to be a channel with eroded banks. The absence of large riparian trees makes it difficult to follow the river's course. The valley floor supports a grassland of low tussocks that may be sacaton and tobosa. The bands of small dark trees flanking the river channel and growing at the base of the foreground hill are probably frost-damaged mesquites. Although the hills are relatively open, they support a prominent shrub stratum, varying greatly in density. The elevation is 4,050 feet.

Plate 48b (1962) The river channel, outlined by cottonwood, willow, and mesquite, is deeply entrenched in the valley just beyond the cultivated fields, which have replaced the former grassland. The hills to the east are covered by shrubs, among which are Chihuahuan whitethorn, creosote bush, and ocotillo.

Plate 48c (1994) The valley floor is much changed. In the former pasture at right are many mesquites. A few tussocks of sacaton are present and, but for the mesquites, might signal a return to the grassland of the previous century. The cultivated field on the left appears to have been abandoned long enough for some mesquites to become established. The trees of the riparian strip are mostly larger and more numerous. This linear community of large, closely spaced trees exemplifies the gallery forest that has come to dominate the San Pedro River Valley during the twentieth century. No detectable change is seen in the shrub cover on the distant hills.

48b

48c

49a

Plate 49a (1880) This camera station is a few miles downstream from the camera station for the preceding plate. The camera points east-southeast toward Bronco Hill and the Gird Dam, 1.5 miles above the old town of Charleston. The dam, built about one year before this photograph was taken, supplied water to the mills at Charleston for processing the ore from the Tombstone mines. The dam was destroyed by a flood on July 26, 1881. Extensively redesigned and rebuilt by March 1882, it was washed away for the last time by floodwaters in July 1887.[23] The tree at the upper left (arrow) is probably an Emory oak. More are scattered across the moister slope on the other side of the ridge.[24] They are at the dry lower edge of their elevational range here. Their presence on slopes away from moist valleys is surprising and can be best explained as a relict from the time when Oak Woodland extended across the San Pedro Valley from the Mule and Dragoon Mountains on the east to the Huachucas and Whetstones on the west. The elevation is 4,000 feet.

Plate 49b (1960) The modern steep-walled channel is obscured by trees. Erosion has claimed nearly all of the ground on which the wing of the dam rested (the left side of the photograph). Cottonwood and Goodding willow line the river; seep willow and Mexican devilweed grow abundantly on the sandy bed. The rocky hill in the foreground supports a distinctive community of Wright lippia and Chihuahuan whitethorn, an association that also dominates the hills in the right midground and the left background. On the slopes of Bronco Hill, center, sotol joins the other two plants. In all locations, the vegetation is appreciably denser than before. Along the floor of the little valley in front of the camera, mesquite and Chihuahuan whitethorn are the dominants. Also visible in the photograph are ocotillo, desert broom, chamiso, soapberry, and sacaton. Except for one survivor, the oaks have vanished. The nearest ones occur perhaps 10 miles away and nearly 1,000 feet higher in elevation. At left center, a short section of track of the Arizona and Southeastern Rail Road is visible. This railroad was built by the Phelps Dodge Corporation in 1889 to connect the mines at Bisbee with an existing railroad line at Fairbank.[25]

Plate 49c (1994) Large woody plants that appeared in 1960 still dominate the scene. The cottonwoods and Goodding willows along the San Pedro River are larger, as are many of the velvet mesquites below the camera. Most of the larger mesquites have dense branch clusters or "witches brooms" that are the result of frost injuries probably sustained during the catastrophic freeze of 1978. More sacaton appears with the mesquites below the camera. This reach of the San Pedro River Valley is now part of the San Pedro National Riparian Conservation Area, which was recently established to protect the new riparian forest and its surroundings. The railroad track is now abandoned.

50a

Plate 50a (ca. 1885) From a hill north of Charleston, the view is south-southeast across the town of Millville and toward Bronco Hill. The San Pedro River is out of the view to the right. The Gird and Corbin Mills (left midground and left background, respectively) got water from Gird Dam, which is shown in the previous plate. The flumes shown here conveyed tailings to the pond depicted in the next plate. Two plants of sotol appear at lower left; the grass across the foreground is either cane beardgrass or Arizona cottontop. Ecologically, the point of most interest is the open, grassy appearance of the rolling country on the far side of town. Across the river, Charleston also lies in grassland. The elevation is 4,050 feet.

Plate 50b (1960) The only structures remaining are the mill foundations and the adobe walls of the old Gird house. The foreground grass has given way to a thicket of Chihuahuan whitethorn, with an understory of tanglehead and sideoats grama. The midground, center and right, is dominated by mesquite and sacaton. The hill behind the Gird house supports Wright lippia and Chihuahuan whitethorn. In the darker area at center and immediately to the right of the old house, the Wright lippia drops out, leaving a pure stand of acacia. In the lighter, more open area to its right are small mesquites, sacaton, and *Aristida glauca*. On the rolling country in back of town, the grassland has been engulfed by a dark tide of Chihuahuan whitethorn studded locally with patches of tarbush and creosote bush. The light area (arrow) at extreme right is perhaps a relict colony of the old grassland; in it occur scattered mesquite, palmilla, Rothrock grama, and *Aristida glauca*. Also abundant are cane cholla and catclaw.

Plate 50c (1994) Except for the loss of the building, changes during the previous one-third century have been slight. Woody vegetation across most of the scene is as dense or more so than before, although at a couple of places, including the grassland remnant, the vegetation is more open. Tanglehead no longer grows in the foreground.

50b

50c

51a

Plate 51a (1880) This photograph is taken from the hill that appears between the camera and Gird Mill in the preceding plate. Looking downslope and southwest toward the Huachuca Mountains, the view depicts Charleston, a town of some 2,000 residents and, for eight years, one of the principal cities of Arizona. The San Pedro River runs from left to right through a channel trench that is well defined. The rectangular barren area in midground is a tailings pond associated with the mill. The foreground hillside is spotted with Chihuahuan whitethorn. One lone cottonwood stands on the riverbank; the slopes of the terrace across the river are open and relatively free from brush. The elevation is 4,000 feet.[26]

Plate 51b (1960) Charleston has vanished. Its demise began when the Tombstone mines struck water and could no longer be worked, making the mills useless. Then came a drop in silver prices, further accelerating the decline of the mill towns. Finally, leveled by an earthquake in 1887, Charleston disappeared into the mesquite. On the foreground slope, Chihuahuan whitethorn has sharply increased and now dominates a sparse flora that includes cane cholla, catclaw, desert zinnia, black grama, sideoats grama, and three-awn. Although tailings ponds are usually considered inhospitable sites for plant growth, the pond seen in the 1880 view is now densely covered by mesquites. Young cottonwoods, perhaps established following the record stream flows during the winter of 1940–41, line the river channel, delineating faithfully its backward-S curve. The channel itself is more deeply entrenched than in 1883. The terrace sides, once grassy, support an almost impenetrable thicket of Chihuahuan whitethorn.

Plate 51c (1994) Little change can be discerned except that the cottonwood and willow trees lining the entrenched channel of the San Pedro River are larger. The mesquite thicket between the camera and the river appears about the same. Many plants have frost injuries, presumably from the December 1978 freeze.

51b

51c

52a

Plate 52a (ca. 1890) With this plate, the photographic progression down the San Pedro River is interrupted, and the camera records upland conditions about 7 miles east of the river. The view is due north toward the Dragoon Mountains across Tombstone in its silver age. The Cochise County courthouse, just constructed, stands at the far left. The vegetation of the hills on the near side of town is unmistakably grassland. Dotting the grassland are leaf-rosette plants such as sotol, agave, and beargrass. In the foreground, Wright lippia, century plant, and sotol can be recognized. The elevation is 4,850 feet.

Plate 52b (1960) The grassland has given way to the vegetation of the Chihuahuan Desert. On the hills in front of town, creosote bush, ocotillo, tarbush, mesquite, and Chihuahuan whitethorn are the principal plants. Ocotillos have increased appreciably in the foreground, and many branches were cut away to clear the view. On the north-facing slope in the foreground, the flora is varied and dense; it includes Wright lippia (the dominant), beargrass, sotol, century plant, tarbush, mariola, mesquite, and a species of *Brickellia*. The bajada behind town is covered by a patchwork of light-colored grassland and darker shrubland. Mortonia is also an abundant bajada plant.

Plate 52c (1994) Comparison of the three photographs in this set suggests that Tombstone has undergone little expansion since its 1880s heyday. The photographs belie the extent of urbanization, however. Most of the land shown is not available for urban development: the foreground and the disturbed area at midground are owned by mines; the land on the far side of town is either Arizona state or federal land. Elsewhere, out of view, many new houses have been built on the outskirts of the old village. Wright lippia, mariola, ocotillo, and Chihuahuan whitethorn are still among the dominant plants visible in this photograph. Palmer's agave is all but absent from the site, a loss that may be related to the increased shrub cover. Several agaves grow nearby on open refuse heaps, where competition from other plants is missing. The arrows point to Walnut Gulch, where the next four sets of photographs were taken.

52b

52c

53a

Plate 53a (ca. 1890) About 6 miles west of Tombstone, from a station part way up the north terrace of Walnut Gulch, the camera faces southeast across the gulch toward the south terrace and the Three Brothers Hills. The flood plain is open and largely free from brush. The two dark shrubs designated by white arrows look like Mexican tea. The clumps of grass are probably sacaton. At left, the black arrow points to what may be a young mesquite. Several cattle can be seen standing in the shade of other mesquites on the arroyo floor. The elevation is 4,000 feet.

Plate 53b (1960) The floor of the wash has undergone a pronounced shrub invasion by both mesquite and Chihuahuan whitethorn. Normally the two plants do not occupy the same habitat, but here they grow side by side, a juxtaposition conveniently illustrated in the foreground, where acacia is in the center and mesquite at the right. A large mesquite occupies the same spot as the small tree identified by a black arrow in plate 53a. The south terrace is shrubbier than before and supports the Chihuahuan whitethorn–creosote bush–tarbush community typical of much of the region. One Mexican crucillo (black arrow) is visible at center on the slopes. A half-dead Arizona walnut tree grows at the right (white arrow). Out of sight behind the mesquites are burrobrush, desert willow, and *Rhus microphylla*. In the midground are Mexican tea, palmilla, and some clumps of sacaton.

Plate 53c (1994) To avoid honey mesquite that completely blocks the view, the camera station has been moved about 40 feet north of the original station. Mesquite and Chihuahuan whitethorn have increased noticeably in size, if not in number. As in 1890 and 1960, scattered sacaton occupies the level floodplain. Also present are whitethorn, little-leaf sumac, creosote bush, chamiso, yellow bird-of-paradise flower, desert zinnia, burrobrush, palmilla, and doveweed. Many snakeweed plants are dead. Mexican tea, present in the previous photographs, is missing. The floodplain shows evidence of sheet flooding. Desert broom borders the wash, which has downcut 10–12 feet. Desert willow and Arizona walnut occur along the wash; tarbush, mariola, and Chihuahuan whitethorn grow on the uplands.

53b

53c

54a

Plate 54a (ca. 1890) This photograph, which partly overlaps the preceding one, is taken from the same camera station as plate 53, but the view is east. Both photographs show the same vegetation. The shrubs at the south base of the little hill in center are young mesquites (see next plate). The abundant dark plant on the plain to the left of the hill may be tarbush. The elevation is 4,000 feet.

Plate 54b (1960) This photograph registers the same invasion by mesquite and Chihuahuan whitethorn shown in plate 53b. The mesquite has come in most densely at the base of the terrace along the course of a small tributary that cannot be seen. The conspicuous bunch grass in the fore- and midground is sacaton (1). The abundant smaller plants are vine mesquite (a grass) and bullnettle. Among the distinctive, easily recognized shrubs near the camera are Chihuahuan whitethorn (2), Mexican tea (3), some small mesquites (4), and chamiso (5), which here is heavily grazed and assumes a low, compact shape.

Plate 54c (1994) There are several dead mesquites near the camera station. Most plants of the two mesquite species in the area show frost damage, probably from the severe freeze of December 1978. Both species have nevertheless increased in importance. Large scattered clumps of sacaton occur across the level floodplain of Walnut Gulch and are about as plentiful as before. Mexican tea, a common plant in 1960, is absent from the same area. Tarbush is still the dominant plant at the mouth of the small canyon at left.

54b

54c

55a

Plate 55a (ca. 1890) From a camera station on the flood plain of Walnut Gulch, this view is northeast toward the small hill that appears in the preceding plate. The mesquites are all young—perhaps five to fifteen years of age—a fact that helps date the invasion and suggests that twenty years ago the hill may have been almost completely free from brush. The elevation is 4,000 feet.

Plate 55b (1962) The mesquites have matured, and the hill is densely crowded with Chihuahuan whitethorn. Beneath them is bush muhly, a grass that commonly is found under shrubs. As is usually the case, mesquite mostly occupies the lowlands; Chihuahuan whitethorn, the rockier uplands. Sacaton, chamiso, and cane cholla also grow on the hillside; sacaton and chamiso, in the foreground on the floor of the wash. The Chihuahuan whitethorn invasion has probably kept pace with that of mesquite. Many of the adult whitethorns in this photograph may have been present in the old one as seedlings.

Plate 55c (1995) The large mesquite in the center is dead; thus, the three views cover its lifespan. Sacaton is more abundant today than in either of the earlier scenes. Chamiso is no longer present.

55b

55c

56a

Plate 56a (ca. 1890) From an island in the middle of Walnut Gulch about a mile below the preceding station, this view looks southwest down the wash toward its junction (right) with the San Pedro River, which is inconspicuous here. The Huachuca Mountains are in the background at the left. As in the preceding plates, the valley floor is open and apparently grassy. The large shrubs, many showing signs of frost damage, are probably mesquite. The elevation is 3,950 feet.

Plate 56b (1960) The replicate view is from a point 25 feet south of the old camera station to avoid a dense mesquite. The wash, having changed its channel, now sweeps to the right of the island. Along the abandoned bed, rabbitbrush, desert willow, burrobrush, and *Carlowrightia* (*linearifolia*?) have become established. The north-facing terrace slope at the far left, much brushier than before, has been invaded by Wright lippia, Chihuahuan whitethorn, tarbush, and especially creosote bush, toward the top of the slope. The valley floor is heavily overgrown with mesquite, rabbitbrush, burrobrush, desert willow, and some very large catclaw trees. Cottonwood trees clearly define the course of the San Pedro River. The water tank in the midground at the left belongs to the railroad station at Fairbank. In front of it and to the right is a small, light-colored hill that looks much as it did ca. 1890. The slope that faces the camera supports a heavy cover of black grama, with some bush muhly near the few shrubs. The far side is dominated by the typical Chihuahuan whitethorn–creosote bush–tarbush community, and there are many dead mesquites.

Plate 56c (1994) A branch of Chihuahuan whitethorn crosses the field of view at left. The mesquites (both honey and velvet mesquites) are larger than before. Branches of adjacent trees often overlap. The channel of Walnut Gulch appears larger, perhaps the result of floods during the previous one-third of a century. Rabbitbrush, a common streambed shrub in 1994, might be represented in the 1890s photograph by small, dark clumps on the floodplain. Scattered desert willows, a few of them dead, also occur along the wash margin. Plants found on the slopes are gray thorn, little-leaf sumac, Wright's lippia, mariola, Chihuahuan whitethorn, desert zinnia, and sacaton. Large cottonwoods, visible in the background, still line the banks of the San Pedro River.

56b

56c

57a

Plate 57a (ca. 1890) From a hill west of Fairbank, the view is southeast across the junction of Babocomari Creek (lower arrow) with the San Pedro River (upper arrow). The Mule Mountains are at the right, the Tombstone Hills at the left. The south terrace of Walnut Gulch, the site of the stations for the preceding four plates, rises in front of them. This rare photograph, lacking though it is in resolution, reveals much about conditions before the onset of arroyo cutting. Neither stream has a distinct channel. Babocomari Creek, actually an irrigation ditch, winds sluggishly through a marshy, grass-filled plain.[27] The course of the San Pedro is almost invisible. Except for the mound in the right midground, the site of the modern railroad station at Fairbank, the valley appears to be treeless. The elevation is 3,800 feet.

Plate 57b (1962) In carving a channel, part of which is visible as it sweeps around the lower left, Babocomari Creek has cut deeply into the hill from which the original photograph was taken. The new camera location is about 100 feet northwest of the old, which cannot be reoccupied. The dike at the left of the photograph forms part of a project to divert the creek to a new channel, visible at the center and about on line with the water in the old photograph. The San Pedro's deeply incised channel is obscured by brush; the junction of the two streams is out of the photograph to the left. Mesquite, cottonwood, and Goodding willow choke the valley floor. Sacaton, the dominant grass, forms a distinct community with mesquite at many places along the San Pedro.

Plate 57c (1994) The open, treeless valley of a century ago has not returned; many of the tall trees seen in 1962 are present after thirty-two years. The forest of cottonwoods, Goodding willows, and mesquites that now occupies the valley floor became established in the 1930s following earlier downcutting and channel widening.[28] This reach of the San Pedro River is now within the San Pedro Riparian National Conservation Area. The 1962 camera station, already moved because erosion destroyed the original site, is now gone, perhaps as the result of the high flood of 1983. The present camera station is at the top of an actively eroding bank and will be lost with a little further erosion. Plants on the floodplain below the camera station include desert broom, saltcedar, chamiso, and sacaton.

57b

57c

58a

Plate 58a (ca. 1890)   For this photograph the camera looks due north and shows some hills lying southwest of St. David near the San Pedro River. Of the shrubs in the foreground, the one farthest to the right is probably chamiso, and the others look like somewhat deformed young mesquites. The elevation is 3,700 feet.

Plate 58b (1962)   For an exact match, the camera should be farther to the left. Mesquite and gray thorn are the large shrubs visible; chamiso and one small palmilla (right foreground) make up the remainder of the vegetation. At the left side of the pyramidal hill in the right of the photograph, a grove of cottonwoods marks the course of the San Pedro River.

Plate 58c (1995)   The severe freezing temperatures in December 1978 damaged most mesquites in this area.[29] On the older large plants, many of the branch tops were killed. Subsequent regrowth produced a persistent "witch's broom" of dead and living branches. Small young plants that were killed to the ground have recovered and now look like the plants in plate 58a. Thus, a growth form that was originally attributed to woodcutting[30] is more likely the result of a hard freeze a few years before the first photograph was taken. Many large mesquites in the vicinity are dead. Plants of the area include palmilla, tarbush, burroweed, zinnia, chamiso, all thorn, Engelmann prickly pear, Christmas cactus, staghorn cactus, plains bristlegrass, bush muhly, and Russian thistle.

59a

Plate 59a (ca. 1890) This photograph, taken about 1 mile southeast of the previous camera station, captures a south-southeasterly view across the San Pedro River toward the Tombstone Hills. Except for palmillas and frost-damaged mesquites, closely cropped grass appears to be the primary cover. Cottonwoods are conspicuously absent from the lowlands along the river. The elevation is 3,700 feet.

Plate 59b (1962) The open plain of the 1890s is densely occupied by woody plants such as mesquite, catclaw, Chihuahuan whitethorn, and tarbush. Sacaton is also present. *Bahia absinthifolia* appears as one of the abundant small herbs in the foreground mixed with desert zinnia, fluffgrass, and burroweed. In the midground, a lower, probably moister, area has a dense cover of mesquite, Chihuahuan whitethorn, sacaton, and bristlegrass. The tall trees in the background are cottonwoods, recently established along the course of the San Pedro River.

Plate 59c (1994) Transformation of the open, grassy plain of a century ago to a dense, mesquite-dominated savanna is virtually complete. Interlocking tree crowns make human passage difficult. A recently cleared field and roadway have destroyed some of the mesquites seen in 1962. Burroweed is the dominant small shrub. Palmilla, an important constituent of the 1890s vegetation, is still apparent; also present are barrel cactus, desert zinnia, Lehmann lovegrass, catclaw, and prickly pear. Cottonwoods, barely visible through the mesquites, are still plentiful along the distant San Pedro River.

59b

59c

Changes in the Grasslands

Our photographs clearly show that grasslands in our region underwent substantial change after the turn of the century. In the most altered locations, grasses became too scarce to suggest their past importance. Our findings agree closely with previous studies.[31] The virtually unanimous conclusion is that, after the turn of the century, woody species proliferated and grasses declined at most grassland locations. The most frequent species were mesquite, ocotillo, Chihuahuan whitethorn, tarbush, Wright lippia, threadleaf groundsel, wait-a-minute, whitethorn, turpentine bush, one-seed juniper, catclaw, and gray thorn. Some of these are also among the increasers in Oak Woodland.

As with the Oak Woodland, mesquite is the species most frequently seen encroaching at our grassland photograph stations. At all of the twenty-eight grassland stations where mesquite could clearly be seen, the plant increased during the period prior to 1962; from that year to the 1990s, increase was observed at eighteen, with no change in the others. It appears that the rate of spread has diminished during recent years.

Chihuahuan whitethorn, a shrub that has increased at grassland sites in the San Pedro River drainage, was seen at nine stations in the early photographs and had increased at all nine by 1962. During the next third of a century, it continued to increase at only three, making no change at the others. Although before 1900 Chihuahuan whitethorn played a minor role at our camera stations, it has since become a dominant in the vegetation of the uplands and the bajadas, a situation comparable to that of mesquite on the bajada around the Santa Rita Range Reserve (plates 18, 19, 38) and of whitethorn on terraces near Tumacacori Mission (plates 40 and 41). Together with tarbush and creosote bush, Chihuahuan whitethorn forms a distinctive community on limestone soils. In company with Wright lippia, it inhabits rocky hillsides. By itself it dominates large stretches of gentle bajada, where it may create nearly impenetrable thickets in areas that a hundred years ago were more open and grassy. Its proliferation must rank as one of the major events in recent vegetative developments in the grassland.[32]

Our Semidesert Grassland photographs have provided insight into one of the region's most intriguing problems: the timing and extent of woody plant encroachment. It is apparent from our glimpses at many stations through time that, with few exceptions, the changes were more profound during the early part of the twentieth century than during the last third of that century. Our photographs suggest that one-seed juniper, unlike the other woody plants, has continued to increase sharply through both time periods. Two leaf succulents (sotol and Palmer agave) and a cactus (prickly pear) have increased more in the later time period of our study than during the earlier.

Before approaching the question of causes, we will turn to a series of photographs from the next lower vegetation area, where landscapes have a different history and potential for change.

Chapter Six The Desert

Judged either by diversity of life forms or by number of species, the Sonoran Desert supports the most complex vegetation of the four desert regions of North America. Where broken terrain, well-developed drainage patterns, and rocky soils produce a variety of microclimates and substrates, a single vicinity may contain plants representing more than half of the twenty-five life forms known from deserts.[1] Although a single community may include only a few species, there are so many communities that the total number of plants growing in the Sonoran Desert is large indeed.

On the basis of life-form and species composition, the desert proper has been subdivided into six regions that can be conveniently referred to as provinces (figure 6.1).[2] (Three of these provinces can be found in our region and are described in more detail below.) In general, broad climatic and topographic factors shape the provincial boundaries, but within a single province the vegetation may range from simple to complex depending on variations in soil and microclimate. Descending the gentle outwash slope, or bajada, which surrounds a desert mountain range, one finds a progressive change toward simplicity in soils and vegetation. The terrain becomes smoother, the soil becomes finer and more uniform, and drainage patterns become less well defined. These factors combine to produce homogeneous soil conditions. Between the upper part of the bajada and the lower, many perennial species drop out and only a few new ones enter. The life forms become fewer in number, with a marked tendency in the lowlands for dominance to be assumed by evergreen plants that are capable of biseasonal growth.[3]

Soil determines the character of desert vegetation by regulating both the quantity of water available and its duration.[4] The relatively level sites at the base of a bajada are more xeric than the slope above because of soil texture differences.[5] Other physical features such as soil depth and, in some places, its salt content also influence the distribution and abundance of plants. Physical condition of soil is in turn related to soil age. Desert soils are not static substrates. Rather, near the surface is a complex mosaic of soils of different ages that have been forming for perhaps millions of years to the present. Vegetation in the Sonoran Desert is patterned upon this mosaic; thus plant communities reflect soil age.[6]

The effect of soil depth may be seen in the reduction of perennial plant cover to values of 5 percent or less where the relatively coarse, absorbent, upper layers of soil are shallowly underlaid by impervious horizons of caliche or of hardpan. Because the downward movement of soil moisture is impeded, much of the water that might otherwise be available to plants is lost by evaporation.

Where the soil surface is dissected and irregular, vegetation is heterogeneous. Where level terrain, undissected by runnels, imposes more uniform soil moisture conditions, vegetation tends to be simpler. Perhaps nowhere in the Sonoran Desert are differences so stark between the plant life of these two kinds of terrain as they are in the Arizona Uplands.

The Arizona Uplands

By reason of the relative abundance of moisture it receives and the wide range of elevations it spans, the Arizona Uplands is the most diverse of the three Sonoran Desert provinces in our region (figure 6.2). It lies along the northeastern edge of the desert region, where true desert vegetation may extend upward to elevations above 3,500 feet on warm, south-facing slopes. Toward the west and south, the plant life of the province extends downward to elevations of about 1,000 feet. Average annual rainfall varies from 7 inches to 12 inches, depending on elevation.[7] Mean annual temperatures range roughly from 64°F to 72°F. This province is set apart from the other Sonoran Desert provinces more by its winters, which are relatively cool and wet, than by its summers, which are hot and moist. The temperature regime clearly reflects isolation from the Pacific Ocean.[8]

Arizona Uplands vegetation is most luxuriant on upper bajadas, where 40 percent or more of the surface may be covered by the crowns of woody and succulent perennials. Vegetative cover may be so dense that it is difficult to see for more than a few hundred feet. Beneath the low shrubs are such small plants as pincushion cactus. Above these, bursage commonly grows in a uniform layer about 1 foot high. Low trees like foothill paloverde and ironwood, their crowns perhaps 20 feet higher still, occur on the low ridges between runnels. And towering over all the others, widely spaced saguaros may often reach heights of 30–40 feet. Between the overstory of low trees and the understory of bursage and/or brittlebush, many other

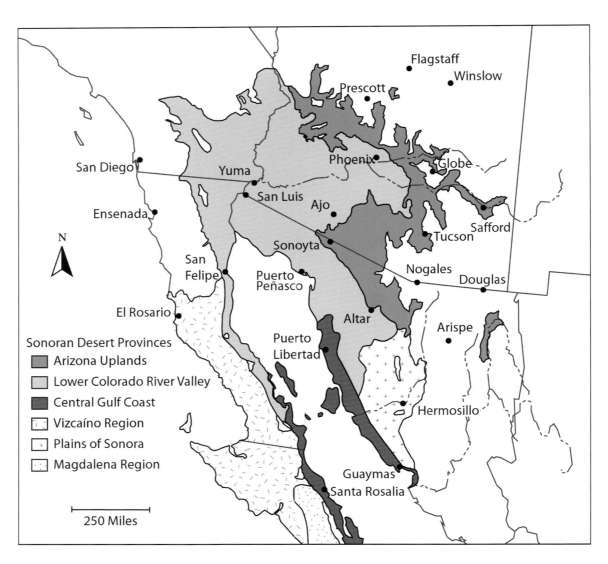

Figure 6.1 The vegetative provinces of the Sonoran Desert. (After Shreve 1964 as modified by Brown and Lowe 1994 and by Turner and Brown 1994.)

plants can be found. They vary so greatly in stature, however, that no layer of intermediate height can be recognized.[9]

At the base of a bajada, many perennials of the upper slope are no longer present. Ironwood, foothill paloverde, bursage, and saguaro, among others, are gradually lost in the descent. Toward their lower limit, the individuals of these species become confined to small drainage ways, and the intervening areas are occupied by creosote bush and white bursage.

In our region, plant species typically respond uniquely to seasonal changes in moisture and temperature. As a result, periods of leafiness might overlap but seldom coincide precisely except at the onset of the summer rains, when the combination of sudden moisture and warm temperatures breaks the long estivation of foresummer. Several species, like limber bush, leaf out only with this coincidence of available moisture and high tem-

perature. Another group, including bursage and creosote bush, produces leaves during all seasons when there is water. Intermediate between these two extremes are whitethorn, ocotillo, and foothill paloverde, which do not produce leaves during the cold winter but may foliate in the spring, summer, or autumn if soil moisture is present in sufficient quantity. These plants may lose and regain their leaves several times in a given season, with ocotillo perhaps holding a plant world record by changing its leaves five or six times during some years.[10] In still another phenological category, the saguaro is able to absorb and accumulate water during the winter and summer, but no growth occurs until the warmer spring and summer periods.[11]

The diversity in habits of growth is even more striking when one considers leaf fall and flowering habits. Ocotillo loses its leaves abruptly with the onset of drought; foothill paloverde retains

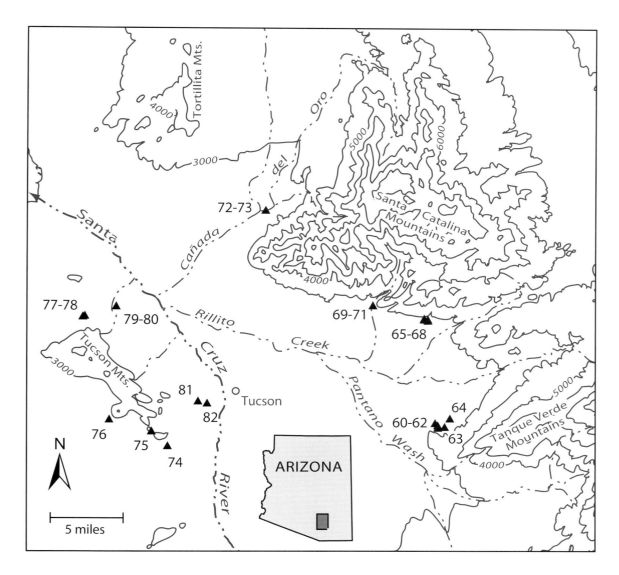

Figure 6.2 Location of photograph stations in the Arizona Uplands.

its foliage over longer periods of dryness. Bursage loses its leaves slowly, the older ones first and the younger ones later, until with prolonged drought only the very youngest leaves, at the stem tips, remain; these too may die if desiccation is extreme.

In the Arizona Uplands, the dominant plants flower in spring, starting with bursage in February and ending with whitethorn in May. The duration and abundance of bloom are sensitive to rainfall amounts, and, after dry winters, foothill paloverde, bursage, brittlebush, and other species flower briefly and sparsely. In very dry years, they might not flower at all. The timing of flowering is controlled by a combination of environmental factors. Warmth, calculated as degree-days above a certain threshold, is required for development of flower buds. The trigger that initiates this process varies among species. The first substantial rain of the cool season is the trigger for spring flowering of bursage, ocotillo, creosote bush, and brittle-

bush. In the case of saguaro, the trigger is a complex combination of day length, solar radiation, and cool-season rain.[12]

A second conspicuous community of the bajadas occurs in narrow, branching ribbons along drainage channels and is analogous to the gallery forests of higher elevations. Whitethorn and catclaw are most typical here, their size depending upon the amount of moisture available. Where minor washes coalesce in a single large channel, these plants reach heights of 15–20 feet and may be joined by mesquite, blue paloverde, desert willow, canyon ragweed, and other species requiring the improved moisture balance of this habitat. Several of these desert riparian species invade nonriparian positions in the Semidesert Grassland and Oak Woodland.

On the plain below the bajada, the plant community is low in stature and simple in composition, a consequence of a less favorable moisture supply. Creosote bush is the principal dominant, occurring as a widely and rather uniformly spaced plant about 2 feet high in the drier habitats and over 6 feet high where moisture is more abundant. Under optimum conditions, it may attain coverage values of 15–20 percent. White bursage, its principal associate, occupies the broad openings among the larger shrubs. It rarely exceeds 2 feet in height and may have coverage values of from 1 percent or less to 10 percent. Total coverage for the community may vary from 15 to 30 percent, depending on the availability of soil moisture.[13] Unlike the paloverde-saguaro association, the creosote bush–white bursage community is not restricted to the Sonoran Desert but occurs along the valleys of the Mohave Desert as well.[14]

The Lower Colorado Valley

Heading west from Tucson, Arizona, to San Luis on the Colorado River in Mexico, one crosses a series of plains that descend to the river like a giant staircase. The steps are separated by mountain ridges that grow lower and less massive toward the west. With each tread, the plant life grows more impoverished: the simple creosote bush–white bursage community is no longer confined to the valleys, and foothill paloverde, bursage, ironwood, and saguaro are restricted almost entirely to drainage ways.

The Lower Colorado Valley, a low, arid province, is drained by 250 miles of the Colorado River and 200 miles of the Gila River. The western edge lies in the Imperial Valley of California. To the south, narrow extensions into Mexico flank the head of the Gulf of California. The eastern boundary occurs near the longitude of Phoenix, Arizona.

Throughout its extent, the province lies mainly below 2,000 feet, and in the Imperial Valley it descends well below sea level. It is at once the hottest and the driest of the desert subdivisions. Mean annual temperatures range from 65°F to 74°F;[15] average annual precipitation ranges from slightly more than 1 inch to 9.5 inches.[16] The rainfall tends to be equally distributed between winter and summer or, in the westernmost reaches, slightly unbalanced in favor of winter.

As one might expect from the degree of aridity, the vegetation is simple, sparse, and relatively uniform. Height and density vary with the amount of moisture available from the local soil. The most common community, dominated by creosote bush and white bursage, covers vast stretches of plains, bajadas, and even volcanic hills. On some volcanic outcrops, creosote bush and bursage are replaced by low, widely spaced foothill paloverde and infrequent saguaros.

The runnel vegetation of the province includes foothill paloverde and ironwood but lacks whitethorn, one of the more conspicuous plants in similar habitats in the Uplands. Along some of the broader washes, mesquite and blue paloverde occur. The smoketree, absent from most of the Arizona Uplands because of its sensitivity to cold, grows here in riparian situations.[17] Desert willow, which has higher water requirements than smoketree, occurs only sparsely below about 1,500 feet,[18] although it is abundant along washes in the Arizona Uplands and in the grasslands.

On the low interfluves between minor washes is a community dominated by desert saltbush, which, much like the creosote bush–white bursage community, also extends into the Mohave Desert.[19] Desert saltbush grows in essentially pure stands on well drained, fine, sandy loam and on other types of soil where surface drainage is impeded.[20] The total plant coverage expresses the relative water balance of each location; coverage values as low as 3.5 percent have been noted in a site where downward percolation is blocked by hardpan. As with many plants of the desert plains, vegetative growth occurs during both rainy seasons; flowering is restricted to autumn.

Expanses of stabilized sand characterize large areas of the Lower Colorado Valley. Big galleta, a perennial grass that grows in pure stands or in association with creosote bush and other woody plants (plates 83, 84, and 85), is often the dominant. The apparent anomaly of a grass-dominated community in this very arid province can be explained by a combination of environmental and biological factors. In deserts, pure sand is often moist compared to nearby substrates, especially those with a high proportion of clay. Grasses, which have higher water requirements than most desert shrubs and trees, can meet their moisture needs on dunes and sand sheets even when the atmospheric climate is quite dry. In areas where sand moves freely, however, most grasses cannot grow fast enough to escape burial and death. Because big galleta, like many desert shrubs, bears perennating buds above the soil surface, it can grow fast enough to remain above the surface as sand accumulates around the plant.[21]

In its physiography, the Lower Colorado Valley is more uniform than the other two provinces considered. Only a few low mountain ranges interrupt the plains. Sand dunes and malpais fields—the latter the product of recent volcanic activity—occur toward the south and support distinctive communities of their own. The Pinacate Mountain region (plates 83–90), where lava flows and dunes exist side by side, is a checkerboard of dark and light substrates. Vegetation, responding sharply to the shifting patterns of soil and albedo, is an intricate mosaic. Pockets of relatively deep sand support big galleta. Sandy flats are dominated by ocotillo or, rarely, saguaro. The

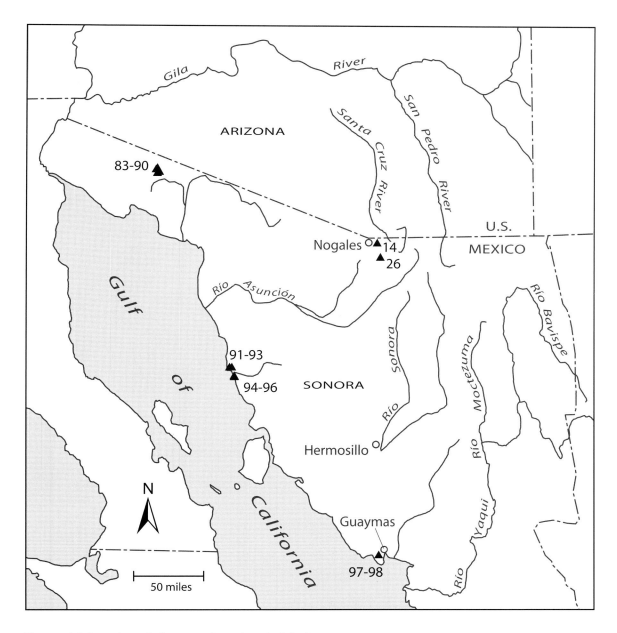

Figure 6.3 Location of photograph stations in Mexico.

volcanic hills maintain a sparse cover of foothill paloverde, ironwood, elephant tree, and *Jatropha cuneata*. The latter two plants have affinities to the south where, with their fleshy-stemmed relatives, they form the characteristic vegetation of the Central Gulf Coast province.

The Central Gulf Coast

The vegetation of the Central Gulf Coast occurs in two coastal strips along opposite sides of the Gulf of California. To the north, both give way to the Lower Colorado Valley and today, if not in the past, the two strips make no contact with each other. The province has been described as the driest in the Sonoran Desert,[22] but from the weather records now available this distinction

clearly belongs to the Lower Colorado Valley.[23] The coastal region seems to be characterized more than anything else by rapidly changing gradients of rainfall. The isohyets are closely crowded, more or less paralleling the coastline, and the amounts decline precipitously as one approaches the gulf from the east. Guaymas, one of the few stations with a long-term record, receives an annual average of about 9.25 inches, with 79 percent falling during the six hot months.[24] This amount is probably near maximum for the province; most parts receive between 4 and 6 inches.

In contrast to the two provinces already described, the vegetation of the Central Gulf Coast lacks the sharp distinction between communities of the plains, on the one hand, and the bajadas and hills, on the other. As the traveler approaches the region from the Lower Colorado Valley, creosote bush and white bursage become sporadic in their occurrence and no longer dominate extensive areas. Hills and intermontane plains, valleys, and bajadas alike are covered by open communities of small trees or shrubs.

Relative to the Arizona Uplands, foothill paloverde loses its dominant position in the vegetation, and ironwood and blue paloverde gain in importance. Elephant tree and *Jatropha cuneata*, plants of limited occurrence in the Lower Colorado Valley, are joined by a host of other fleshy-stemmed, sometimes aromatic plants, including *Bursera hindsiana*, *Jatropha cinerea*, and *Euphorbia misera*. In most situations, none of the perennials asserts dominance to the degree that foothill paloverde, for example, does in the Arizona Uplands. The saguaro is of minor importance, its place in the vegetation being occupied now by the cardón, or by other columnar cacti such as organpipe cactus and sinita.

The cirio, another fleshy-stemmed species (plates 94 and 96), is found at only one location on the Mexican mainland but occurs abundantly there. Demographic studies of this unusual plant give widely varying estimates of its maximum longevity. One study notes the possibility that plants may reach ages of 500–600 years.[25] Another places the maximum age at about 250 years.[26] The mainland occurrence of this bizarre tree extends from the vicinity of Puerto Libertad (Punta de Kino or Punta de Cirio) south through the Sierra Bacha nearly as far south as Desemboque.[27] The disjunct distribution of the cirio from its present center of occurrence in Baja California to minor outposts on Isla Angel de la Guarda and the mainland remains one of the many intriguing problems involving the flora of the Sonoran Desert.

An important community, found also in other provinces of the desert with coastal contacts, is dominated by *Frankenia palmeri* (plate 93). Often associated with *Atriplex barclayana*, it dominates a low, windswept community that typically is found just inland from the strand. On sandy soils the monotonous appearance of the association may be broken by taller plants such as *Jatropha cuneata*, *J. cinerea*, *Euphorbia misera*, jojoba, and teddybear cholla (plate 95).

The photographs that follow sample only two localities within the Central Gulf Coast (figure 6.3). Plates 91–96 show the vegetation around Puerto Libertad, Sonora, and, in general, reveal remarkably little change during the past thirty years. Neither the density of most species nor the size of individual plants has appreciably altered. The exceptions to this generalization are found most notably among such columnar plants as cardón and cirio, which have increased in size and, in the case of cardón, in density as well. In the much-altered landscape of the Islas Mellizas in Guaymas Bay, the second of the two places in the Central Gulf Coast province for which photographs are available (plates 97 and 98), cardón is similarly involved.

60a

Plate 60a (ca. 1935) This camera station is located on a hill within the newly established Saguaro National Monument near Tucson. The view is northeast toward Agua Caliente Hill and looks down a gentle incline that drains into Tanque Verde Creek. The saguaro stand pictured here is a particularly memorable one because of the many large individuals and the absence of small ones. It is an example of the Arizona Uplands at its upper colder elevation limit on level to rolling terrain.[28] The dominant plants here are brittlebush, foothill paloverde, velvet mesquite, and saguaro. In the distance, on the floodplain of Tanque Verde Creek, saguaros become infrequent and mesquite predominates. On the banks of Tanque Verde Creek at right, the large trees are cottonwood and Goodding willow. The elevation is 3,050 feet.

Plate 60b (1960) Through natural attrition, the saguaro population has undergone a reduction of about one-third. Some workers predict that, at this rate, the population will be zero by 1998.[29]

Plate 60c (1998) Saguaro National Monument is now Saguaro National Park. The decline in saguaros has continued but at a rate slightly less steep than predicted. Many of the scattered survivors appear unhealthy and soon will be gone. Meanwhile, the regeneration of this forest has already begun, as shown by findings from a long-term study-plot located at left midground. There has been a sharp increase in establishment of new cacti, especially during the 1970s.[30] These new plants, mostly 1–3 feet tall and hidden by the trees, will become visible in another decade or so. Given time, the entire flat might once again look as it did in the 1930s. An increase in both size and number of velvet mesquite, foothill paloverde, and other woody plants makes the overstory noticeably denser than in 1960. Brittlebush is still the dominant small shrub in the foreground.[31] Beyond the park boundary, construction of houses is removing saguaro habitat at a rapid rate and will doubtless have an impact on regeneration trends there. Note the increase in large trees along Tanque Verde Creek.

60b

60c

61a

Plate 61a (ca. 1935) This photograph, taken about half a mile south of plate 60 and about 200 feet southwest of plate 62, looks north-northwest toward the Santa Catalina Mountains. Some of the saguaros are 35 feet tall, or about 150 years old. The hole in the tall stem at the right has probably been used as a bird's nest. Such entries are usually sealed off from the interior of the plant by a hard, ligneous material.[32] The elevation is 2,950 feet.

Plate 61b (1962) Only seven individuals—a quarter of the earlier population—remain. The decline results not only from the death of old individuals but from the failure of young plants to become established in sufficient numbers to replace the old. In the foreground are staghorn cholla, desert hackberry, and creosote bush. The trees are foothill paloverde and velvet mesquite. Whitethorn, Mexican crucillo, gray thorn, and catclaw constitute the principal shrubs in the midground; desert zinnia, paperdaisy, burroweed, and brittlebush are the more important low perennials. Most of the mesquites bear old axe marks.

Plate 61c (2000) Loss of large saguaros is complete although there are now five small saguaros in the area where the large ones once stood.[33] The new saguaro in the foreground, established in a low-lying relatively moist area, is 14–15 feet tall and became established in about 1936. Woody species such as velvet mesquite, desert hackberry, Mormon tea, Mexican crucillo, and whitethorn have increased in biomass, as has prickly pear. Several nearby creosote bushes have died.

61b

61c

62a

Plate 62a (1935) This view is south-southwest at a hillside about 100 yards from the preceding location. The flora of the wash in front of the hill includes velvet mesquite, catclaw, creosote bush, and desert hackberry. The coarse, gravelly slopes support the usual dominants, foothill paloverde, saguaro, staghorn cholla (arrows), and bisnaga. A variety of shrubs are present, including Mexican crucillo, creosote bush, mesquite, and ocotillo. The elevation is 2,950 feet.

Plate 62b (1960) The hillside strewn with dead plants shows that, as saguaros have declined, so has much of its associated community. In this site, the number of saguaros is about 40 percent less than in 1935. In the stand of chollas marked by arrows in the 1935 photograph, the decline has been even steeper. More than a quarter of the paloverdes on the hill are dead.[34] The small perennials have also undergone a severe attrition.

Plate 62c (2000) Because the foreground mesquite now completely blocks the view, the camera has been moved back about 20 feet and slightly to the right.[35] As at other nearby sites, saguaro numbers have declined still further. The small portion of the midground slope that is visible in the photograph fails to convey the magnitude of paloverde mortality. In 1962, eighty-seven foothill paloverdes were alive on the entire slope. When surveyed in 1996, the population stood at fifty-seven, a decline of 65 percent. This decline acquires relative significance from a statement made by Forrest Shreve in 1911: "I have had a great many thousands of Palo Verdes come under my observation within a radius of 75 miles of Tucson, and have seen only two dead trees of full size."[36] Prickly pear cactus has increased, as have the small woody perennials of the area. In the foreground, creosote bush biomass has declined, perhaps the result of shading by the growth of larger plants since 1962.

62b

62c

63a

Plate 63a (1935) This photograph, taken in the vicinity of the two preceding plates, looks north-northwest toward the Santa Catalina Mountains and provides a good look at some plants that go unrecognized in many of the photographs. Four cacti are present: saguaro (1); staghorn cholla (2); jumping cholla (3); and bisnaga (4). Desert zinnia (5) is the common low perennial in flower; burroweed (6) can be recognized among the desert zinnias. *Ambrosia confertiflora* occurs as a tiny herb, and (7) may be paperdaisy. The elevation is 2,950 feet.

Plate 63b (1962) One of the old saguaros, recently fallen, lies on the ground at the right. In the twenty-seven years separating the photographs, the ranks of the giant cactus have been decimated. The shrubs—foothill paloverde, whitethorn, mesquite, gray thorn, and desert hackberry—have fared almost as badly. The fortunes of small perennials such as brittlebush and burroweed fluctuate with short-term variations in rainfall. Just two years before this photograph was taken, their numbers were very low. Since then, small perennials in general have registered a comeback, and brittlebush in particular, almost gone in the 1960s, has made rapid gains.

Plate 63c (1995) Because the jumping cholla at extreme lower right now blocks the view, the camera has been shifted to the left. No large saguaros can be seen, and close inspection of the area within 200 feet of the camera revealed no small plants for replacement. Because all saguaros present at this site in 1935 and 1962 were already old, probably dating from the mid-1800s, their absence today means that saguaro establishment has failed for 100 years or more. In contrast, other succulents, such as prickly pear (2 species), have increased substantially. The dark shrub in the center at midground is desert hackberry, much increased in size since 1962. Velvet mesquite, many with old axe marks, has increased in volume if not number, as have gray thorn and Mexican crucillo. Brittlebush, desert zinnia, and burroweed are still common. There is more grass today than earlier, the dominant being red three-awn. The once common bisnaga is still absent. Pincushion cacti are common, especially beneath the crowns of burroweed.

63b

63c

64a

Plate 64a (1932) The camera is located on a ridge about half a mile northeast of the station for plate 60. The view is north-northwest toward the Santa Catalina Mountains. The dense saguaro stand is populated mostly by large, branched individuals. Prickly pears are conspicuous. Mesquites and foothill paloverdes are the dominant trees. The small shrubs are probably burroweeds. The elevation is 2,900 feet.

Plate 64b (1962) Small perennials have almost vanished from the landscape. The prickly pears are gone. Death among the saguaros has been widespread, and mortality of the shrubs is nearly as great. With some exceptions, the conditions shown here prevail throughout the whole lower part of Saguaro National Monument, a section characterized by rolling terrain and soils that are coarse, well drained, and relatively homogeneous. Elsewhere in the Monument, as in the rocky foothills and along the lower slopes of the Tanque Verde Mountains, the saguaro seems to be repopulating, and the plant communities appear to be stable. Whatever factors are responsible for the widespread mortality at lower elevations in the 1960s, they affected many species, not merely one.

Plate 64c (1995) Shrubs near the camera block more of the view than before. The biomass of trees and large shrubs appears to have increased, as has that of the intervening small shrubs, reversing a trend noted in plate 64b. The decline in saguaro numbers, however, has continued unabated from the time of the earliest photograph. Grasses such as bush muhly and red brome, a non-native annual, are abundant beneath shrubs and in openings on the flat terrain at midground. Several creosote bushes have died on the flat, and several prickly pears are visible in the background. Urbanization of the Santa Catalina Mountains foothills has notably increased during the past thirty years.

64b

64c

65a

Plate 65a (1915) This and the next three sets of photographs were taken around the mouth of Soldier Canyon at the foot of the Santa Catalina Mountains near Tucson, Arizona. This view, which looks north-northeast into the canyon toward the mountains, shows teddybear cholla, foothill paloverde, brittlebush, ocotillo, and saguaro. The elevation is 2,950 feet.

Plate 65b (1960s) A few teddybear chollas have disappeared; the ocotillo at the left has added some stems; and the small saguaro in the foreground at left center has grown to a respectable size. Several mesquites have gained prominence in the right foreground. The Mt. Lemmon highway has taken its toll on the native vegetation, but, except for the mesquites, there has been little significant change. Trees and shrubs are mostly foothill paloverde, mesquite, and wolfberry. Brittlebush is the principal semishrub.

Plate 65c (1993) Teddybear cholla has continued to decline in importance. Foothill paloverde and velvet mesquite have increased in size and now block much of the midground view. The status of saguaros on the far slope is difficult to evaluate but is probably little changed. Several saguaros seen on the right in the 1960s are now gone and a new one, near the camera, has grown into view, as has a small 1-foot-tall plant just right of center in the foreground. Engelmann prickly pear is present but not abundant on this rocky slope. The ocotillo at left is gone, but the species is still well represented at the site. Fairyduster and brittlebush are the dominant small shrubs; other shrubs and subshrubs include staghorn cholla (center foreground), desert hackberry, wolfberry, Arizona carlowrightia, and *Janusia gracilis*.

65b

65c

66a

Plate 66a (1908) This photograph was taken seven years earlier than plate 65, but the camera station is only a few yards away. As in plate 65, the view is northeast—away from the plain and toward the rocky slopes of the Santa Catalina Mountains. Brittlebush blankets the foreground. Saguaro, ocotillo, mesquite, and foothill paloverde can also be identified. The elevation is 2,900 feet.

Plate 66b (1960) Of the four saguaros closest to the camera in 1908, one has died. Otherwise, in contrast to conditions on the bajada (plate 65), the saguaro population here, 100 yards away in the foothills, has been stable for half a century. The amount of brittlebush in the foreground has declined, and the foothill paloverde at center is gone. In addition to the plants that can be identified in the 1908 view, catclaw, staghorn cholla, mesquite, and Engelmann prickly pear are present.

Plate 66c (1993) Turnover among the foothill paloverdes is conspicuous; some were lost but have been replaced. Of the four nearby saguaros seen in 1908, only two remain, and both show frost damage. The velvet mesquite that barely extends into the photograph at extreme left in 1908 has continued to increase in size. The status of Engelmann prickly pear, seen in the foreground here and common in the 1960s view, is difficult to ascertain because of the dense foreground foliage but probably has increased during the past thirty-three years. Common species here include fairyduster, brittlebush, wolfberry, limberbush, *Janusia gracilis*, pincushion cactus, bush muhly, slender poreleaf, and a species of three-awn.

66b

66c

67a

Plate 67a (1915)   The view is turned south-southeast from the preceding view and now looks across the bajada toward present-day Saguaro National Park, located on the outwash slope on the other side of Tanque Verde Creek. The Santa Rita Mountains are at the right, and the Tanque Verde Mountains are at the left. Compared to its density in plate 66a, the vegetation here looks relatively sparse. The absence of young saguaros is a sign that the population is in decline. The elevation is 2,900 feet.

Plate 67b (1962)   In contrast to conditions on the mountainside, the vegetation here has been greatly altered. Of the score or more saguaros in the midground in 1915, none stands now. The area appears to be brushier, although it is not clear whether this is because there are more individuals or merely because their average size is larger. Many plants occur in both the 1915 and 1962 photographs: desert hackberry (1); mesquite (2); and foothill paloverde (3). The Engelmann prickly pear (4) in the foreground has come in recently. Also visible are brittlebush (5); blue paloverde (6); wolfberry (7); and catclaw (8).

Plate 67c (1993)   Several small saguaros have become established on a rocky slope nearby. On level ground, the decline noted in 1962 has continued. The greater biomass of woody plants can be attributed both to growth of existing plants and establishment of new ones. The small mesquite at middle foreground in the 1915 view is dead. A mesquite closer to the camera (number 2 in plate 65b) now blocks much of the foreground. Other plants that persist from 1915 are desert hackberry (at extreme left, midground) and blue paloverde (number 6 in plate 65b, obscured by foreground velvet mesquite in 1996). The foothill paloverde seen on the extreme right in all three views is dead in 1996. Engelmann prickly pear is more prominent; fifty-six plants grow within a 23-foot radius of the camera. Other species present include fairyduster, limberbush, brittlebush, staghorn cholla, pincushion cactus, Arizona cottontop, and tanglehead.

68a

Plate 68a (1908) Here the camera station is on the bajada about one-fourth to half a mile southwest of the three preceding stations. It faces east toward the Santa Catalina Mountains (left) and Agua Caliente Hill (far right). Individuals that carry over into the 1962 view are foothill paloverde (1); three mesquites (2), two of which are already mature adults; a desert hackberry (3); and two saguaros (4). The lower story is relatively dense and probably contains burroweed, paperdaisy, Mexican tea, and brittlebush. The elevation is 2,850 feet.

Plate 68b (1962) Like plate 67, which also shows vegetation of the bajada, this one depicts a decline in the number of saguaros and coverage of subshrubs. There has been a general increase in woody vegetation. Of the two saguaros seen at center in 1908, only one remains, and it is more dead than alive.

Plate 68c (1994) The low perennials, which declined between 1908 and 1962, are more numerous. The taller trees have increased in size and in number. Large saguaros have declined, although a cursory examination of the area around the camera station revealed many small seedlings. Engelmann prickly pear is much more numerous than before. Three of the prominent trees, a mesquite and two paloverdes, were present in 1908. The desert hackberry at left foreground has also persisted since 1908. Saguaro establishment is vigorous here. The dominant small shrubs today are burroweed, bursage, brittlebush, and turpentinebush. Many cacti are found here, including pincushion cactus, barrel cactus, staghorn cholla, prickly pear, and jumping cholla. This site, in a subdivision named Sunrise Ranch, will soon be developed. Indeed, a new house in a neighboring development lies perhaps 75 yards away, just out of view to the left.

69a

Plate 69a (1911)   This is a view of the hill at the mouth of Sabino Canyon, Santa Catalina Mountains, east of Tucson. Readily visible on this southeast-facing slope are typical Sonoran Desert plants such as saguaro, foothill paloverde, prickly pear, and teddybear cholla. The elevation is 2,900 feet.

Plate 69b (1961)   A 3-acre, permanent saguaro study plot established on this slope in 1961 contains 225 saguaros. The field of view is smaller than in the 1911 photograph because the camera lens is narrower. The leaning saguaro at left foreground can be seen in the earlier photograph. The old road along which this photograph was taken has been widened, removing a swath that stops just short of this plant and of the lower edge of the photograph. Careful study will show other saguaros that persist from 1911.

Plate 69c (2000)   A wider lens was used, and the view now includes the roadway and the vertical roadcut. Teddybear cholla appears less abundant now than in 1911, prickly pear more abundant. When the plot was last examined in 1989, saguaro numbers had fallen to 172. Fifty-six plants have died, and only three are new. A study of the age distribution in 1989 revealed that, from about 1775 to 1950, the population was maintaining itself. Only since 1950 has the decline been abrupt.[37] The large clumps of light-colored grass seen at mid-slope and above (near center) are fountain grass, an introduction from Africa widely used in ornamental plantings.

69b

69c

70a

Plate 70a (1890) Plates 70–71 are views across Sabino Creek, an ephemeral stream where water flows several months of the year and the rest of the time lies near the surface. The creek drains the south side of the Santa Catalina Mountains near Tucson. This early view looks west across the creek toward the entrance to the canyon and the location of the previous plate. Two distinct plant communities and habitats are juxtaposed: on rocky slopes, a foothill paloverde–saguaro community, and at stream side, a gallery forest. The treetops of that forest are just visible along the lower edge of the photograph. The elevation is 2,850 feet.

Plate 70b (1962) The saguaro study plot (plate 69) is located on the hillside at upper right. On the slopes, foothill paloverde is more abundant everywhere. Although saguaros are less numerous in a few locations, overall their total number is somewhat greater, especially on the lower hillside. The road has been widened and paved. At right center on the hill above the road, teddybear cholla is more conspicuous but not necessarily more abundant; the fuzziness of the old photograph makes comparison difficult. Mesquite has increased along the runnel that parallels the highway. Fairyduster, catclaw, limber bush, cockroach plant, brittlebush, hopbush, and wolfberry occupy spaces among the dominants. On the lower hillside, just above the gallery forest, are desert hackberry and coursetia.

Plate 70c (1995) The roadside embankment has been colonized by trees and shrubs, especially foothill paloverde and velvet mesquite. Throughout the view, canopy coverage of paloverde has increased as individual plants have grown. Prickly pear has proliferated on midground slopes. On the far slope at right, as in the permanent plot, saguaros appear to have declined. The trees along Sabino Creek are much taller and denser than in 1890.

70b

70c

71a

Plate 71a (1890) From near the same location as plate 70, the view is northeast up Sabino Canyon. Oaks, sycamore, and velvet ash dominate the scene. An ecologist describing the locality in 1915 noted the importance of these species in the gallery forest but noted that cottonwood was only "occasional" and failed to mention mesquite at all.[38] His description agrees well with the conditions portrayed here. The elevation is 2,850 feet.

Plate 71b (1962) The dense growth on the canyon floor is largely the result of the increase in velvet mesquite. A count of saguaros visible on the hill behind the bosque shows that the population is slightly larger than in 1890. Toward the top of the hill, there are fewer individuals; toward the bottom and in the valley, there are more. Foothill paloverde dominates the hillside along with saguaro. Other species present there include desert hackberry, catclaw, ocotillo, wolfberry, limber bush, and teddybear cholla. On the foreground slope are mesquite, catclaw, and foothill paloverde.

Plate 71c (1995) View of the foreground slope is now almost fully blocked by catclaw and foothill paloverde that have enlarged with time. Despite the blocked view, it is evident that prickly pear is more abundant now than earlier. The ring-like scars on the large saguaro at left are typical of frost damage. The number of annual constrictions above the rings shows how many years have passed since the damage occurred. The thirty to thirty-five annual constrictions on this plant suggest that the damage almost certainly happened during the catastrophic freeze of the night of January 11–12, 1962.[39] If this assumption is correct, the saguaro seen in plate 71b had been subjected to the damaging low temperatures about nine months earlier but did not yet show freeze symptoms. (The photograph in plate 71b was taken on October 2, 1962.) The canyon-bottom habitat of this saguaro is prone to freezes because of cold-air drainage. Saguaros growing on nearby slopes were probably unaffected by the January 1962 freeze. This same brief freeze killed 35–75 percent of saguaros at higher, colder sites nearby.[40]

1b

71c

72a

Plate 72a (1913) This and the next set of photographs show a riparian habitat greatly different from Sabino Canyon. Although this ephemeral wash, Cañada del Oro, carries surface water only after a heavy rain, infiltration through its sandy bed is sufficient to ensure an adequate supply of groundwater at all times for perennials with moderately deep roots. The vegetation correspondingly is more xeric than that on the floor of Sabino Canyon, but more mesic than that of the surrounding desert. The view is east-southeast toward Pusch Ridge on the west side of the Santa Catalina Mountains near Tucson. A dark outpost of pine forest is visible on the peak at right center. The elevation is 2,650 feet.

Plate 72b (1962) Burrobrush, crownbeard, and a species of stickleaf occupy the foreground and the channel. The thicket on the opposite bank is denser than in 1913. Mesquite, desert willow, catclaw, whitethorn, and pencil cholla are its main elements. Increased mesquite biomass apparently is responsible for the greater density. On south-facing slopes in the hills, foothill paloverde and saguaro are the dominants. The paloverde, at least, has undergone a substantial increase in numbers.

Plate 72c (1993) The wash floor in the foreground is densely covered with burrobrush, a sign that this part of the water course has not carried heavy flows for perhaps a decade. The meandering waterway has shifted to the base of the far hill. The hydrology of this channel was altered for a few years following construction of Golder Dam, 8 miles upstream from this site. The bed of the proposed lake failed to hold water, and the dam was finally breached by court order. A section of the opposite bank has been stabilized by a soil-cement coating to protect the approach to the highway bridge spanning the Cañada del Oro. This approach was washed away during the large flow of October 1983.

72b

72c

73a

Plate 73a (1913) This view, taken from the same station as the preceding plate, faces southeast toward slopes with northerly aspect. The left margin overlaps slightly with plate 71, and the composition of the bank and channel vegetation is continuous between the two photographs. The elevation is 2,650 feet.

Plate 73b (1962) These moist, north-facing hillsides approach the upper edge of the desert and their vegetation has many elements in common with the Semidesert Grassland. In general, three communities can be distinguished. On steeper slopes, trees are few. Desert hackberry, catclaw, and whitethorn are the dominant shrubs, and sotol, desert honeysuckle, fairyduster, wait-a-minute, hopbush, burroweed, turpentine bush, and species of *Opuntia* are the secondary plants. Along the gentler gradients that prevail over most of the hillside, ocotillo joins these same species, forming an understory to blue and foothill paloverde. On warm, dry, south-facing slopes, where many of the species typical of the grassland drop out, blue paloverde becomes rare and saguaro gains codominance with foothill paloverde. The apparent increase in paloverde is confirmed by analysis of mortality and relative age in 1962. A hillside study plot in the middle of plate 72 has 108 paloverdes of both species, none dead. Of the thirty foothill paloverdes present, 53 percent are under 3 feet in height. Twenty-eight percent of the trees are foothill paloverde, and 72 percent are blue paloverde.

Plate 73c (1993) Now, approximately the same area supports 98 paloverdes, 65 percent of which are foothill paloverdes, 31 percent blue paloverdes, and 4 percent hybrids between these two species (a category not recognized earlier). In addition, the proportion of small foothill paloverdes has declined from 53 percent to 23 percent.

73b

73c

74a

Plate 74a (1915) This view in the Tucson Mountains is northwest across Robles Pass toward the southeast face of Cat Mountain. Like the rest of the Arizona Uplands, the area is floristically varied. Dominants on the rocky, north-facing hillside in the foreground are foothill paloverde, whitethorn, gray thorn, and wolfberry. The understory likely comprises bursage, white ratany, range ratany, fairyduster, cockroach plant, paperdaisy, cholla, Engelmann prickly pear, tobosa, *Janusia gracilis*, *Ayenia pusilla*, Coulter hibiscus, and desert hibiscus. The elevation is 2,650 feet.

Plate 74b (1962) The highway is higher than the old road and follows a different alignment. In addition to the hillside, which has already been described, five other habitats can be distinguished. The area on the near side of the pass and sloping northward is dominated by foothill paloverde with a few saguaro, bursage, creosote bush, and mesquite. The area along a runnel has no foothill paloverde, and its cover is predominantly creosote bush and whitethorn. The area on the opposite side of the pass and sloping gently to the south is similar to the near side of the pass but, as might be expected of a warmer exposure, has an abundance of ironwood and saguaro as well as paloverde. The dominant small shrub is triangle leaf bursage. On the rocky lower slope of the mountain, brittlebush replaces bursage in the understory, and ironwood drops out to leave foothill paloverde and saguaro as the large dominants. Above the escarpment line, foothill paloverde and saguaro diminish in importance, and a group of unidentified shrubs, probably including jojoba, is taking over. The most important change between 1915 and 1962 is the great increase in the number of foothill paloverdes, a change that is most evident on the lower slopes.

Plate 74c (1994) The large plants in the foreground are whitethorn and foothill paloverde. Species composition of the understory is much the same as in 1962. The most notable change, aside from the appearance of scattered houses, has been a striking increase in prickly pear, which is now a codominant. On most of these plants, the lowest stem segment is triangular in cross-section, which indicates that the plant grew from seed rather than from a fallen stem segment. Thus, the search for causes of prickly pear proliferation at this site should be centered on life-cycle events involving seed production and germination.

74b

74c

75a

Plate 75a (1914) The camera station is located at the foot of Cat Mountain about 1.5 miles northwest of the station for plate 74. The view is south-southeast toward Beehive Peak. Floristically, the scene is similar to the lower slopes in plate 74. The abundant foreground shrub is triangle leaf bursage. Foothill paloverde and saguaro are the chief large dominants. Ocotillo, staghorn cholla, and several other species of cacti, including Engelmann prickly pear, are evident. The elevation is 2,750 feet.

Plate 75b (1962) Bursage is less abundant in the foreground—a change that may or may not be significant since picnickers have had a heavy impact on the area. Jojoba plants are about as numerous as bursage. Cholla, whitethorn, creosote bush, staghorn cholla, and prickly pear also occur. The most notable change is the extent to which foothill paloverde has increased, resulting in a less open view.

Plate 75c (1994) After the decline noted in 1962, bursage is again about as abundant as in 1914. The same may be so for prickly pear. The increase in foothill paloverde has not continued, and its density has apparently remained stable since 1962. Mature saguaros have continued a steady decline, and their replacements, if any, are still too small to be seen. Indeed, within a 50-foot radius of the camera station, only one small saguaro was found. It was roughly 15 inches tall, or about twenty-five years old. Establishment is spotty for this species, however, because within half a mile of this station we noted many young plants beneath paloverdes.

75b

75c

76a

Plate 76a (1915) The third set of photographs in a circuit of the Tucson Mountains, this view is from a station 3 miles west of the preceding one and looks west-northwest toward Brown Mountain. Here, as in plates 74 and 75, the most notable difference over the years is the extent to which the plants, particularly foothill paloverde, have increased in coverage. The arrows mark two saguaros that carry over into the new photograph and that can be readily identified. The elevation is 2,650 feet.

Plate 76b (1962) The buildings of Old Tucson, built after 1915, are visible at the far right. Foothill paloverde (denser than before), and the giant cactus (less dense) are the dominant plants. Other species are creosote bush (1), ocotillo (2), staghorn cholla (3), Engelmann prickly pear (4), limber bush (5), whitethorn (6), bush muhly (7), and *Opuntia phaeacantha* (8). Present but not labeled are fairyduster, catclaw, a species of grama grass, coldenia, and desert hibiscus. The common foreground shrub is bursage.

Plate 76c (1994) The most notable changes have been the progressive closing of the view by growth of paloverdes and the continuous decline through time in saguaro numbers. Three saguaros—7, 15, and 20 inches in height— are growing within a few yards of the camera but are obscured by other plants.

76b

76c

77a

Plate 77a (1916) This view is north-northwest toward Safford Peak (left of center) from a ridge on the west side of the Tucson Mountains. Saguaro, ironwood, foothill paloverde, and ocotillo dominate the midground; brittlebush dominates the foreground. The smaller plants likely include whitethorn, creosote bush, white ratany, bursage, senna, staghorn cholla, jumping cholla, prickly pear, *Dyssodia porophylloides*, wolfberry, and *Janusia gracilis*. The elevation is 2,900 feet.

Plate 77b (1962) Brittlebush in the foreground has greatly decreased, as have at least two of the large dominants, ironwood and foothill paloverde. Within a 100-foot radius of the camera, 20 percent of the foothill paloverdes and 40 percent of the ironwoods are dead. Although mortality like this is abnormally high for these long-lived species, the condition seems to be local.

Plate 77c (1988) The die-off of ironwood and foothill paloverde has not continued; near the camera station, no unusual amount of mortality is evident. Brittlebush, which is frequent in the foreground, has increased somewhat since 1962 and is approaching its abundance in 1916.

77b

77c

78a

Plate 78a (1916) The camera station is a few hundred feet northeast of the station for plate 77, and the view is north-northwest toward Panther Peak (left center). The hill in front of the camera is at the far right in the preceding plate. Some of the same plants are in both views. This photograph shows the Arizona Uplands in an advanced stage of development. Foothill paloverde and ironwood constitute the principal arborescent element. Mesquite, omnipresent in the valleys, plays a minor part in this hilly country; the few individuals only grow along runnels. By virtue of their size and shape, saguaros and ocotillos assume a visual importance out of proportion to their actual numbers. The elevation is 2,900 feet.

Plate 78b (1962) The foothill paloverde on the right and one ocotillo on the left still frame the view, and the foreground looks as cluttered as before. The first hill has as many saguaros as in 1916 but fewer paloverde trees. The apparent decline in paloverde may be due in part to seasonal differences in foliation. In the earlier photograph, leaves are present; in this November view, they are absent.

Plate 78c (1988) Populations of the most prominent species, foothill paloverde and saguaro, appear relatively static since 1916, nor has prickly pear shown the dramatic increase evident elsewhere in the Tucson basin. Ocotillo is more evident now than at either of the earlier times. The most prominent change is the man-made overburden dump (far left) associated with a limestone mine that serves a nearby cement manufacturing plant.

78b

78c

79a

Plate 79a (1908) In this northwestern view of a rocky cliff near Picture Rocks in the Tucson Mountains, the camera station is a few miles northeast of the preceding one. This photograph forms a panorama with the next plate, and the two record some of the floristic detail of the Uplands. Foothill paloverde and saguaro dominate the scene; jojoba and creosote bush are important secondary shrubs; limber bush, brittlebush, triangle leaf bursage, and white ratany constitute the dense lower story; and four cacti—teddybear cholla, staghorn cholla, jumping cholla, and bisnaga—lend additional diversity to the landscape. Tanglehead and *Janusia gracilis* complete the list of the major species. The elevation is 2,400 feet.

Plate 79b (1962) A high degree of stability characterizes the vegetation at this site. Of thirty-six saguaros in the 1908 view, twenty-three are still alive. Although thirteen saguaros died, fourteen new ones have appeared, and the population remains almost exactly the same at the end of fifty-four years. The saguaro at center has lost its arms; the woody rods forming the structural framework still protrude from the stumps.

Plate 79c (1996) The high degree of stability seen between 1908 and 1962 has not continued into 1996. Some creosote bush have died; many others are shrunken remnants of the original robust plants. The greatest overall change is in the foreground, where jojoba and foothill paloverde have increased since 1908. Brittlebush is now largely confined to the upper half of the slope. Bursage is scattered in the foreground, common on the lower slope. Less obvious but nonetheless significant changes in saguaro and paloverde are summarized in the caption for plate 80c.

79b

79c

80a

Plate 80a (1908) This view is north-northwest from the same station near Picture Rocks. The young saguaro at the left—perhaps thirty years old—appears also in plate 79a. Pale lycium, whitethorn, and Engelmann prickly pear occur here, but not in plate 79. In general, the steep, rocky, upper slope constitutes one habitat; the more gentle, gravelly, lower slope another. Brittlebush, the principal small shrub on the upper slope, gives way to bursage on the lower. In a similar fashion, the stubby teddybear cholla yields to the larger jumping cholla, and foothill paloverde to creosote bush. Jojoba seems equally at home in both habitats. The elevation is 2,400 feet.

Plate 80b (1962) Only the teddybear cholla has radically changed in number over the years, but its increase is quite striking. Creosote bush proves to be the most stable of the species present; at least three-fourths of the individuals visible in this view can also be seen in the older one. Foothill paloverde shows about the same degree of stability as the saguaro (see plate 79). Of fifty paloverde trees in the earlier view, thirty have persisted, twenty are new. The annual mortality rate of 0.7 percent is too high to agree well with an earlier suggestion that foothill paloverde lives 300–400 years.[41]

Plate 80c (1996) The foreground has undergone the most change. As in the preceding plate, brittlebush has retreated up-slope, its place taken by jojoba, teddybear cholla, and jumping cholla. Bursage remains the dominant low shrub. The sixty-one foothill paloverdes on the slope represent an increase of 20 percent since 1962. The number of saguaros has risen only slightly from thirty-six to thirty-nine, suggesting a large degree of stability. However, an examination of the population age composition shows that 8 percent of the thirty-nine became established before 1900, 20 percent between 1900 and 1950, and 11 percent since 1950.[42] The relatively low number of recent additions is striking. A stable population would have several times more plants in the youngest group than in the next older, leading to the conclusion that this population is in decline, at least temporarily, and that fifty years from now few large plants will be growing here, producing a scene similar to that in 1908. The large flat rock at center foreground provides a reference from which to judge soil erosion. The above-ground part, long exposed to the elements, is coated with dark "desert varnish" in the 1908 view. By 1962, erosional losses exposed a lighter unvarnished surface below, which also is visible in the 1996 photograph. The ruler leaning against the rock is 16 inches long, showing that soil loss (most of which had occurred by 1962) amounts to about 3–3.5 inches. A shallow runnel has developed between the rock and the camera station.

80b

80c

81a

Plate 81a (1928) The Tucson Mountains as seen from the Carnegie Institution of Washington's Desert Botanical Laboratory, west of Tucson. The laboratory preserve was fenced in 1907 to exclude livestock. Roughly a millennium before this photograph was taken, the area was the site of intensive dry farming, remnants of which are still visible throughout the area.[43] The photograph, looking along an old road, was taken to document the recovery of that roadway. Creosote bush is the dominant plant. One foothill paloverde stands at the right. The small shrubs appear to be seedlings of bursage. The elevation is 2,650 feet.

Plate 81b (1985) Most creosote bushes are gone. Their dead root crowns can be found through close inspection. Bursage and ratany are now the clear dominants. One ocotillo has died and is lying on the ground, several others have become established, and the one at right midground has grown larger. Foothill paloverde has increased, as have saguaros. Fairyduster is common. The small tufted grass is fluffgrass. The area is now surrounded by the city of Tucson. The power lines crossing the background at left serve houses that are just out of view in the distance.

Plate 81c (1998) Creosote bush has not yet staged a revival although three or four old plants, a small fraction of the original number, have survived. This is surprisingly high mortality for a plant that can live for thousands of years.[44] Unlike creosote bush, bursage, fairyduster, and range ratany increased between 1928 and the mid-1950s.[45] These three species (especially the last two) are palatable to livestock, and their comeback might reflect cessation of grazing in 1907. The high mortality of creosote bush might be related to luxuriant growth of range ratany, a root parasite, after cattle were excluded from the Desert Laboratory grounds.[46] The loss of creosote bush at the Desert Laboratory is confined to an area of perhaps 100 acres where the soil is highly calcareous. Nearby populations have thrived on different soils and with different combinations of associated plants.

81b

81c

82a

Plate 82a (1906) The Carnegie Desert Botanical Laboratory as seen from the laboratory grounds west of Tumamoc Hill. The laboratory is just over three years old, and the laboratory grounds will not be fenced for another year. Numerous minor roads, constructed by teamsters for collecting rocks, cross the hill up to about midslope.[47] The rocks were hauled to nearby Tucson, where they were used in constructing house foundations and walls. The activity of the "rock pickers" was a source of concern to the scientists and may have had an impact on saguaro establishment.[48] The dominant tree on the hill slopes is foothill paloverde. The elevation is 2,600 feet.

Plate 82b (1968) Remnants of the rock-hauling roads are still visible on the lower slopes of Tumamoc Hill. The new road crossing the scene follows a power line. That foothill paloverdes on the slopes have increased in size is evident from the trees silhouetted against the sky along the ridgetop. The creosote bush at left midground appears to be dead. The small tufted grass in the foreground is fluffgrass, a short-lived perennial.

Plate 82c (1999) The old rock-hauling roads have almost completely healed. The road along the utility easement receives infrequent use and is still evident. The light patches on the hill slopes are grasses. The patch downslope to the right of the buildings is buffelgrass, an aggressive perennial from Africa; the others are native grasses, mainly tobosa and three-awns. The increase in cacti, especially prickly pear and jumping cholla, is consistent with observations at several of our photograph stations from desert to grassland. About 11 percent of the foothill paloverdes on the rocky, west-facing slope died within the past six years. Mortality was highest among large trees on steep slopes. It seems likely that drought in the 1990s interacted with natural senescence of an aging population, weakening old trees and hastening their deaths.[49]

82b

82c

83a

Plate 83a (1907) This view looks north-northwest toward the United States from a spur of the Hornaday Mountains at MacDougal Pass in the Pinacate Mountain area of Sonora, 45 miles west of Sonoita. This hot, sandy plain, flanked partly by volcanic intrusions and partly by dunes drifting inland from the Gulf of California, is near the upper edge of the arid Lower Colorado Valley province, the site of plates 83 through 88. A positive identification is impossible, but the dominant tree in this photograph appears to be blue paloverde. In the distance, several trees with dark crowns are probably mesquites. This flat, referred to as "a fine meadow of galleta grass" at the time of this photograph,[50] is dominated by big galleta, a coarse, semiwoody grass. The elevation is about 900 feet.

Plate 83b (1962) Striking change in the vegetation is not entirely apparent from the photograph, since dead plants clutter the scene and give an impression of greater plant density than is really the case. The tree dominants of the earlier view have declined sharply, partly because of woodcutting and partly, perhaps, from drought.[51] A study plot occupies the front two-thirds of this and the adjacent view (plate 84). Of the 219 mesquite trees on the plot, woodcutters chopped down 43. Ninety-one mesquites, most of them small, are dead from natural causes, and 27 others are nearly dead. Only 58 mesquites are alive and healthy. Among the blue paloverdes, a tree that is not suitable for fuel, only one has been chopped. Of the remaining 55, 30 are dead and only 25 are alive. Clumps of big galleta—the dominant grass—are also markedly fewer in number.

Plate 83c (1997) Ninety years are spanned by these three photographs, and within that relatively short period three distinct out-of-phase pulses are recorded by the dominant plants: saguaros have steadily increased through time; big galleta has steadily decreased; and paloverde and mesquite have declined during the first fifty-five years and then rebounded during the last thirty-five years. The small saguaro first seen in the foreground in 1962 died before April 1992.

83b

83c

84a

Plate 84a (1907) The photograph overlaps with plate 83 and is from the same camera station looking north-northeast. This "fine meadow" of big galleta grass supports many mesquite and paloverde trees, some of which shelter young saguaros. The small saguaro at center foreground stands next to the remains of the tree under which it became established. The crater at left is probably a product of geologically recent volcanic activity in the Pinacate region. The elevation is about 900 feet.

Plate 84b (1959) As in plate 83, mortality among mesquite and blue paloverde has been high. Two species of wolfberry, one of them *Lycium macrodon*, have also undergone a substantial reduction in numbers: twenty-eight of ninety-two plants in the study plot are dead. The increase in saguaro numbers stands in marked contrast to the many dead blue paloverde, mesquite, wolfberry, and big galleta. Creosote bush is common in the study plot, although much of it is also dead. Bursage is rare, white bursage infrequent. A few ironwood trees can be seen. The foreground saguaro noted in the 1907 photograph is dead, although the basal portion is still standing.

Plate 84c (1997) The asynchronous pulses of plant biomass noted at the adjacent site shown in plate 83 are seen here as well. Minor erosion channels are visible for the first time. A small cattle ranch was established about 2 miles from here in the mid-1960s, and the area receives some impact from these animals.[52]

84b

84c

85a

Plate 85a (1907) View to the northwest from a small rocky outcrop a few hundred yards south of the preceding site. This photograph shows the expedition from the Carnegie Desert Botanical Laboratory traversing MacDougal Pass, one of several geographic features named by this group.[53] Only three or four saguaros can be seen. The scarcity of large cacti is noteworthy. Branches of two paloverdes appear at left and right in the immediate foreground. The elevation is about 900 feet.

Plate 85b (1958) Many saguaros now punctuate the scene. A large segment of this population was undoubtedly present, but hidden from view, a half century earlier. Ironwood and paloverde are the trees; the dominant small plants are big galleta. Drought has not yet produced conspicuous dieback of woody plants (see plates 83 and 84).

Plate 85c (1997) While the increase in saguaros is most obvious, the loss of big galleta is remarkable, as well. The prominent saguaro at right foreground in the earlier views is now dead. The paloverde and ironwood have increased in size.

85b

85c

86a

Plate 86a (1907) Looking east-northeast through Macias Pass in the Hornaday Mountains, this view shows the sandy plain also shown in plates 83, 84, and 85. The large patch of cactus crossing the foreground is devil's club cholla; the shrub below the camera is bursage. Beyond can be seen teddybear cholla, creosote bush, ironwood, and saguaro. This site is described as "a wonderful desert botanical garden." Young saguaros, mistakenly thought to be mature individuals of a dwarf race, are abundant.[54] The elevation is about 900 feet.

Plate 86b (1962) Big galleta, not bursage, now dominates the immediate foreground. Most of the bursage is dead, although many dead plants still stand. The same is true of creosote bush; only two large living clumps can be seen (1). The foreground cluster of devil's club cholla is still present, but hardly thriving. The density of teddybear cholla and saguaro seems about the same as before. Two ironwoods (2) and a foothill paloverde (3) grow on the sandy floor. Foothill paloverde, elephant tree, and *Jatropha cuneata* are present on the rocky slopes of the Hornaday Mountains.

Plate 86c (1997) The devil's club cholla growing across the foreground has changed little in ninety years. Other species, including teddybear cholla, creosote bush, and bursage, have declined, although the timing of the losses differs. Dieback of creosote bush and bursage dates from the 1907–62 period, whereas mortality of teddybear cholla and saguaro postdates 1962. The tree component at this "desert garden" site has increased over the past ninety years. The big galleta has come and gone.

86b

86c

87a

Plate 87a (1907) The view is west toward the sands of the Gran Desierto, from the eastern rim of MacDougal Crater in the Pinacate region about 2 miles from the preceding stations. The crater—a caldera in geological parlance—is about a mile in diameter and 450 feet deep. Its floor, largely free from the ubiquitous influence of humans, provides a unique opportunity for studying vegetation change. The isolation of the locality, the escarpments ringing the crater, and its steep talus sides all combine to make human entry difficult and, equally important, to keep out cattle. The elevation is about 900 feet.

Plate 87b (1959) Even from a distance of half a mile, the vegetation of the floor presents a strikingly altered appearance. The playa in the center appears desiccated, and the crowns of most of the shrubs have shrunken. Large mesquites at the center of the playa have many dead branches. Encircling the playa and scattered throughout it are large, dead creosote bushes. Death is obviously widespread among the plants both on the playa and along the runnels leading into it (plates 83 and 84). Foothill paloverde, saguaro, creosote bush, big galleta, brittlebush, and a few mesquites and few ironwoods make up the impoverished perennial vegetation of the crater floor. Elephant tree (primarily on north-facing slopes), pygmy cedar, *Jatropha cuneata*, and ocotillo grow on the sides, none of them in abundance.

Plate 87c (1997) Most of the mesquites at the center of the crater died back to the ground and then resprouted. Death among the large creosote bushes within and around the mesquite patch reached almost 100 percent. Replacing the creosote bushes are numerous additional mesquites that became established after rare tropical storms in 1970 and 1972 drenched the area. (See plate 89.) The large paloverde and mesquite trees on the crater floor have made a slight recovery.[55]

87b

87c

88a

Plate 88a (1907) In this enlargement of the valley floor seen at extreme left in plate 87a, most of the plants are saguaros, foothill paloverdes (the large trees), creosote bush (the small regularly spaced shrubs), and big galleta. There are roughly twenty paloverdes. The saguaro population, including some plants nearly hidden by nurse plants, stands at approximately ninety-five. The elevation is 900 feet.

Plate 88b (1959) Since 1907, sixty-seven saguaros have died and fifty-one new ones have appeared, leaving a total population of seventy-nine plants. The 17 percent net loss in saguaros was accompanied by an even more serious loss among the large desert trees and creosote bush. Foothill paloverde has declined by about 50 percent, and creosote bush has declined even more.

Plate 88c (1997) Of the ninety-five saguaros present ninety years ago, only eight persist today. The fifty-nine saguaros carrying over from 1959 are joined by twenty new ones, so the population has suffered no net loss during the previous thirty-eight years. The number of trees has remained roughly the same since 1959, although big galleta has declined during the same period. Except in better watered situations, the creosote bush population remains low. Data from permanent plots established in the 1960s show that the creosote bush population has since declined by 91 percent and that no new plants became established.[56] The rapid pre-1959 losses in major plant species occurred during what was probably the most serious drought since A.D. 1700.[57]

88b

88c

89a

Plate 89a (1959) This plate provides close examination of drought effects on the crater floor. The camera station is on the south side of the crater, looking northwest across the area viewed from the rim in the preceding plate. The larger dead plants are mesquites and paloverdes, including foothill, blue, and their hybrids; the smaller ones, creosote bush. Here, at the margin of the central playa, creosote bush is growing robustly along runnels and dying in the interfluvial spaces. The grass, big galleta, is mostly dead. Although saguaro seems to be maintaining itself for now, its seedlings require shade, and death of the large nurse plants ultimately spells decline for it, too. The white paper bands on many of the saguaros mark them as subjects in a long-term study of growth and survival.[58] The elevation is about 400 feet.

Plate 89b (1984) Twenty-five years later and toward the end of a relatively moist period, this view accentuates the quick ascendance of the short-lived brittlebush (light-colored shrubs with hemispheric crowns) and the rejuvenation of many stressed creosote bushes that had undergone severe drought-pruning. Several saguaros and many creosote bushes have died, presumably victims of the drought.

Plate 89c (1998) After fourteen years, brittlebush coverage is lower. The creosote bush and saguaro numbers are little changed. There is no sign of a comeback among the large nurse plants, most of which died forty years ago.

89b

89c

90a

Plate 90a (1959) This is the view north across the floor of MacDougal Crater from a point toward the outer edge of the dark central area seen in plate 87a. Sand has drifted across the crater floor, which is bare from drought, and accumulated on the lee (eastern) side of the shrubs. The vegetation is composed almost wholly of creosote bushes with a scattering of taller mesquites. Most branches of the bushes are dead. The elevation is about 400 feet.

Plate 90b (1974) Most of the creosote bushes survived in this area, which lies just beyond the center of the crater where mortality was close to 100 percent. The survivors have regained normal stature and are joined by numerous seedlings of western honey mesquite, each marked with a white paper flag. Seventy-five to eighty mesquite seedlings are in the group in front of the camera. These dense concentrations of seedlings appear to be confined to low areas that were, presumably, temporarily flooded by Hurricane Joanne in October 1972. No seedlings of creosote bush are found in the area.

Plate 90c (1998) Mesquite survival was great enough that the earlier open shrubland is now an almost impenetrable thicket. In a nearby study plot, the number of mesquites increased from 2 in the 1960s to a high of 413 in 1974, then declined to 249 in 1984 and to 80 in March 2000.[59] The number of creosote bush fell from five to two through the 1970s and 1980s but was back up to five by March 2000. Note the dead creosote bush at right foreground and the two new ones at mid-foreground. As the new, dense mesquite thicket dies back and new creosote bushes come in, perhaps the playa will return to a creosote bush scrubland.

90b

90c

91a

Plate 91a (1932) The view is west toward hills about 4 miles northwest of Puerto Libertad, Sonora. The common occurrence here of fleshy-stemmed perennials like *Jatropha cuneata* (1), elephant tree (2), and cliff spurge (3) marks this vegetation as part of Shreve's Central Gulf Coast subdivision of the Sonoran Desert, the setting for the remaining sets of photographs. Other plants visible in the photograph are white bursage, bursage, brittlebush, jojoba, creosote bush, ocotillo, and *Frankenia palmeri*. The large columnar cactus is cardón, and the two multistemmed species in the background are sinita and organpipe cactus. The elevation is 450 feet.

Plate 91b (1965) On first glance, there seems to have been remarkably little change in thirty-three years. Many of the shrubs on the plain below are relics, and most of these have not changed in size. The ocotillo in the left foreground has apparently produced no new stems since 1932, and the existing stems have not perceptibly elongated. Only the cardón has increased in height. "The slowness of growth, great longevity and low rate of establishment among the perennials gives the desert an extremely stable character."[60] Despite the general effect of suspended animation among the larger species, a count shows that the smaller shrubs have diminished in number.

Plate 91c (1989) The cardón that stood right of center has died. Otherwise, most large plants have remained the same for fifty-seven years. The ocotillo at left foreground has still changed little since 1932, and the plants on the distant bajada show remarkably little change as well.

91b

91c

92a

Plate 92a (1932) The station is about 40 feet away from the preceding one, and the camera looks southeast toward Puerto Libertad, a few buildings of which are visible where the bay curves behind the hills in the midground at the right. In the foreground are ocotillo, brittlebush, *Jatropha cuneata*, and teddybear cholla. The elevation is 450 feet.

Plate 92b (1965) Passage of thirty-three years has resulted in little significant vegetation change. The relative openness of the valley at the base of the foreground hill continues as before. Most of the plants carry over from 1932. The ocotillo at left foreground has lost some branches. In the foreground, gains and losses in the limberbush and teddybear cholla populations cancel each other out. Among the species visible in the midground are ironwood, foothill paloverde, jojoba, elephant tree, cardón, sinita, cliff spurge, and species of *Lycium*.

Plate 92c (1989) Changes in the view have been largely cultural. Puerto Libertad has burgeoned as the community serving a large oil-fueled power plant. A new paved road connects Puerto Libertad with communities to the south. Many of the buildings, radio towers, and smokestacks are also new. These three plates, like the preceding ones, seem like a study in suspended animation, especially on the marine-derived soil of the valley floor. The next plates, taken from a position visible at the left edge of this photograph, show the nature of these sites in close detail.

92b

92c

93a

Plate 93a (1932) The view, from a camera station about 2 miles north-northwest of Puerto Libertad, is to the west. The photograph looks up the bajada toward the small hills at center where the stations for plates 91 and 92 are located. The poverty of the vegetation on marine deposits of this part of the Central Gulf Coast is evident. Such xeric plants as ocotillo, bursage, and creosote bush occupy the runnels. *Frankenia palmeri* dominates the ridges between the runnels. The elevation is 200 feet.

Plate 93b (1965) Thirty-three years later, most of the same ocotillos are present. The net loss of bursage, *Jatropha cuneata*, and cliff spurge makes the landscape appear more arid. *Frankenia palmeri*, the dominant plant directly in front of the camera, has also experienced a net loss.

Plate 93c (1989) The *Frankenia* population has remained relatively static during the previous twenty-four years. The overall impression is that more change occurred from 1932 to 1965 than from 1965 to 1989.

93b

93c

94a

Plate 94a (1932) The camera station is on Punta de Cirio, and the view is west toward the Gulf of California. In upland areas like this, the vegetation of the Central Gulf Coast reaches its maximum development, typically displaying a floristic wealth that rivals the Arizona Uplands. Cirios, some of them 35 feet high and several centuries old, dominate the scene. This and the adjacent region are the only places where this tree occurs outside of Baja California. In the foreground are elephant tree (1), *Jatropha cuneata* (2), ocotillo (3), white bursage (4), brittlebush (5), jojoba (6), and a small cardón that is about 4.5 feet tall (7). The elevation is 150 feet.

Plate 94b (1965) The elephant tree at the right has died, as has the *Jatropha*. Close scrutiny reveals the loss of one large cirio near the center of the view. The ocotillo has acquired a few more stems. A certain amount of leisurely growth has been enjoyed by the cirios. The young cardón at left center is now more than 13 feet tall. A small cardón grows to its left. The two photographs might otherwise have been taken one year apart. Teddybear cholla, desert lavender, creosote bush, sinita, triangle leaf bursage, stickweed, chuparosa, *Solanum hindsianum*, *Bursera hindsiana*, and species of barrel cactus and wolfberry may also be found in the vicinity.

Plate 94c (1989) The ocotillo has changed little in twenty-four years, but the small cardón has died, and the view is missing at least four large cirios. The persistent cardón is now 16.5 feet tall. There is little change in other aspects of this view.

94b

94c

95a

Plate 95a (1932) Between Punta de Cirio and Puerto Libertad, this view looks southwest across a mesa toward the Gulf of California. Cardones dominate the scene. Like saguaro in the Pinacate Mountain region, cardón may forgo a rocky habitat for a sandy one in very arid sites. The middle story consists of elephant tree, *Bursera hindsiana*, *Jatropha cuneata*, *J. cinerea*, jojoba, and teddybear cholla. A large plant of *J. cuneata* protrudes into the left edge of the picture. The elevation is 100 feet.

Plate 95b (1962) The fleshy-stemmed jatrophas and burseras have changed very little. The grasses are fewer. As elsewhere in the Central Gulf Coast province, cardón has proliferated and teddybear cholla has declined. Several short-lived plants of desert mallow appear in the foreground.

Plate 95c (1997) The large shrub with stout branches in the middle of the picture is *Jatropha cinerea* and persists from 1932, as does the elephant tree to the right. No vestige remains of the *J. cuneata* that previously extended into the left foreground. Cardones are more numerous than before. No plants of teddybear cholla are visible in this view.

95b

95c

96a

Plate 96a (1932) This near-shore community, located where the canyon shown in plate 92 debouches into the Gulf, is atypically dense and varied. *Frankenia palmeri* and *Atriplex barclayana* form the low matrix in which *Jatropha*, *Bursera*, cliff spurge, and teddybear cholla are embedded. The view is southeast toward north-facing slopes on which all of the plants mentioned in plate 90 are to be found, plus *Fagonia californica*, desert hibiscus, *Echeveria pulverulenta*, and desert mallow. The elevation is 50 feet.

Plate 96b (1965) Many plants persist from 1932. The relatively large shrub at right, just a few yards from the camera, is cliff spurge. The shrub to its left in 1932 has died recently and was probably the same species. The openings between the shrubs appear larger now than previously, due in part to the loss of teddybear cholla, which has markedly declined, but also due to a general decrease in size of the dominant *Frankenia palmeri*.

Plate 96c (1997) The number of teddybear cholla stabilized after 1965. The large cardón at left midground is now gone, although the total number of cardones is greater than in 1932. The population of cirios on the slopes appears little changed.[61] A new fence, seen along the cleared strip on the hill at right, surrounds a large part of the Sierra Bacha (Sierra de Cirio) and delineates the Rancho Punta de Cirios. The plants and animals of this privately owned preserve are monitored by the federal government in an effort to protect this unique habitat.

96b

96c

97a

Plate 97a (1903) One of the Islas Mellizas in the bay at Guaymas, Sonora. The photograph is taken from the southernmost of the two islands and looks northeast toward the outskirts of the city. Here, as at Puerto Libertad and Punto Cirio, the desert reaches down to the water's edge. Columnar cacti are incongruously juxtaposed with mangrove thickets. A sparse stand of old cardones is scattered over the island. Chuparosa, *Euphorbia tomentulosa*, and *Zizyphus amole* occur among the rocks, black or red mangrove in the water. The elevation is about sea level.

Plate 97b (1961) A striking increase has taken place in the number of cardones. A census shows that there are 5,836 plants on this small island, or about 3,200 per acre. The same sort of increase has occurred on other islands off the coast.[62]

Plate 97c (1996) During the previous thirty-five years, the number of cardones seems to have declined little if at all. The plants have grown in size, and their biomass is many times greater than it was in 1903 or even in 1961. As human population has increased on the mainland, so have human incursions on this little island. Many small plants were "decapitated" and a sizeable area burned at the time of a visit in 1984.[63]

97b

97c

98a

Plate 98a (1903) The camera station is located on the island shown in plate 97. The view is southwest toward the second of the Islas Mellizas, the location of the camera station for plate 97. In the background at the left is the peninsula that extends east to Punta Baja and forms the southern perimeter of Guaymas Bay. The cardón population is striking because it is composed of large individuals only, suggesting that successful establishment has been lacking for several centuries—that is, during the life of this population. The elevation is about sea level.

Plate 98b (1961) The cardón population is perhaps three or four times larger now than in 1903. The Islas Mellizas, like Mac-Dougal Crater, represent a partially controlled situation, and grazing can be dismissed as an ecological factor. The islands are near enough to the shore to be reached by swimmers, and near enough to Guaymas to ensure that they are, in fact, frequently visited. Nevertheless, they should be among the more stable desert habitats: the temperature and humidity are controlled within narrow limits by the water, and animal interference with the plant life is minimal. A major fluctuation in the population of a long-lived perennial is difficult to explain. On the adjacent mainland, seemingly identical habitats support few cardones.

Plate 98c (1996) The cardón population is as dense as before. The many small plants have now reached heights of 6–12 feet. Many of these plants have attained sexual maturity, providing a rich source of food for those animals capable of flying from the mainland to seek out the pollen, nectar, fruits, and seeds when seasonally available. Rodents are not residents on these two islands, although remains of black rats (*Rattus rattus*) have been found here. The large shrubs on the slope are *Zizyphus amole* and appear little changed in size or number since 1903. The town of Guaymas has spread to the hills on the adjacent mainland.

98b

98c

Local Variability of Plant Populations

During the first half of this century, vegetation change in the desert was neither so striking nor so consistent as in Semidesert Grassland and Oak Woodland. In some localities shrubs decreased, in others they increased. Numbers of mature saguaros at Saguaro National Monument had plummeted by the 1960s; in MacDougal Pass the reverse was true. Teddybear cholla thrived at one location; at another it was dying. Remarkably little change occurred between the 1930s and the 1960s in the Central Gulf Coast province, where, with the exception of cardón and cirio, most plants did not even increase perceptibly in size. The abundance of most species stayed about the same, except for cardón, which became much denser on the Islas Mellizas.

Apparent Trends in the Vegetation

As we examine the photographs from the desert, we question whether we should expect the vegetation of this arid region to change through time in a manner different from that of the more moist zones we examined earlier. It is not clear the extent to which one ought to expect random and unsystematic fluctuations in the vegetation of an arid region as a concomitant of the high spatial variability in rainfall. On the one hand, we might argue that in the drier reaches of the desert, some localities, even over a period of many years, will receive more precipitation than adjacent ones and that the vegetation will reflect this unequal distribution by exhibiting a mosaic of stable, decadent, and thriving patches.

On the other hand, it is equally possible to argue that the present vegetation must be adapted to wide variation in rainfall, or it could not exist where it now does. Extremes resulting from spatial inequalities, even if continued from one year to the next over a period of many years, and reinforced by temporal variability, should mean little to a long-lived perennial, once it has become established. During the first few years of its existence, before it has developed an adequate root system or, in the case of the succulents, an adequate capacity for water storage, a plant may, indeed, be vulnerable. But to the saguaro, the paloverdes, the cardón, and possibly the mesquite —perennials that go on producing seed year after year for periods of well over a century—even the loss of many consecutive crops of plants should matter little. Over their lifetimes, local mosaics of abundance and paucity in rainfall surely must average out into smooth mean amounts along well-defined gradients.

It is not clear which of these two viewpoints ought to prevail. The problem, then, is which of the changes shown in the plates represent ran-

dom, local fluctuations and which, if any, indicate general conditions from which one might infer long-term variations in climate. All of the observations that follow have to be qualified in light of this basic uncertainty.

Although saguaros underwent dramatic population reductions at a few of our stations, especially those at Saguaro National Park (plates 60–64), this pattern was not upheld over the range of conditions sampled by all of our desert photographs. Of the twenty-six stations at which saguaros are seen, 48 percent showed an increase, 26 percent a decrease, and 26 percent no change during the period from 1915 (or earlier) to 1962. This net change of +22 percent is to be compared with a net change of –14 percent during the period from 1962 to the 1990s (based on values of 24 percent increase, 38 percent decline, and 38 percent no change).

For foothill paloverde at eighteen stations, increases during the earlier period were found at 72 percent, declines at 17 percent, and no change at 11 percent (net change of +55 percent). During the later period, there were no stations where the tree declined, but it increased at 64 percent of twenty-two stations and remained constant at the other 36 percent (net change of +64 percent).

Creosote bush is seen at ten stations where increases, decreases, and no change values were 40 percent, 30 percent, and 30 percent, respectively, for the first time period, and 30 percent, 30 percent, and 40 percent, respectively, for the second, providing net change values of +10 percent and 0 percent for the first and second time periods, respectively.

Mesquite, often associated more with grasslands than desert, registered an increase in three of four stations during the first period and then in all four stations during the second.

One of the most prominent members of the Chihuahuan Desert flora sampled by our photographs is Chihuahuan whitethorn. At the nine stations where this plant is seen, 100 percent reveal an increase during the early period, with 33 percent increasing and 67 percent remaining unchanged during the more recent period.

Cardón is seen at only two stations with early photographs taken during the pre-1915 era, and there the plant increased dramatically during both time periods.

Changes during the two time periods for the major long-lived species can be summarized as follows: for saguaro, there was a shift from stability to a slight decline; for foothill paloverde, a net increase during both time periods; for creosote bush, a slight net increase at first followed by no change; for mesquite, the net change was upward during both time periods but was greater during the second period than for the first; for Chihuahuan whitethorn, an increase at all stations in the first time period, followed by only a slight net increase later; and for cardón, a continuous upward shift during both slices of time. We will address possible reasons for the patterns of change among these species in chapter 7 and 8, where we will augment the limited photographic analysis provided in these last three chapters with a broadened analysis based upon additional matched photographs in our extensive archive.

Chapter Seven The Pattern of Change

In the three preceding chapters, we have described changes seen at the ninety-eight photograph stations featured in this book, using only the information available from those limited views. In this chapter we draw upon those stations and about 200 additional sets of photographs. Altogether, they represent 260 camera stations within the region discussed in this book.[1] Instead of examining the changes by vegetation zone as in the previous chapter, we will consider trends for individual species across all vegetation zones. In addition, we will review findings by others that may help explain the changes seen here. Of the seventy-two species identified in our photographs, only those reviewed below appeared in enough photographs to provide useful information. The net percentage change cited below is calculated by adding the percentage of photographs in which increased biomass occurred, subtracting the percentage in which a decrease was registered, and assigning a value of zero to the percentage showing no change in biomass (table 1).

Mesquite

The two species of mesquite occurred at more of our photograph stations than any other plant.[2] Mesquite was found from sea level to 5,500 feet and from desert to woodland. For the pre-1962 period, mesquite showed a net increase in biomass in 94 percent of the 148 stations at which it occurred; during the post-1962 period, there was a net increase of 70 percent at 159 stations (table 1). Our photography shows that, though mesquite increase slowed during the last one-third of the twentieth century, it has not stopped. The apparent slackening of pace shown by our photographs is partly the result of the 1978 freeze, which caused dieback of mesquites at several of our camera stations.[3] At some sites, mesquite has probably reached its maximum density and cover, preventing further gains.

The fruits of mesquite are eaten by livestock and by wildlife such as coyotes and deer. The seeds are covered by a hard coat that must be broken before germination will occur.[4] Seed coat perforation may be accomplished by passage through the alimentary canal of an animal, through scarification in floodways, or through bacterial/fungal activity. Seeds are disseminated by the animals that consume mesquite fruits and by flood waters along waterways.[5] The hard seed coat contributes to the seeds' long-term viability, which may ex-

ceed twenty years.[6] Grass fires kill mesquite seeds that remain on the soil surface. Burial at depths as shallow as 0.78 inches will prevent damage to seeds.[7]

Seed germination of mesquite probably occurs mainly during the summer rainy period, this timing the result of high temperature requirements for both germination and subsequent seedling growth.[8] The search for climatic ties to mesquite encroachment might, therefore, be directed at shifts in summer rainfall.[9]

Mesquite invasion is commonly thought to have begun just before the turn of the nineteenth century.[10] The proliferation continued at mid-century, and speculation that it would continue into the last part of the century has proven correct.[11] Without intervention, it is likely that mesquite will continue to gain in biomass at elevations lower than 5,500 feet, its approximate upper limit,[12] until it reaches some biomass density beyond which no increase can occur. As noted in chapters 4 and 5, the increase in mesquite is not confined to the Semidesert Grassland, as others have suggested,[13] but has also occurred in the Oak Woodland and the desert from the late 1800s onward.

At its upper, frost-induced limit (plates 1, 2, and 3) and along the upper San Pedro River Valley (plates 58 and 59), the tree is exposed to infrequent severe freezes that limit its growth and may actually kill some plants.[14] Many mesquites along the San Pedro River Valley are deformed as a result of frost, which can kill all branches back to the ground or, in larger plants, back to larger branches.[15] If the plant survives, numerous new branches are produced from living tissue just below the anatomical frost line, producing plants with a characteristic pollarded growth form (figure 7.1).

At its lower, drought-induced limit (plates 83 through 90), mesquite has undergone wide fluctuations during the past half-century. In a long-term study plot on the floor of MacDougal Crater, the density of mesquite in a 26-foot x 197-foot (8 meter x 60 meter) plot increased from 2 plants during the 1960–66 period to 413 plants in 1974, after which the number plummeted to only 80 plants in the year 2000 (figure 7.2).[16]

Saguaro

This large cactus was found at our photograph stations from sea level to almost 4,000 feet. We were

Table 1 Changes in Selected Sonoran Desert Species as Depicted in Repeat Photography

Species	1st Photo Dates			2nd Photo Dates			3rd Photo Dates			No. Photo Pairs		% Change	
	No.	Median	Range	No.	Median	Range	No.	Median	Range	1st Period	2nd Period	1st Period	2nd Period
Brittlebush	21	1932	1907–1935	23	1962	1958–1965	21	1990	1984–1998	21	21	+12	+5
Burroweed	9	1910	1890–1932	14	1962	1962–1968	13	1995	1993–1996	9	13	+66	−7
Chihuahuan whitethorn	19	1890	1880–1915	21	1962	1960–1962	18	1994	1994–1998	19	18	+95	+21
Cardón	18	1932	1903–1935	18	1962	1961–1965	18	1994	1984–1997	18	18	+56	+61
Cottonwood	30	1890	1883–1915	30	1962	1960–1971	29	1994	1994–1998	30	28	+60	+54
Creosotebush	20	1908	1890–1959	24	1962	1933–1985	24	1994	1984–1998	20	24	−11	+8
Foothill paloverde	68	1914	1890–1959	82	1962	1933–1985	77	1994	1984–2000	68	77	+61	+53
Goodding willow	8	1890	1889–1909	8	1962	1962–1968	7	1994	1994–1996	8	7	+75	+71
Jumping cholla	17	1928	1906–1956	17	1963	1960–1985	19	1995	1988–1998	17	17	+23	+35
Mescal	7	1890	1884–1915	10	1962	1960–1975	10	1994	1994–1998	7	10	−43	+70
Mesquite	148	1895	1883–1935	172	1962	1956–1975	161	1994	1985–1998	148	159	+94	+70
Oaks	58	1891	1887–1937	60	1962	1960–1973	49	1994	1989–1996	58	46	−33	+10
Ocotillo	58	1909	1880–1941	64	1962	1922–1985	65	1994	1984–1998	58	63	+55	+22
One-seed juniper	20	1902	1887–1925	24	1962	1956–1973	22	1994	1987–1996	20	22	+100	+100
Prickly pear	38	1915	1890–1935	47	1962	1960–1985	46	1994	1984–2000	38	46	+19	+70
Saguaro	82	1915	1890–1915	98	1962	1928–1985	91	1994	1984–2000	82	91	−10	−31
Sotol	9	1891	1887–1911	9	1962	1962	10	1994	1989–1995	9	9	+22	+45
Triangle leaf bursage	11	1928	1906–1941	13	1980	1928–1985	13	1994	1988–1998	11	12	+36	+42

able to evaluate biomass changes at eighty-two station pairs for the pre-1962 period and ninety-one for the post-1962 period. For the earlier period, there was a net change of −10 percent; for the latter period, the value was −31 percent (table 1).[17] Populations of this large cactus seemed to decline slightly during the first period, somewhat more strongly during the second. Diminishing biomass of saguaros is consistent with findings that numbers of individuals are decreasing throughout much of the range. This decline has been sharp since about 1970 at Tumamoc Hill[18] and since about 1960 at Organ Pipe Cactus National Monument and MacDougal Crater.[19] The decline seen in our photographs is apparently not a local phenomenon but a reaction to some persistent, regional influence that, if continued, could reduce the prominence of saguaro across the northern part of the plant's range.

Saguaro is highly dependent upon summer rainfall. Its short-lived seeds ripen and are dispersed with the onset of summer rains in July. They require warm temperatures for germination[20] and any that remain in the soil through the winter lose their viability before the return of warm, moist conditions.[21] Protection by rocks or "nurse" plants enhances survival among saguaros,[22] although not all species serve equally well in this role.[23] The life expectancy of individual plants is roughly 175 years.[24]

In our 1962 photographs of the Pinacate region, saguaro stands out as the only dominant plant that did not decline before 1962. At Mac-Dougal Pass, it increased notably (plates 83, 84, and 85) and in MacDougal Crater maintained itself (plates 86 and 87). Trees such as mesquite and paloverde experienced sharp losses at the same time. Although the saguaro population was not in decline in the crater or at the pass, there was speculation at that time that a decline was im-

Figure 7.1 Mesquite tree along San Pedro River Valley near site for plates 58 and 59 (1998). The plant's pollarded appearance almost certainly results from freeze damage during the period Dec. 6–10, 1978.

minent because woody plants, which provide the protection necessary for seedling establishment, were disappearing. From this it was posited that the disappearance of the saguaro would merely lag the disappearance of its shade-providing nurse plants, and that the current crop of small saguaros would be followed by far fewer plants in the future.[25]

A third of a century later, the trees at MacDougal Pass returned (plates 83–85), while in MacDougal Crater they did not (plates 87–90). To determine what happened to saguaros during this time, we estimated ages of saguaros in MacDougal Crater (in 1993) and at MacDougal Pass (in 1997).[26] Our sample at each site was a subset of the population. Saguaro establishment at MacDougal Pass (figure 7.3a) showed a marked decline during the decades between 1940 and 1970, followed by an abrupt increase in the 1980s and 1990s, a pattern similar to that of woody plants at the same site.[27] At MacDougal Crater, where the decline in nurse plants continued after the mid-century drought, the saguaro population also continued downward (figure 7.3b). The close parallel between saguaro and nurse-plant survivorship at these two sites may result from the interaction between nurse plants and saguaros. The possibility exists, as well, that woody plants and saguaros are responding similarly to local environmental conditions.

Cardón

This massive cactus, known only from northwest Mexico, is far more restricted in its mainland dis-

tribution than the saguaro.[28] Our photographs of cardón are all from the Gulf Coast. Seed germination probably depends on warm-season rainfall.

In eighteen triplicate photograph sets, initially taken between 1903 and 1935, cardón showed a 56 percent net increase to the mid-1960s, followed by a net increase of 61 percent afterward. Unlike saguaro, cardón appears to be increasing (table 1).[29]

In 1964, we measured the heights of all 5,836 cardones on one of the Islas Mellizas, the island shown in plate 97. Their density was more than 3,200 plants/acre. Of the 5,538 unbroken and uninjured plants, 94 percent were 5 feet tall or shorter. The few taller plants were mostly no more than fifty or sixty years of age (figure 7.4). This recent increase is duplicated on other nearby islands in Guaymas Bay and elsewhere in the Gulf of California, but not on the mainland. We do not know what caused the difference in cardón reproduction on these islands versus the mainland. Because they are free of most direct or indirect human impacts, they should provide natural laboratories for the study of vegetation change in the absence of human interference.

Foothill Paloverde

This widespread Sonoran Desert endemic was found at sixty-eight stations documented from 1914 to 1962 and at seventy-seven stations documented from 1962 to 1994.[30] During the earlier period, the net change was +61 percent. During the latter period, the net change was +53 percent (table 1). Thus, there appears to be a net increase in biomass of foothill paloverde during both periods, although in later years the change upward has diminished.[31]

Foothill paloverde flowers in the spring, the seeds then ripening just before the onset of summer rains.[32] For germination to occur, the hard seed coats must be broken, but until this occurs the seeds retain viability for several years in the soil.[33] They germinate only in the summer in response to moist, warm conditions.[34]

Death of foothill paloverde has been noted only rarely over the years. In 1911, Forrest Shreve wrote that "I have had a great many thousands of Palo Verdes come under my observation within a radius of 75 miles of Tucson, and have seen only two dead trees of full size." He noted that during ordinary drought the trees shed branches, a tactic that preserves the life of the individual. Since Shreve's time, it has become apparent that mature foothill paloverde does succumb to extreme drought. At Tumamoc Hill in Tucson, mortality was found to be episodic and more likely to affect senescent trees during moderate to extreme deficits in winter or summer rain.[35] The effects of the physical environment on seedling survival

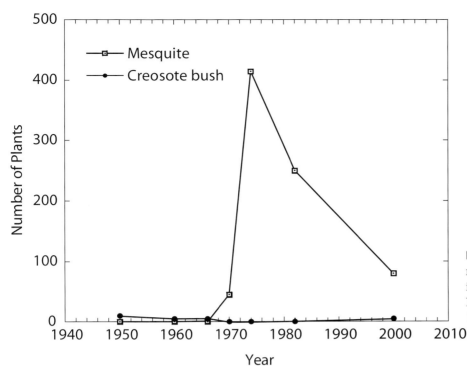

Figure 7.2 Number of mesquites and creosote bushes in a permanent plot in MacDougal Crater, 1950–2000.

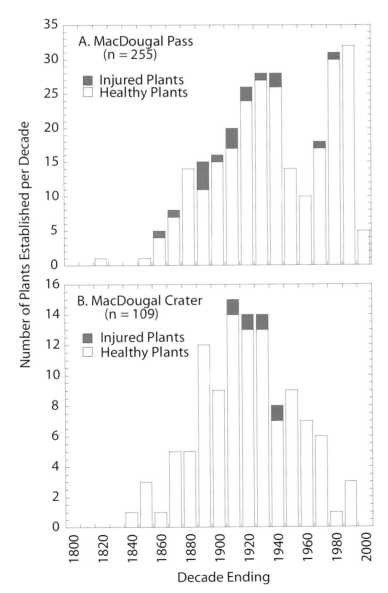

Figure 7.3 (a) Saguaro population age structure at four photograph stations in MacDougal Pass, 1997. (b) Saguaro population age structure in a permanent plot in MacDougal Crater, 1993.

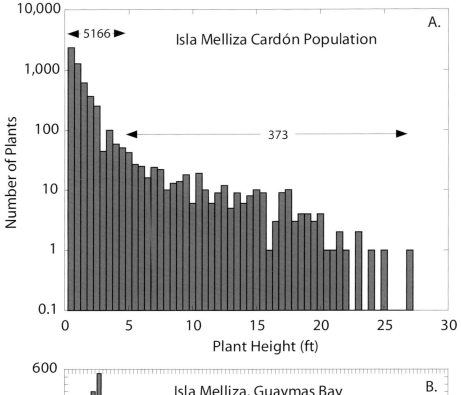

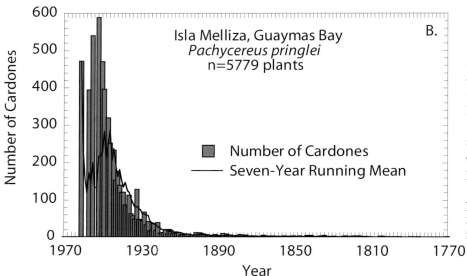

Figure 7.4 (a) Height distribution of all cardones (5,779) on one of the Islas Mellizas in Guaymas Bay in 1964. (Note the logarithmic scale for the *y* axis.) (b) Age distribution of the same population of cardones in 1964. Each bar represents a year; the tic marks along the x-axis define 2-year intervals.

were swamped by effects of the biological environment.[36]

In MacDougal Crater, a site that has been neither grazed by livestock nor chopped by woodcutters, dead plants outnumbered living ones in the 1960s (plates 87–89). The larger trees were probably 100–200 years of age when they died.[37] Their deaths were almost certainly a consequence of prolonged drought from the 1940s to the 1960s. The plants lost in the 1960s were not replaced by new individuals. In contrast, at a nearby station outside the crater, mesquite, foothill paloverde, and blue paloverde all suffered huge losses before 1962, but, by the 1990s, all three had rebounded to their 1907 levels (plates 83 and 84).[38]

More often than would be expected by ran-

dom chance, paloverdes grow in close proximity to saguaros. This nonrandom pattern presumably arises because paloverdes provide favorable conditions for saguaro seed germination and seedling survival, and because the birds that disperse saguaro seeds tend to perch and roost in paloverdes.[39] After many years, the close relationship between the two plants often turns lethal for the paloverde, apparently because the saguaro is more efficient at extracting moisture from the soil.[40]

Prickly Pear Cactus

This common cactus, found at our desert and grassland stations, has a history of change markedly different from that of the other plants noted so far. Prickly pears were depicted at thirty-eight stations during the decades before the early 1970s, where its net change was +19 percent. After the early 1970s, prickly pear appeared in forty-six views at which net change was +70 percent. In the set of recent photographs, mesquite and prickly pear increased at about the same rate, but in earlier views, mesquite outpaced prickly pear by a factor of five (table 1).

Prickly pear increased at grazed and ungrazed sites; thus, livestock grazing does not appear to be responsible.[41] Establishment of prickly pear probably depends on a sequence of favorable events, including abundant production of pads, followed a year or two later by enough winter rain for good fruit set, then, in another year or two, an unusually wet summer.[42] The plant is sensitive to freezing temperatures, and infrequent catastrophic freezes will deplete populations in frost-prone areas. Some stockmen burn off the spines with a torch to provide additional food for their animals during periods of drought.

Ocotillo

Seen at fifty-eight stations in the period ending around 1962, and at sixty-three stations after that, ocotillo, like several other plants in our photographs, increased during both periods with greatest increase before 1962. During the earlier period, the change was a 55 percent net increase; during the later period, the net increase was 22 percent (table 1).

As with prickly pear, establishment depends on a sequence of favorable conditions. To promote flower production, winter rains are probably a requirement. The seeds, which need warmth and moisture for germination,[43] disperse during the arid foresummer and are quickly consumed by animals. For many seeds to escape predation, summer rains must arrive early. Ocotillo may live to an age greater than seventy-eight years.[44] The plants are highly flammable and very sensitive to fire,[45] a fact of some significance in that part of their range lies within the grasslands of the Sonoran Desert region where fires burned frequently in the past.

Chihuahuan Whitethorn

This shrub appeared in photographs from within the Chihuahuan Desert or from stations nearby. Of the nineteen stations from the first time interval, Chihuahuan whitethorn showed a net increase of 95 percent. During the post-1962 period, the plant appeared at eighteen stations, where it registered a net increase of 21 percent. Although several other woody species in our study increased more during the first period than during the second, Chihuahuan whitethorn stands out among

these for the much greater increase during the earlier period than during the later one (table 1).

The seeds, which are long-lived, probably germinate only in response to summer rainfall. Wild and domestic animals aid in its dissemination and establishment.[46] Plants probably respond to fire in much the same way as whitethorn, a closely related species, which suffered 68 percent and 31 percent mortality, respectively, following two wildfires.[47]

Creosote Bush

This species registered an 11 percent net decline at twenty stations between 1908 and 1962, followed by an 8 percent net increase at twenty-four stations photographed mainly during the 1990s (table 1).

Within our region, creosote bush grows at elevations up to 3,500 feet, and it grows up to 5,000 feet in the Chihuahuan Desert.[48] Seeds germinate only in late summer or early autumn.[49] Flower production is impaired by excessive soil moisture, thus limiting seed production to relatively dry times and places.[50] Its resinous leaves make the plants highly flammable. Because they sprout only weakly when burned, the plants are extremely susceptible to fire damage. However, fires in the desert are uncommon because there is little fine fuel to carry them.[51] Creosote bush typically produces expanding clonal rings that persist for centuries, and some individuals of the plant may be among the oldest living individual organisms.[52] This species has expanded its range significantly during the past century in southern New Mexico.[53]

Establishment is an exceptionally slow process on new geomorphic surfaces and may not occur until the new surface has been exposed for roughly 1,000 years.[54] The rate of encroachment into already established plant assemblages is an order of magnitude less.[55] Adaptation of creosote bush to extreme aridity involves, in part, its ability to maintain net photosynthesis when leaf water content is exceptionally low.[56] Thus, in soil dominated by creosote bush roots, moisture removal continues to levels well below those proving lethal for most associated species. Ratany has circumvented this competitive advantage by parasitizing the roots of creosote bush,[57] an ability that may help explain the remarkable loss of creosote bush shown in plate 81.

One-seed Juniper

The one-seed juniper[58] is found above the desert in Semidesert Grassland and Oak Woodland. Of the twenty stations with photographs spanning the initial period and the twenty-two stations spanning the later period, it increased at 100 percent of the stations during both times, making it unique in our study (table 1). In the Oak Woodland zone, one-seed juniper often becomes established beneath living or dead oaks, presumably because its seeds are dispersed by perching birds. The matched photographs show a steady increase within grassland communities in central Arizona during the twentieth century.[59]

Jumping Cholla

From 1928 until 1963, seventeen stations showing jumping cholla recorded a net increase of 23 percent; during the later period, from 1963 to 1995, the net increase was 35 percent at seventeen stations (table 1). Although jumping cholla produces flowers, it seeds are mostly sterile, hence reproduction is largely asexual. Fruits and stem segments can root after falling to the ground, especially those falling from late winter to midsummer.[60] The lifespan, about forty to eighty years, is moderately long.[61]

Oaks

To obtain a large sample for analysis, we grouped all species of arboreal oaks together. Our pooled data set includes fifty-eight sets of photographs spanning the period from 1891 to 1962 and forty-six sets in the second period, ending in 1994. The net change in oaks was –33 percent during the first interval, +10 percent in the second period (table 1). These results show that whatever caused the loss of oaks during the first part of the twentieth century did not continue into the last part.

Our initial photographs were taken after the period of massive oak harvesting, hence the subsequent decline in oaks cannot be attributed to that spate of woodcutting.[62] The oaks in our region are adapted to fire and sprout after being burned, so we doubt that fire is responsible for the decline, either. In our photographs, oak dieback occurred primarily at the lower edge of Oak Woodland—that is, at the arid lower limit of oaks in our region.

Cottonwood

Cottonwood occurred at thirty stations in the period from 1890 to 1962. During this period, the tree experienced a net increase in 60 percent of the photograph pairs. For the second time interval (1962–94), twenty-eight photograph pairs recorded a net increase of 54 percent (table 1).

Growing beside perennial and ephemeral streams or in sites with shallow groundwater, cottonwood reproduces primarily by seed and also by sprouting after stumps and branches are buried by flood deposits. The ability to stump-sprout de-

clines with age.[63] The seeds, which ripen in late winter and early spring, remain viable for only a few weeks[64] and germinate almost exclusively on the freshly exposed alluvium deposited as spring flood waters recede.[65] Its ability to withstand fires is dependent upon time of year and fire intensity. Cattle eat the seedlings and may prevent successful seedling regeneration.[66] Our finding that recruitment is affected more by flooding and the creation of suitable habitat than by grazing pressure was anticipated by earlier work.[67]

Goodding Willow

A shrub or tree to 60 feet tall, Goodding willow grows near perennial or ephemeral streams and other locations where soil surfaces are flooded seasonally. The seeds, which are shed in the spring, are viable for only a few days and germinate best on freshly deposited alluvium.[68] Goodding willow sprouts vigorously from the root crown[69] and probably sprouts following fires.

Comparison of eight photograph pairs spanning the period from 1890 to 1962 shows a net increase in biomass of 75 percent. For the period ending in 1994, seven pairs record a net increase of 71 percent.

Mescal

This agave was seen in seven photograph pairs spanning the first period (1890–1962) and ten pairs covering the later period (1962–94). There was a net change in biomass of –43 percent during the first period and +70 percent during the second (table 1).

The claim that numbers of mescal were reduced following the 1950s as the result of a decline in nectar-feeding bats has been challenged.[70] Analysis of the few photograph pairs in which this plant is seen suggests that a decline did occur prior to the 1960s but that there has been a sharp increase subsequently. The plant reproduces mainly by seeds; after the seeds are produced, the plant dies.[71] The seeds are apparently long-lived, remaining in the soil seed bank at least through the first winter after maturation.[72] The plant's partial resistance to fire is probably related to habitat conditions; the scant fuel supply in its preferred rocky habitats does not promote intense fires.[73] Establishment may be enhanced by disturbance.[74] Some stockmen cut the flowering stalks to augment livestock feed during periods of drought.

Brittlebush

Brittlebush is a low shrub, typically with a compact hemispheric crown. Its dry-season leaves are densely covered by white hairs, whereas leaves produced during moist periods are green and nearly glabrous. This shrub was found at twenty-one camera stations and changed little during the two time periods, first with a net increase of 12 percent, and then of 5 percent (table 1).

The main period of flowering is in the spring, although the plants may flower as early as November.[75] The seeds remain viable in the soil for a year or more, and they typically germinate from October to April.[76] This is a relatively short-lived species, with maximum longevity in one study found to be thirty-two years, although few individuals survived longer than seven years.[77] Unlike many species of woody plants in our matched photographs, individuals of this short-lived species probably do not persist from one photograph to the next. Brittlebush is frost sensitive, a trait contributing to its short-term, sporadic occurrence at some sites.[78]

Burroweed

A small, short-lived semiwoody shrub, burroweed is found in the Sonoran Desert and the higher grasslands. The seeds of this plant are shed in the fall and germinate in response to cool season rainfall. It is known to increase on depleted rangelands. Burroweed registered a net increase of 66 percent in nine photograph pairs spanning the period from 1910 to 1962. It declined by 7 percent in thirteen pairs taken between 1962 and 1995 (table 1).

Triangle Leaf Bursage

Found at only eleven and twelve camera stations representing the first and second time periods covered by our analysis, this plant showed a net increase of 36 percent and 42 percent, respectively, at those stations (table 1). This small shrub, an important member of the paloverde-saguaro plant community, is found along the eastern and northern portions of the Sonoran Desert. Triangle leaf bursage has a shallow root system, and its leaves are rapidly lost during seasonal dry periods. New leaves are produced when rains return. Triangle leaf bursage is confined to those parts of the Sonoran Desert where rainfall biseasonality is the norm. It flowers mainly from February to April,

and its seeds germinate following heavy autumn or late winter rains. Although not a desirable forage plant for cattle, it is readily used by horses and other equids.[79]

Sotol

This leaf succulent appeared in only nine photograph pairs from each time period and had a net increase of 22 percent and 45 percent during the first and second periods, respectively. This plant, often called "desert spoon," may develop a short trunk upon which the whorl of narrow leaves is borne. The shoot is sensitive to fire and will die back to the soil if burned. Plants sprout readily after fire. Where fires are frequent, sotol takes the form of low-growing leaf whorls and can be confused with beargrass when in the background of photographs. This possible confusion resulted in our omitting many photograph pairs from the analysis.

Summary of Changes Shown by Extended Photograph Series

Of the eighteen species shown in table 1, fourteen species showed a net increase in biomass during the first time period and four species showed a net decrease. During the second time period, the values were sixteen and two species, respectively. The foregoing tabulation shows that, in those instances where increases were gained during both time periods, in seven cases out of thirteen, the increase was greater during the earlier period; in five cases, the increase was greater in the later period; and in one case, there was no change. Of the four species that declined during the earlier period, all but one, saguaro, increased during the later one; saguaro's decline was more pronounced in more recent times. Just as saguaro stands out as the only species to decline during both periods, one-seed juniper stands out as the only species to increase at 100 percent of the photograph stations during both periods. The reduction in the pace of biomass increase in most of the species may be the result of habitat saturation or it may be caused by other factors, such as change in herbivore pressures or shift in climate. These topics will be examined in the following chapter.

Chapter Eight **Change and Cause**

As shown in the original *Changing Mile*, vegetation change in the Sonoran Desert region before the 1960s was almost startling in its extent. We have now found that changes through the next third of a century occurred with only slightly slackened pace, if the pace changed at all. Although these changes, most of which involved increase in biomass of native plants, were most pronounced at elevations above the desert, the changes were surprisingly large at many desert sites as well. Among the most thoroughly sampled species, the increase through time was often at different rates during the two time periods sampled by our photographs. In a few instances, the species declined through time. That so many different species with vastly different autecological traits were caught up in these changes illustrates the complexity of the situation. For example, population fluxes may have been tied to shifts in rainfall during the critical season for seed germination, or to rare events such as dissipating tropical cyclones.[1] Some plants may respond to the recurrence interval of catastrophic freezes. Physiological differences between C_3 and C_4 plants (see chapter 1) may be another source of complexity. The possible role of global warming must be examined as a force behind some of the changes.

Our sequence of photographs depicts deaths among oaks, mesquites, saguaros, and ocotillos. In some cases, only scattered plants have died; in others, massive die-offs have been revealed. Whether death has resulted from natural attrition, climatic extremes, or human activities is a key question in our search for causes of vegetation change.

Although death is inevitable, replacement of dead plants with new ones is not. We suspect that differences in germination requirements may dictate much of the variability in establishment. Different species germinate in response to different climatic cues, and seed longevity determines whether germination will occur when favorable conditions do arrive. Examples include saguaro, which has relatively short-lived seeds that require substantial summer rains for germination, and brittlebush, which has longer-lived seeds that germinate in response to good winter rains. Our photographs show pulses of establishment in both these species. Their episodic pattern of recruitment may reflect interannual variation in climate, especially rainfall.

The degree of complexity suggests that no single factor should be sought as the ultimate driving force behind all the changes observed. Because the effects of one force may be much like those of another and because several forces may have reached a critical intensity at about the same time, determining specific causes for vegetation change is difficult, if not impossible. Furthermore, networks of interactions may amplify any force, affecting the entire ecosystem and causing counterintuitive changes such as local extinction of ant and rodent species following abundant cool-season rainfall.[2]

Thus, the increase of woody plants in the grasslands could stem from livestock utilization, from decreased fire frequency, from increased atmospheric CO_2, from changes in rodent numbers, from some climatic shift such as more winter rain, or from these forces acting together. Of the many possible agents of vegetation change in the desert region during the twentieth century, seven—woodcutting, rodents and other wildlife, increased atmospheric CO_2, domestic livestock, fire, soil erosion, and climate—have been assigned major roles by different researchers. Some of these potential forces have relatively local effects; others embrace the entire region. We consider each in the following pages.

The Effect of Woodcutting

Fuelwood cutting may once have had a major impact upon the evergreen woodlands and mesquite thickets near major mines of our region.[3] According to one study, much of the lower Oak Woodland in southeastern Arizona was cut over for fuelwood by the 1900s, and harvesting continued well into the twentieth century.[4] Tombstone's bonanza period (1879–86) was the time of most intense timber harvest from that town's "woodshed." Several of our photograph stations lie within this woodshed near Tombstone's mines.

Had photographs been taken of the Oak Woodland prior to the peak woodcutting period, we could compare a relatively pristine woodland with one that subsequently had been exploited. Unfortunately, by the time of our earliest photographs, the Oak Woodland had already been heavily harvested, and no such comparison is possible.

Fuelwood cutting within hauling distance of mines clearly had a potential for a great effect.[5] Much of the impact might have occurred at sites for which we have no photographic documentation. The photographs we have seen suggest that woodcutting was neither uniform nor universal. In plate 2, loss of oaks apparently did result from fuelwood cutting. However, two plates show oaks growing near active mines where ricks of firewood are piled nearby for fueling the mill's furnaces.[6] At one of these stations, many of the standing trees seen in the earliest view were still alive in the 1990s. At the other station, their fallen remains have persisted. The presence of living trees close to large wood-consuming operations suggests that the fuel source was not necessarily local. Apparently the trees growing near wood-burning furnaces were not always targets of the woodcutter's axe.[7]

Several of our photographs have shown Oak Woodland declining (plates 15, 16, 17, 22, 23, 24, 25, and 26). These declines are not always associated with previous mining operations. They are, however, consistently found at the lower, arid edge of the woodland, and, in most instances, the remains of the dead oaks are still lying on the ground. The pre-1960s losses have been reversed only slightly during the succeeding third of a century. These observations strongly support the view that woodcutting is not closely associated with the oak decline.[8]

Several of our photographic sequences illustrate the expansion of riparian woodlands into areas that were mainly grassy valley bottoms at the time of the earliest photographs (plates 6, 7, 10, 27, 42, 43, 46, 47, 48, 50, 51, 56, and 57). The former lack of trees along these valley bottoms has been attributed to fuelwood cutting.[9] We see no stumps or other signs of woodcutting activity in the early photographs, although many of these date from the period shortly after woodcutting began. We also see no sign of minor forest elements, such as shrubby forest underlings. In addition, many of the historical accounts describing conditions before heavy woodcutting began leave a clear record of predominantly open, treeless expanses on valley floors. Undoubtedly some cottonwoods were harvested for window lintels, and trees such as mesquite were burned as fuel. This does not mean that there was available as a wood source anything comparable to the essentially continuous riparian forest that became established later.

In summary, fuelwood cutting was undoubtedly severe in some areas, but its influence was local and transient.

The Effect of Rodents and Other Animals

Evidence from a variety of sources suggests that small mammals such as rabbits, mice, prairie dogs, and kangaroo rats play a role in the depletion of grass cover and in the establishment and elimination of woody species.[10] Although some early work indicated that the effect of rodents and jackrabbits is greatest on heavily grazed ranges,[11] more recent studies suggest that small herbivores can dramatically alter ecological systems in the absence of livestock.

Researchers have long recognized that the activities of black-tailed jackrabbits (*Lepus californicus*), kangaroo rats (*Dipodomys* spp.), and white-throated woodrats (*Neotoma albigula*) promote shrubs at the expense of grasses. Not only do they eat many seeds, but rodents and rabbits also consume as much as 40 percent of the grass forage produced. In one study, removal of three species of kangaroo rats from cattle-free grassland allowed grasses to increase significantly; at the same time, rodent species typical of grassland also increased.[12]

By harvesting, distributing, and planting seeds, rodents further encourage the spread of shrubs such as mesquite and whitethorn acacia.[13] Some researchers have argued that native seed dispersers such as rodents were ineffective in promoting encroachment of mesquite onto grasslands. Thus, "prior to the introduction of domestic livestock, limited seed dispersal may have been the primary constraint to *Prosopis* invasion of grasslands."[14] Other evidence, however, indicates that rodents, particularly Merriam's kangaroo rat (*Dipodomys merriami*), may in fact be more efficient at seed dispersal than cattle, and that cattle act primarily as seed predators rather than disseminators.[15]

Rodents are common in the desert as well, where they devour saguaro seedlings and even the succulent tissues of mature saguaros.[16] Woodrats[17] may be alone among rodents of the Sonoran Desert in their ability to consume the outer parenchyma of saguaro without suffering ill effects from oxalates present in the issue. Saguaro stems are consumed mainly in habitats lacking either prickly pear or cholla, which are apparently preferred water sources.[18]

The beaver *(Castor canadensis),* another rodent that can have considerable impact on vegetation, was once abundant along the rivers of southern Arizona and northern Sonora, including the Santa Cruz and San Pedro.[19] These animals were heavily trapped in the early part of the nineteenth century and for a while disappeared from Arizona rivers. According to some views, the extirpation of beaver had two major impacts: downcutting of stream channels,[20] and development of riparian forests. Because beaver populations rebounded by the 1850s,[21] several decades before downcutting began, the first view seems untenable. The second view also seems unlikely. To flourish, beaver require cottonwood and willow for food and for building dams. Both of these riparian trees grow rapidly after cutting; thus the impacts of beaver on woody riparian vegetation should be transient and local. If extensive riparian forests had existed at the time of our earliest photographs, therefore, we should see the evidence in numerous views. It seems likely that the low incidence of cottonwood and willow before the 1890s reflected not suppression by beaver but spread of fire from adjacent grasslands and perhaps other causes.[22] Prior to downcutting, beaver dams in the valleys would have promoted shallow water tables that helped exclude such woody plants as mesquite.

Beaver have recently been returned to the upper San Pedro River through the efforts of the Bureau of Land Management.[23] Elsewhere in our region, they have again disappeared as a result of reduced stream flow, conflicts with farmers near rivers, and other causes.

Some of the changes seen along the San Pedro drainage may be related to recent extirpation of black-tailed prairie dogs (*Cynomys ludovicianus arizonensis*), which once flourished there. Through the early part of the twentieth century, this ground squirrel was abundant along the San Pedro Valley from Benson south to the Mexican border and beyond.[24] In 1892, it was present in the vicinity of Monument 98 on the U.S.-Mexico border (plates 42 and 43).[25] Active pursuit by government exterminators has completely extirpated prairie dogs from the San Pedro River Valley in the United States.[26] Prairie dogs remove woody plants from grassland sites,[27] and their disappearance from the San Pedro River Valley may have contributed to the increase of mesquite and other shrubs. Changes along the Santa Cruz River Valley and elsewhere to the west and south cannot be attributed to prairie dogs, which were known only as far west as a site northwest of the Huachuca Mountains.[28]

The unique pattern of increase shown by oneseed juniper invites speculation. Its seeds are disseminated by birds that transport the berry-like cones to perches in other trees or shrubs, scattering some seeds along the way and depositing others beneath the perches.[29] As woody plants spread into former grasslands, perching birds would have followed, carrying juniper seeds with them and distributing them from new perches. The spread of one-seed juniper would thus have waited upon establishment of forerunner shrubs and trees. In any case, its delayed and continuing rapid spread suggests that its encroachment has not yet begun to abate.

Except for a very few species, populations of most rodents and other animals have been relatively stable over the period of change depicted in our photographs. Their role in mediating changes is therefore secondary. In the one example of complete rodent extirpation, the loss of prairie dogs from the San Pedro River Valley is probably not wholly responsible for the increase of woody plants in grasslands: a parallel increase occurred nearby in grasslands where prairie dogs were unknown.

The Effect of Domestic Livestock

Plates 22, 43, and 44 show grasslands in southeastern Arizona as they appeared in 1890 and 1891. These views, recorded just before or at the beginning of a lengthy drought, illustrate the impact that cattle can have on grass-dominated terrain. When grazing animals are removed and drought ends, however, such grassland can quickly respond, like an abandoned closely trimmed lawn, until the former biomass is restored. Regrowth no doubt occurred in our region during good years; nevertheless, abundant historical evidence indicates that the net effect was not in the direction of recovery. At the end of the nineteenth century and beyond, especially in the vicinity of water sources, the grasslands were virtually denuded of their cover and remained in this condition for many years.

The rapid expansion of the cattle industry in southeastern Arizona in the 1880s happened at about the same time as the onset of vegetation change. The chronological association invites the conclusion that one event is causally related to the other. Certain experimental work suggests mechanisms that might be involved, such as scarification and dispersal of seeds by livestock.[30] Plate 8 shows what is apparently an example of this joint effect near old Fort Crittenden, where a marked mesquite invasion has taken place inside an old corral. Clearly, there is often a close interaction between livestock and mesquite. However, as certain photographs from MacDougal Crater demonstrate (plates 87 and 90), mesquites may reproduce profusely and spread from site to site in the complete absence of livestock.[31]

Domestic grazing animals may contribute to the spread of woody plants in several different ways. In one important role, they act as disseminators. As many as 1,617 undigested mesquite seeds have been found in a single cow "chip," for instance.[32] The seeds of mesquite and other leguminous invaders (catclaw, whitethorn, and Chihuahuan whitethorn, in particular) have hard, impervious coats that must be scarified before water can enter and the seeds can germinate. The passage of these seeds along the alimentary tracts of livestock, such as cattle, and wildlife, such as deer and coyotes, provides the necessary scarification.[33] Mesquite-seed scarification in MacDougal Crater must be accomplished by some nonbovine agent such as bacterial/fungal activity, rodents nicking seed coats, flood-water transport, or wetting/drying cycles.[34]

That cattle can and do disperse seeds of woody invaders is not in question. Whether or not seed dispersal by cattle was a major force in converting grassland to shrubland is another matter. In the millennia between extinction of large native herbivores[35] and introduction of cattle, grassland in our region was not invaded by shrubs even though small native herbivores were present and were presumably spreading seeds of woody plants.[36] The persistence of grassland then suggests that the local distribution of woody plants was not limited primarily by seed dispersal. If so, seed dissemination once cattle were introduced could not have been a primary cause of grassland conversion.

It has been noted that livestock can hinder woody plant establishment by consuming the seedlings of mesquite and other woody plants[37] or by consuming flowers and herbage of mature plants.[38] Countering this argument is the *de facto* evidence that, despite livestock grazing, seedling survival has been high enough to greatly increase the number of woody plants on southern Arizona rangelands and riparian areas since the turn of the century.

Another way in which cattle were thought to contribute to shrub encroachment was by opening up grassland communities that in the past, because of their closed nature, were able to repel shrub establishment.[39] Recent work, however, suggests that this intuitively logical model may not describe reality; survival of mesquite seedlings on grass tussocks may be comparable to that of seedlings grown on bare interspaces between the grass plants.[40] Furthermore, observations in many historically cattle-free exclosures have shown that vigorous stands of grass do not preclude the establishment of mesquite[41] and that woody plant biomass may increase at a significantly greater rate in livestock-free pastures than in adjacent actively grazed pastures.[42] One recent exclosure study in southeastern Arizona reported a three-fold increase in woody plants and major reorganizations of the associated biota since 1977.[43] These exclosure studies fell short, however, of providing a clear lesson about the role of cattle, because fire, an important natural force that would decline in frequency under cattle grazing, was not part of the experimental design.

Our time-lapse photographs show that, during the early part of the twentieth century, woody plants increased as livestock grazing became prevalent across the Southwest and that they continued to spread after the number of cattle declined in the 1960s.[44] This suggests that grazing may be somewhat disengaged from woody plant encroachment; if so, grazing may not be the prox-imate cause of shrub invasion. The relationship between grazing and vegetation change is evidently not direct. Other factors also play important, if not more important, roles.

In the earlier edition of this book, we argued that climate, not livestock grazing, was behind the expansion of woody plants into grasslands.[45] We reasoned that, because intensive livestock grazing in Sonora in the 1820s and 1830s had produced no long-lasting result, then a comparable level of use after the 1880s would not necessarily have caused the conversion of grasslands into shrublands.[46] Recent findings suggest, however, that ranching in Mexico in the early nineteenth century was a transient and modest enterprise,[47] not at all comparable in intensity to that of the late nineteenth century. What was once considered a boost to the "climate change" hypothesis no longer appears relevant.[48]

In summary, cattle do not seem to be the proximate cause of the most common changes we have observed, although their influence is undeniable. The presence of livestock would have resulted in increased soil erosion and reduced fire frequency, for example—two important forces behind those changes.

The Effect of Fire

Fire is a natural force whose past frequency and intensity has helped shape the vegetation of our area as we know it. Anglo-American efforts to suppress fire appreciably lowered the incidence and extent of burning in our region, especially in the Semidesert Grassland and Oak Woodland. Widespread grazing by domestic animals, principally cattle, during the past century also reduced fine fuels over much of our region, limiting the spread of fires in low density shrublands or woodlands. Although a significant proportion of fires is started by humans, the potential for ignitions derives to a larger extent from natural forces such as "dry-lightning" strikes unrelated to man's overt actions. Furthermore, there is strong evidence that fire frequency is closely tied to large-scale climatic fluctuations.[49] However, once the fires are started, their confinement to relatively small areas often follows from elimination of fine fuel by grazing livestock, from active suppression measures, or from interception by artificial firebreaks such as roads.

At lower elevations, increased production of fine fuel resulting from encroachment by exotic plant species has been noted in several areas of the Sonoran Desert.[50] These new establishments have been on a minor scale and have not been responsible for many large fires. Perhaps with time the influence of fires in the desert will increase as exotics become more widespread. In our region, then, fire as a factor in vegetation dynamics should be most important in the Oak Woodland and Semidesert Grassland, where grasses provide abundant fuel, rather than in the desert proper, where fuel is usually insufficient to carry fire.[51]

Consideration of fire as a force for change involves two basic questions. The first has to do with the effectiveness of fire in controlling woody plants: does recurrent burning suppress them? The second concerns the historical frequency of burning: before Anglo-Americans arrived, how frequently did fires occur in the Semidesert Grassland and Oak Woodland?

To stem encroachment by woody species, fires must be frequent enough and hot enough to eliminate seedlings and saplings, many of which have a strong propensity to sprout from the root crown when the shoots are killed. The effectiveness of fire in maintaining shrub-free grasslands is supported by experimental results and observations in the wild.[52] For instance, mortality of burroweed, a small, unpalatable shrub common on deteriorated ranges, is typically high during dry-season fires.[53] Small mesquites (stems less than 6 inches in diameter) are highly vulnerable to fire; although larger plants can be more resistant,[54] flowering on new stems of these larger plants is suppressed for at least one season following fire.[55] A further important influence of fire is that surface-sown seeds of velvet mesquite and white-thorn acacia failed to germinate after being subjected to a grassland burn, whereas seeds buried as if by rodents were unaffected.[56] Obviously, fire and woody plants coexisted at some level throughout our region because there had to have been a seed source for the species that ultimately spread across the grassland. Such seed sources were preserved in areas of consistently low fuel load and wherever fire failed to reach.

Recent historical research and dendrochronological studies of forests bordering grasslands have revealed that the Semidesert Grassland did indeed burn frequently before the arrival of Anglo-Americans.[57] Some fires were undoubtedly set by humans; many more originated in lightning strikes, which occur across the landscape with high frequency. It appears, therefore, that the fire hypothesis is confirmed; even if the Semidesert Grasslands of this region do not owe their origin to fire, they were once maintained (at least in part) by frequent and extensive burning.

The paradox of Oak Woodland existing adjacent to Plains Grassland before the arrival of Anglo-Americans has been raised.[58] One is forced to reckon with the problem of two life zones, intimately intermingled along their border. One zone was kept free of woody species through burning; the chief characteristic of the other was its woody dominants; and the two were linked by a continuous grass matrix capable of conducting fire. This apparent paradox is resolved by recent findings for two important Oak Woodland species, Emory oak and Mexican blue oak. Both sprout vigorously following fire.[59] Large oaks can withstand ground fires; moreover, seedling establishment of oaks improves after fire. Ground fires burning through Oak Woodland thus did not destroy it. In grassland, however, fire was much more effective in removing mesquites, which sprout less vigorously than oak and are not as well protected by large size.[60] The difference may explain how woodland and grassland persisted side by side during the early period of frequent fires. Certainly, during recent decades, when fire frequency has been greatly reduced, both zones have become favored habitats for mesquite.

Demographic evidence also helps to explain the paradox. To start, the dominant oak species in the woodland have average lifespans of 200–250 years.[61] In addition, their establishment is apparently a regular and constant process.[62] In terms of fire, a newly germinated cohort requires a recurrence interval long enough that the seedlings can grow to fire-resistant size. If wildfire sweeps through the woodland before this happens, the cohort will be wiped out. But if the period between fires is long enough, the juvenile plants can grow into a stand of mature trees. Given the long lifespan of oaks and their regular recruitment, fire could recur rather frequently and still permit the existence of Oak Woodland.

Data from several different sources indicate that this scenario closely reflects the actual situation. In lower Oak Woodland (and probably in Semidesert Grassland, as well), fires historically occurred about every 10–20 years.[63] Given a diametric growth rate of 0.14 inch per year,[64] a ten-year-old Emory oak 1.38 inches in diameter could be large enough to resprout after being top-killed by fire, and a twenty-year-old Emory oak 2.76 inches in diameter could resist even topkilling.[65] Thus, the reported interval between fires is consistent with the time required for young oaks to reach fire-resistant size.

In the Semidesert Grassland, fire frequency was probably about the same as in the Oak Woodland, but the dispersal of woody plants in grasslands was somewhat different. Before the advent of cattle-raising, mesquite in much of our grassland was mostly confined to drainage channels. Plants that became established away from the water courses were killed by frequent fires, while those located on bare mineral soil along runnels survived. Scarification in turbulent floodwaters no doubt enhanced germination of many woody plants along runnels.

After about 1880, grassland fires decreased in frequency at roughly the same time that heavy grazing on Arizona rangelands began.[66] The timing suggests that cattle and other livestock, by consuming grasses and forbs, removed the fuel source that once carried fire across vast areas. In addition, the U.S. Forest Service policy of fire suppression, instituted in the early 1900s, lessened the number and size of fires in grasslands. Together, heavy grazing and fire suppression could constitute strong forces that allowed woody species to encroach upon areas formerly closed to them. Although the pace of grass-to-shrub conversion in fire-free parts of our area during recent decades appears to have been hastened by increased winter rainfall, grazing by cattle may have actually slowed this conversion. In areas protected from both grazing and fire, it is projected that grasses will be eliminated or greatly reduced from much of the system as the importance of woody components expands.[67]

The increase of one-seed juniper is of special interest with regard to fire suppression. Its net change during both periods covered by our photographs was +100 percent (table 1). No other species showed such a consistent, large increase. Given the plant's known sensitivity to fire, its recent spread may be entirely the result of diminished fire frequency during the past century.

To sum up, reduced fire frequency is probably the primary reason why woody plants spread from watercourses onto adjacent grasslands after 1880. The open grasslands typical of earlier times were able to persist only as long as frequent fires acted on a fuel-rich landscape free of close-cropping herbivores. With the introduction of domestic livestock, reduction of fine fuel was combined with more efficient seed dispersal, simultaneously reducing the risk of fire and increasing the spread of woody plants. Although our photographs offer no rigorous experimental evidence to support this conclusion, abundant historical evidence shows that fire was much more frequent in the past than is now the case.[68] A recent study on cattle-free pastures compares mesquite and grass biomass on sites with varying fire frequencies. On a site burned less than one time per decade over a thirty-year period, the ratio of mesquite to grass biomass was 116:1; on the site with four fires per decade, the ratio was 1:2.[69] Whether or not increasing fire frequency will return former grasslands to their original state will probably depend on the amount of soil erosion that has occurred during the fire suppression period. If the earlier soil condition has been heavily altered, then return to the original shrub-free state may no longer be possible.

The Effect of Soil Erosion

Although the impact of livestock grazing on soil characteristics has received considerable attention, few studies provide rigorous comparisons of soil traits between grazed and ungrazed pastures.[70] Soil erosion studies in our region, utilizing manually clipped plots to simulate grazing, have shown that removal of plant cover has only a small influence on erosion; reduction of the rock veneer on the soil surface has a greater effect.[71] If accelerated erosion occurs, surface horizons that contain seedbanks and nutrients could be removed, making germination and establishment of new plants more difficult and potentially changing the types of species that could occupy the site.

Any attempt to return landscapes to previous plant assemblages must come to terms with the issue of soil erosion since the advent of domestic livestock grazing. It is likely that, in many instances, soil physical and chemical traits may have been changed too much for a return to some previous condition. The pace of soil formation is measured on a millennial scale; that of erosion by times spanning only decades. If soil structure—in particular, the surface A horizon (rich in nutrients and organic material)—has been removed, new types of plant assemblages may irreversibly occupy the sites.

Arroyo Formation

The twenty-year period from 1875 to 1895 saw the inauguration of arroyo cutting in Arizona, New Mexico, Sonora, and Utah. Anglo-American settlement in these states took place at various times from 1598 (in New Mexico) to the 1870s. Grazing commenced at equally diverse times, although, as noted in chapter 3, channel erosion occurred along many valleys of our region in the late 1800s at the same time that grazing impacts were at their peak.[72] It is clear, however, that arroyo formation is not dependent on livestock grazing: since about 4,000 years ago, when downcutting was initiated during an earlier era along the San Pedro and Santa Cruz Rivers, there have been five cattle-free episodes of arroyo formation that correlate well between these two valleys.[73] The heterogeneous history of the region makes it difficult to link settlement, and therefore cultural factors, with arroyo cutting; the relatively synchronous onset of erosion, however, points to the operation of a broad, regional factor such as a climatic fluctuation.[74]

No matter what the cause, downcutting has had a huge impact on the region's biota. The encroachment of woody plants into riparian areas follows much the same timing as their encroachment into adjacent grasslands. Before the turn of the century, channels of the San Pedro River and other regional streams, such as Cienega Creek, were largely free of trees and shrubs, covered instead by grasses such as sacaton and tobosa. Prior to the floods of August 1890, the Santa Cruz River above and below Tucson supported several ciénegas, much grass, and some forests.[75] Several early photographs of these valley-bottom grasslands show them to be free of stumps, suggesting that there had been no woods for woodcutters to cut. By mid-twentieth century, these channels were densely covered by forests of mesquites, willows, and cottonwoods[76] and, in general, remain densely covered today.

Oddly, several authors have argued that riparian woodland is disappearing in our region.[77] This perception—that the health and vigor of rivers has declined—has been supported by wildly inaccurate reports. In an account about the San Pedro River, one can read that "steamboats once navigated all the way to Charleston to supply Tombstone and Bisbee with goods."[78] In another, one reads: "When Anglo-Americans first came to the Southwest, much of the Gila River was navigable [by steamboat]."[79] Undeniably, dam construction, water diversion, groundwater pumping, and other forces have altered the hydrologic regimes of many of the region's rivers, resulting in dramatic losses of fish and other aquatic species and in the encroachment of exotic plants.[80] Vegetation along such rivers as the Lower Gila and Lower Colorado, where dam impacts are pronounced, has undergone dramatic changes not at all similar to those of our region's rivers. The changes below major dams should not serve as models for the changes observed elsewhere. Along the San Pedro and Santa Cruz Rivers, both of which are free of dams and their dramatic impacts, our photographs show not a *loss* of vegetation but a major *change* in types of vegetation and an increase in woody biomass. For example, the San Pedro Riparian National Conservation Area was established to maintain riparian forests that, as our pictures show, have replaced former grasslands only during the twentieth century.

The sequence of changes leading to riparian forests apparently was initiated on valley floors where swards of sacaton underwent seasonal flooding. Conditions of alternating wetting and drying on these low-lying, heavy soils promoted the growth of grasses and the exclusion of woody plants.[81] With downcutting came a drop in the water table, accompanied by an end to seasonal flooding and to alternating anaerobic and desiccated conditions. Woody plants such as mesquite were able to occupy the newly created terraces. The deepened channel provided stretches of open mineral soil ideal for the establishment of cottonwood and willow. With subsequent channel widening,[82] the bands of trees broadened and became multiaged, gallery forests.[83]

Dramatic changes in bird life have accompanied the alteration of grassland to forest. The grasslands were prime habitat for many sparrows, including Baird's and Botteri's, both of which are now "species of concern":

> Until about 1878 [Baird's sparrow was] an abundant transient and doubtless winter resident in the grasslands of southeastern Arizona . . . ; until 1920 decidedly uncommon but still a winter resident about the bases of the Chiricahua and Huachuca Mountains. Now apparently much rarer. . . . [Botteri's sparrow is a] rather uncommon summer resident, . . . [u]sually in giant sacaton grass. . . . Formerly much more common, especially before 1895.[84]

The forests of cottonwood and willow that replaced sacaton grasslands support a broad array of new species, including the southwestern willow flycatcher, a federally listed endangered species. The change from grassland to forest came at a time when livestock grazing was heavy, and, although cattle have been shown to consume cottonwood and willow seedlings,[85] apparently sufficient seedlings were left behind to promote establishment of the grand forests that are the subject of much well-deserved interest and concern today.

The Effect of Climate Change

The aspects of climate that are possible forces in driving vegetation change are increased carbon dioxide (CO_2), temperature change, and changes in precipitation. Temperature and precipitation may be under partial control of shifts in ocean temperature and circulation and can be examined together in this context.

Increased Atmospheric CO_2

Increases in atmospheric CO_2 can have profound effects on plant growth, changing the ability of species to utilize environmental resources, particularly limited resources such as water.[86] The possibility that increasing CO_2 has contributed to woody plant encroachment into grasslands has received recent support.[87] For example, one study documented increased nitrogen fixation, increased root biomass, and increased water-use efficiencies in greenhouse-grown mesquite seedlings exposed to high CO_2 concentrations.[88] In contrast, mesquite seedlings grown with a C_4 perennial grass had none of these improvements, indicating that competition with grasses may negate the beneficial effects of elevated CO_2.

Models have been developed to help anticipate biotic changes in various parts of the world as a result of doubling of atmospheric CO_2.[89] One such model shows the retreat and advance of potential growth areas for saguaro and creosote

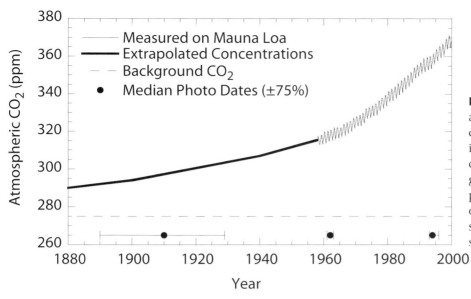

Figure 8.1 Changes in atmospheric carbon dioxide concentrations on Mauna Loa in Hawaii. This station is widely considered representative of global CO_2 concentrations. Box plots indicate the dates of the original photographs, the first set of matches, and the second set of matches.

bush as CO_2 reaches a value twice that of its 1960 level by about the year 2050.[90] These models suggest that the increases in atmospheric CO_2 may have profound effects on the distribution of vegetation in the Sonoran Desert, and that many of these effects are yet to come.[91]

The primary data showing the twentieth-century increases in CO_2 are shown in figure 8.1.[92] The 34 percent increase in CO_2 over the past 200 years has had a preferential fertilizing effect on woody plants, most of which have the C_3 photosynthetic pathway. In contrast, C_4 plants, which in our region are mostly grasses, benefit from lower CO_2 concentrations. As shown in figure 8.1, the period between the first set of photographs (median date = 1910) and second set (median date = 1962) experienced only a slight increase in atmospheric CO_2 (ca. 20 ppm) compared to the interval between the second and third sets (median date = 1994; increase = ca. 50 ppm). From this information, increases in atmospheric CO_2 concentrations would more likely have affected plant growth in the latter half of the twentieth century than earlier. Our photographs also show a large increase in plant biomass during the earlier period; thus, the increase in atmospheric CO_2 and the onset of woody plant expansion were not closely correlated in time. For this and other reasons, the likelihood seems remote that the vegetation changes in our photographs derive from this source.[93] We should look elsewhere for the basic causes of vegetation change.

Temperature Change

Temperature change can affect vegetation in many ways. Shifts in the frequency, duration, or value of daily winter minima have the potential for shaping some segment of the region's biotic communities. A change in the length of the growing season or in the value of summer maxima can also be expected to influence some component of this complex system. Saguaros, mesquites, and other frost-sensitive species will be killed or damaged by lowered temperatures. Oaks may be negatively affected by increased summer highs, and these and other C_3 species may benefit from a longer growing season.

The possibility that global temperatures are increasing is a topic receiving abundant attention from both scientists and politicians. Although most models predict increases in temperature, they do not agree on likely trends in precipitation. Winter temperatures are generally expected to increase more than summer temperatures, reducing the frequency of catastrophic frost and greatly lengthening the growing season. At least one model anticipates more precipitation in all seasons, which greatly complicates any prediction of ecological responses in our region.[94] Fluctuating temperature, as an important variable of our environment, is not merely a recent public obsession. During earlier years of the nineteenth century, climate was thought to be stable. Then it became apparent during the latter half of the nineteenth century in the higher latitudes of the Northern Hemisphere that glaciers were retreating, that timberline was advancing upward and northward, and that the waters of the North Atlantic were becoming warmer.[95] During the first part of the twentieth century, from 1910 to the 1950s, a warming trend was again in effect.[96] There followed, during the period from 1950 to 1970, an interlude of cooling. Since the 1980s, warming of global magnitude has been detected.[97]

There is no question that average annual temperatures are fluctuating at scales from the local to the hemispheric. Figure 8.2 shows that the 1880s

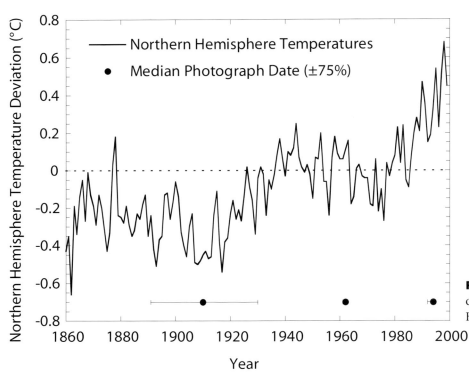

Figure 8.2 Temperature deviations in the Northern Hemisphere.

were relatively warm; this is when the "Little Ice Age" is generally thought to have ended.[98] Temperatures dipped again just before and after 1910. By the mid-1920s, temperatures began climbing to a plateau that was reached in the early 1940s; after twenty years of relatively high values, temperatures retreated slightly, during the period from about 1960 to the mid-1970s; they then climbed again by about 1.1°F to the mid-1990s. According to this evaluation, which minimizes urban heat effects, temperatures in the 1990s were nearly 2°F higher than in the 1860s. Whether the temperature increase is related to the increase in atmospheric CO_2 and other "greenhouse" gases is fiercely debated at present with little in the way of conclusive scientific evidence.[99]

Temperatures in southeastern Arizona have also changed, but the trend more resembles a roller coaster than a directional increase (figure 8.3). As in the Northern Hemisphere, warming has been the trend since about 1975, but southeast Arizona differs in that temperatures were at similar levels around the turn of the century, in the early 1940s, and in the 1990s.[100] The amplitude of the fluctuation is more than 9°F (figure 8.3). The photographs we use were taken at three key points along this temperature curve. The first set, with a median date of 1910, is near the end of a cooling trend following the nineteenth-century high. The second set, with a median date of 1962, is toward the end of another cooling trend following the mid-century temperature high. The last set, with a median date of 1994, is near the maximum of the recent upturn in temperatures.

Nighttime temperatures have increased more than daily highs, even when corrected for nighttime heating effects associated with urban environments.[101] In the rural environment of Tombstone, maximum temperatures have changed little over the twentieth century, but minimum temperatures have increased, particularly in January (figure 8.4). The increase in minimum temperature between the 1950s and 1990s is about 8°F. One of the obvious manifestations of this increase is a decrease in the frequency of severe freezing.[102] In the early part of the twentieth century, severe freezes occurred in southern Arizona about every two to three years. The pace slackened somewhat during mid-century, and during the past forty years severe freezes have been rare, occurring at intervals of a decade or more (figure 8.5).[103]

The higher temperatures—particularly increases in daily low temperatures—are expected to have profound biotic effects. Decreased frequency of severe freezes will increase seedling establishment as well as increase the survival of frost-sensitive species. A less obvious effect of the warming trend is the increase in growing season, which will favor C_3 species, provided cool-season moisture is not limiting.

Moreover, given the spatial relation between temperature and elevation, an increase of 3°F would have the same effect as dropping about 1,000 feet in elevation. Since one can go from the Oak Woodland across the Semidesert Grassland to the upper edge of the Desert in this vertical distance, it appears that the observed temperature

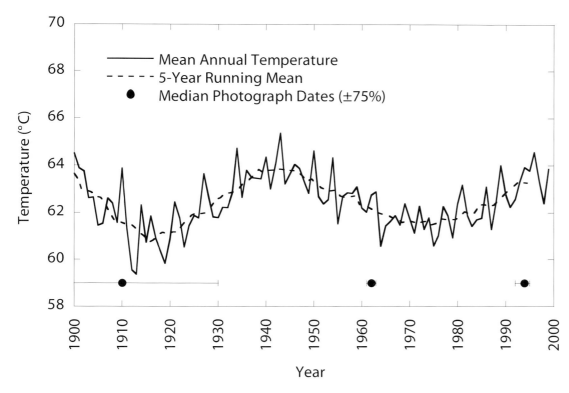

Figure 8.3 Temporal variation in mean annual temperatures in southeastern Arizona.

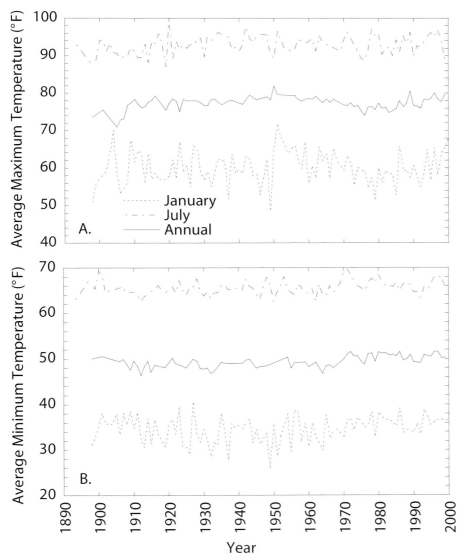

Figure 8.4 Temperature variations at Tombstone, Arizona. (a) Monthly maximum temperature. (b) Monthly minimum temperature.

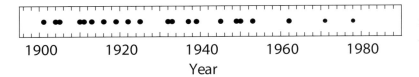

Figure 8.5 Occurrence of severe freezing temperature in Arizona. (Modified from Webb 1996.)

fluctuations since the nineteenth century might have significant effects on plants, particularly those on the edges of their elevational distributions.

Terrestrial bird and mammal populations have clearly responded to the increased temperatures. A recent review of range shifts among mammalian and avian species in the southwestern United States has shown that more than 70 percent of the fifty-five new arrivals since 1890 have come from the Sierra Madre or the scrublands of Sonora and Sinaloa.[104] The remainder are derived from six other mainly temperate biotic realms. The increased density of woody vegetation in both the grasslands and woodlands of the region has undoubtedly fostered the advances of some of these species. Yet, whether or not the increased vegetation density is the result of climatic change, we are left with the possibility that climate change alone acted directly on these animal species.

Precipitation Change

Changes in climate in the southwestern United States during the twentieth century have been documented and disputed.[105] Much of the disagreement appears to center over subtle changes, such as seasonal shifts in precipitation or temperature, rather than large changes in annual conditions. Even so, many observers have documented extreme droughts, although the effects were not long-lasting enough to constitute "climate change." For example, using dendrochronology, E. Schulman documented a severe drought in the middle of the twentieth century in the southern Gila River Basin and inferred effects on native plants:

> It appears highly likely, in view of the general parallelism with the chronologies in Colorado and Utah, that this is the most severe drought since the late 1200s. If this is correct we have, then, a direct climatic explanation for the high mortality which has been observed in recent years in *Carnegie [sic] gigantea* (sahuaro) and in the low-level, woodland

pines *Pinus leiophylla* var. *chihuahuana* and *P. engelmanni*.[106]

The earliest analyses of climate focused on systematic changes in annual precipitation, and none were found.[107] In the early records for Santa Fe and three other New Mexico stations, L. B. Leopold noted a shift toward fewer small rains and more large ones with no significant changes in annual precipitation between about 1880 and 1940. Discussing purported vegetation changes caused by this change in rainfall frequency, and in reference to downcutting of arroyos in the region, Leopold stated that:

> Such a circumstance must have been conducive to a weak vegetal cover and relatively great incidence of erosion. That the modern epicycle of erosion began in the Southwest about 1885 is well established. . . . We see, then, that not only was grazing tending to promote erosion at that time, but meteorological conditions were more conducive to erosion than during the period of the present generation. Thus there is established concrete evidence of a climatic factor operating at the time of initiation of southwestern erosion which no doubt helped to promote the initiation of that erosion.[108]

Leopold's analysis for New Mexico has been criticized[109] because the trends at essentially one station are extrapolated to an entire region. However, R. U. Cooke and R. W. Reeves, in an analysis of precipitation in Tucson, agreed with Leopold's findings.[110]

In *The Changing Mile*, an analysis of the rainfall records since 1895 for five stations in southeastern Arizona showed similarities to Leopold's results. In particular, winter precipitation after 1920 greatly decreased while summer precipitation showed a smaller decline. In an analysis of precipitation between 1898 and 1959 at eighteen stations in Arizona and western New Mexico, W. D. Sellers concluded that "the decrease has occurred almost entirely during the winter months, although there has also been a slight decrease in

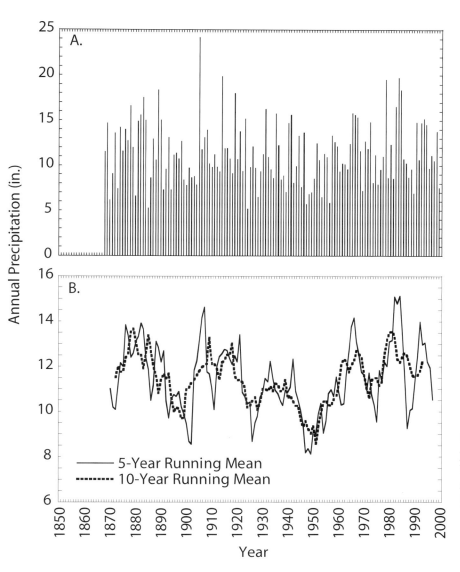

Figure 8.6 Annual precipitation at the Tucson University of Arizona climate station. (a) Annual data. (b) Running means.

summer. None of the trends is statistically significant."[111] Between the early 1920s and 1959, the twenty-year average annual precipitation in Arizona and western New Mexico decreased by about 25 percent. Other researchers in adjacent regions came to similar conclusions.[112]

To update *The Changing Mile*, we analyzed rainfall and temperature data from forty-nine climate stations in southern Arizona and Sonora (appendix 1). We grouped these climate stations into the regions depicted in the plates: Oak Woodland, Semidesert Grassland, Arizona Uplands, Lower Colorado Valley, and Central Gulf Coast. Monthly and annual precipitation was converted to standardized seasonal values,[113] which eliminates problems arising when comparing stations with markedly different total amounts of rainfall. We combined monthly values into two seasons: summer (July–September) and winter (November–March). These two seasons represent precipitation for the growing season of C_3 plants (winter and summer) and C_4 plants (summer).

To evaluate annual changes, we chose the long rainfall record for Tucson University of Arizona (1868–present) as being reasonably representative of the region as a whole. Annual precipitation over the 132-year record has ranged from about 5 to 24 inches with only a few sustained periods of either high or low rainfall (figure 8.6a). A graph of the five- and ten-year running means shows some short, sustained periods of drought and above-average precipitation. If an annual drought in Tucson can be defined as precipitation below 10 inches averaged over five years, five distinct droughts—centered on 1902, 1926, 1949, 1975, and 1987—appear in the record (figure 8.6b). Similarly, if wet periods are defined as precipitation greater than 13.5 inches averaged over five years, six wet periods—centered on 1876, 1882, 1907, 1966, 1983, and 1993—appear in the record. From the ten-year averages, two droughts —late 1890s and early 1950s—and three wet periods—1880s, 1910s, and 1970s–80s—are obvious.

The standardized precipitation graphs for the

five regions (figures 8.7 to 8.16) show complex shifts in annual and seasonal precipitation over the 110 years between 1890, at about the median time when the first set of photographs was taken, and 2000, just after the last set of photographs was taken. The graphs of standardized annual precipitation show a pronounced drought from the early 1890s through 1904–05 in all regions except the Gulf Coast, where no rainfall data were collected in this period. Less severe droughts occurred in the 1920s, from the mid-1940s through the mid-1950s, in the mid-1970s, and in the mid-1990s. In contrast, the wet periods revealed in the long Tucson record are mostly present in the regional records. Because the median date for the first set of photographs is 1910, many of our earliest photographs document highly variable but mostly drought-related conditions. The middle set of photographs, with a median date of 1962, documents sustained drought conditions typical of the mid-twentieth century. Our final set of photographs documents conditions after a sustained wet period between 1977 and the early 1990s.

Seasonal precipitation varies considerably year to year. Considering just the Semidesert Grassland (figure 8.10), winter or summer precipitation (but not both) was above average for fifty-one years between 1900 and 1998; of these, twenty-eight years had above-average winter precipitation with below-average summer precipitation.

Fluctuations in winter precipitation appear to be the source of much of the variability in annual precipitation. For example, in the Oak Woodland, the droughts defined by annual precipitation (figure 8.7) are mirrored by winter precipitation deficiencies (figure 8.8), not summer, with the exception of some years in the 1950s. In the Arizona Uplands, the 1891–1906 drought was the result of consistent summer and winter deficiencies; the 1920s drought was ameliorated by several years of above average rainfall during either summer or winter; and the 1940s–50s drought was broken by above-average amounts of summer moisture in the late 1950s and 1960s. In general, winter precipitation follows a less random pattern than does summer precipitation. In all, these graphs support the idea that means of seasonal precipitation in our region have not been stable through time; in other words, climate is nonstationary.[114]

In a twenty-one year study immediately to the east of our region, the increased shrub cover resulting from a series of unusually wet winters has had a cascading effect on the rest of the system.[115] The second-most abundant rodent, a grassland kangaroo rat (*Dipodomys spectabilis*), went locally extinct. At the same time, a shrubland rodent (*Chaetodipus bayleii*), not seen at the site earlier, has become common. Two species of harvester ants (*Pogonomyrmex rugosus* and *Aphaenogaster cockerellii*) have become locally extinct. In the absence of mound-building rodents, Mojave green rattlesnakes (*Crotalus scutulatus*) and burrowing owls (*Athene cunicularia*) have declined throughout the surrounding area. Thus, through intensive study of the many elements in an ecosystem, it is becoming apparent that certain species hold specific functional roles which maintain the continuity of the system and that short-term climatic shifts can have profound effects, providing a preview of the effects of long-term changes.

The intricate manner in which rainfall controls seedling establishment is shown in a study of black grama during the warming period in the early part of the twentieth century. Only seven years between 1915 and 1968 produced seedlings of this native grass at a site in southwestern New Mexico.[116] The black grama "seedling years" were all within the warming period from 1915 to 1940 and consistently "occurred during periods of winter drought, sandwiched between periods of winter wet, but only when the dry winter was followed by high summer rainfall."[117]

Changes in the seasonality of precipitation would have differential effects on C_3 versus C_4 species, both in terms of growth of existing plants and germination of new ones. Many shrubs and trees, which mostly have C_3 photosynthetic pathways, favor winter precipitation for seed germination and growth, whereas grasses, which mostly have C_4 photosynthetic pathways, are more responsive to summer rainfall. Although the seeds of many shrubs in our region germinate in the winter, velvet mesquite and creosote bush, two of the most common plants of our region, are not among these.[118] An increase in winter precipitation would favor germination and establishment of certain small shrubs such as brittlebush and burroweed. Growth of velvet mesquite might be stimulated when winter soil moisture carries over into spring periods with temperatures high enough for this plant to respond.[119]

Although there has been a shift in rainfall seasonality in our region that corresponds weakly with increases in mesquite biomass, a cause-and-effect relationship seems remote. The period between 1895 and 1917 is a case in point. The first eleven years of this period included only three years of above-average precipitation, and those occurred only during the summer. During the following eleven years, ten had above-average values, all from winter precipitation. This latter period coincides with the first observed range expansion of mesquite. Twelve years between 1980 and 1998 had above-average winter precipitation,

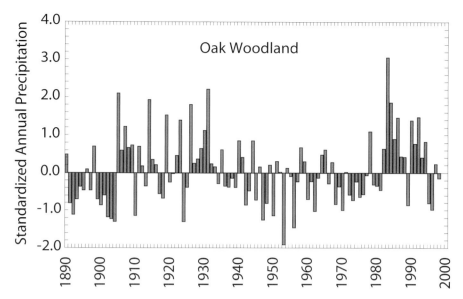

Figure 8.7 Standardized annual precipitation in the Oak Woodland.

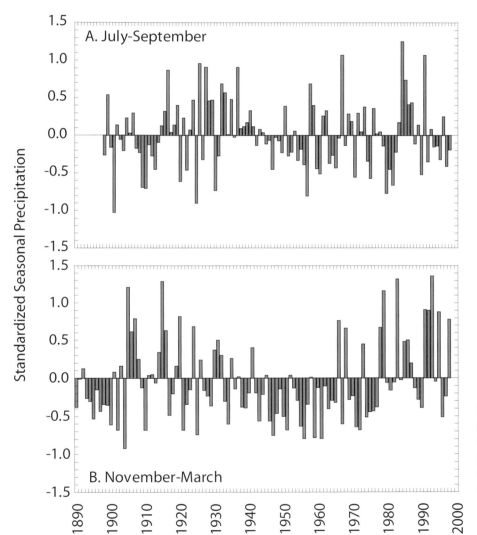

Figure 8.8 Standardized seasonal precipitation in the Oak Woodland. (a) Summer precipitation (July–September). (b) Winter precipitation (November–March).

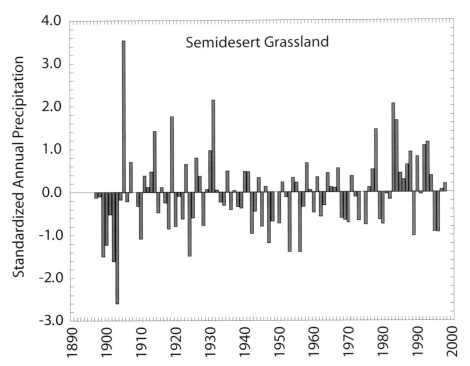

Figure 8.9 Standardized annual precipitation in the Semidesert Grassland.

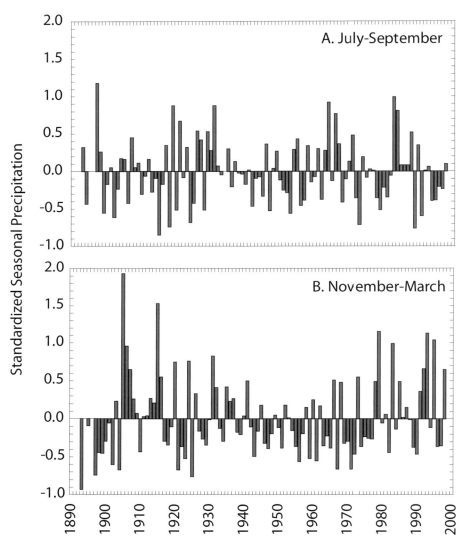

Figure 8.10 Standardized seasonal precipitation in the Semidesert Grassland. (a) Summer precipitation (July–September). (b) Winter precipitation (November–March).

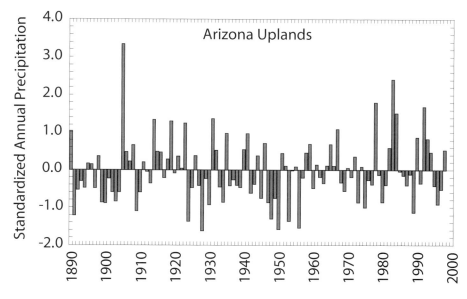

Figure **8.11** Standardized annual precipitation in the Arizona Uplands.

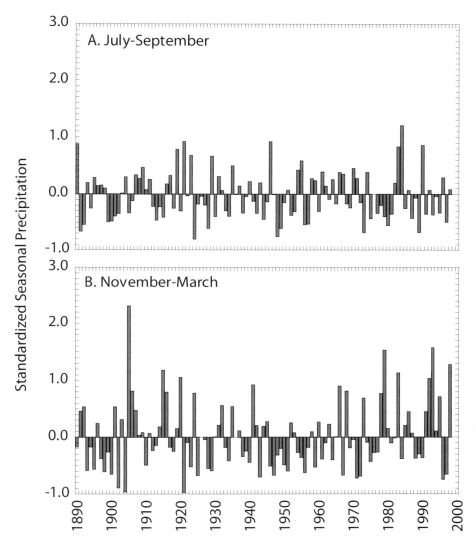

Figure **8.12** Standardized seasonal precipitation in the Arizona Uplands. (a) Summer precipitation (July–September). (b) Winter precipitation (November–March).

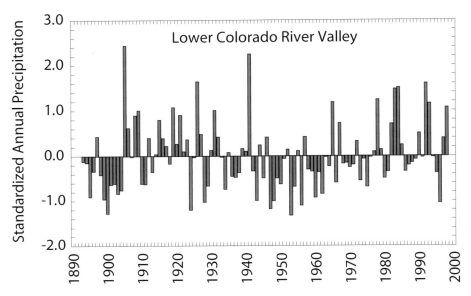

Figure 8.13 Standardized annual precipitation in the Lower Colorado River Valley.

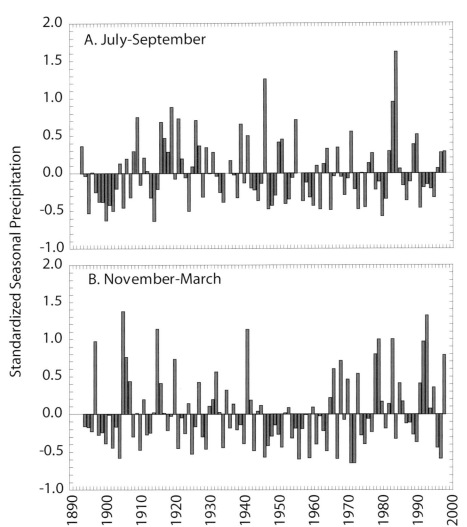

Figure 8.14 Standardized seasonal precipitation in the Lower Colorado Valley. (a) Summer precipitation (July–September). (b) Winter precipitation (November–March).

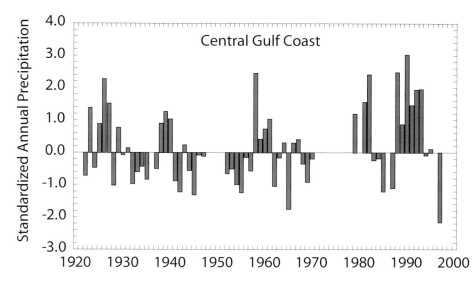

Figure 8.15 Standardized annual precipitation in the Central Gulf Coast.

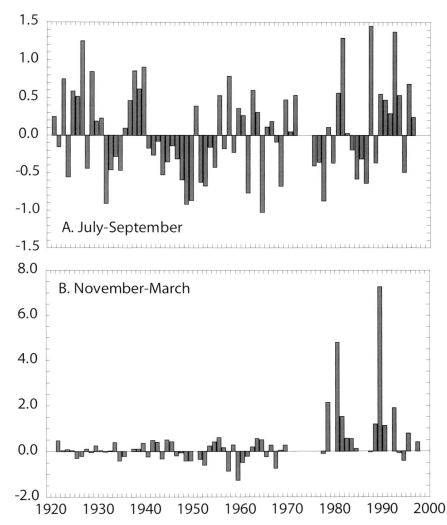

A. July-September

B. November-March

Figure 8.16 Standardized seasonal precipitation in the Central Gulf Coast. (a) Summer precipitation (July–September). (b) Winter precipitation (November–March).

compared with only twelve years of above-average winter precipitation during the previous four decades. The recent wet winters might have promoted growth of certain shrubs, but probably not the growth of mesquite that is seen in our photographs. It would appear that changes in seasonal precipitation are not the proximate cause of rangeland conversion from grasses to mesquite.

The observed loss of oaks came near the end of a drought that exceeded in length, but not depth, the drought at the beginning of the 1900s. Assuming that oak losses seen in the 1962 photographs actually resulted from the 1950s drought, one wonders why the earlier drought did not also take its toll on the oak population. It may be that drought duration is a more telling adversary for the oaks than drought intensity. Also, drought might have more severe effects in an old population than in a young one, and our field observations indicate that many oaks which died during the 1950s were old plants. Another possible explanation is that oaks lost out to mesquites and other woody plants in the competition for scarce water resources, especially at the lower edge of the Oak Woodland, where mesquites had greatly increased in number and biomass in the preceding decades.[120] If this is what actually occurred, then interspecies competition brought on by grazing and fire suppression must be given consideration, along with climate change.

Indices of Regional and Global Climate

Fluctuations in the Pacific Decadal Oscillation (PDO), an index that reflects changes in sea-surface temperatures in the North Pacific Ocean,[121] have been linked to long-term variability in drought and wet cycles in the United States. The effects are particularly pronounced in our region.[122] As shown in figure 8.17, the pattern of fluctuations during the twentieth century is not unlike the temperature curves for southeastern Arizona (figure 8.3). Low values of the PDO indicate reduced sea-surface temperatures, and these are correlated with drought in the southwestern United States. High values of the PDO indicate a higher potential for increased rainfall in our region. Our first set of photographs was taken during a period of positive PDO, in accord with higher precipitation at that time; the second set was taken after a period of sustained low PDO, which corresponds to the mid-century drought; and the third set was taken during the highest period of PDO in the twentieth century (compare figures 8.6 and 8.17).

The Southern Oscillation Index (SOI), the most commonly used indicator of El Niño/La Niña conditions in the equatorial Pacific Ocean, reflects short-term climatic conditions. Positive values of the SOI indicate La Niña conditions, which gen-

erally herald drought in Arizona, whereas negative values of the SOI indicate El Niño conditions, which result in variable but typically wet conditions in the state.[123] El Niño conditions tend to increase winter precipitation in southern Arizona,[124] though summer precipitation is not significantly affected. Some researchers have pointed out interactive relationships between the PDO and the SOI.[125]

Although the frequency of El Niño conditions has not changed significantly in the twentieth century,[126] the amplitude of the SOI has varied considerably (figure 8.18). The SOI was highly variable in the period between 1880 and 1920, when only a fragmentary record was assembled. The highest positive value of the SOI was associated with the intense drought of the mid-1890s, for example, and the 1905–6 El Niño was one of the strongest of the twentieth century. Similarly, the period between about 1960 and 2000 was also highly variable, with pronounced droughts in the mid-1970s, mid-1980s, and mid-1990s and strong El Niño in 1982–83, 1993, and 1997–98. The middle part of the twentieth century had relatively low fluctuations in the SOI.

The effects of the SOI on precipitation were lower between 1920 and 1950 and higher afterward.[127] Prevailing thought holds that the combination of negative SOI (El Niño conditions) and positive PDO (high sea-surface temperatures) yields increased rainfall in the Southwest.[128] Figure 8.18 shows that the dates of our photographs correspond to changes in patterns of the SOI over the past 120 years, and therefore the photographs appear to document conditions at the end of distinct climatic periods in the Sonoran Desert.

The Effect of Urbanization

Several of our photograph sets bear witness to an expansion of direct human influence on the landscape, typically as distant houses on background hills or plains. In at least one instance, however, the original camera station had to be abandoned when human influence approached more closely. This happened with the original plate 95, which in 1932 and 1965 was on the outskirts of the village of Puerto Libertad. By the 1990s, the town had expanded and the camera position was engulfed by houses. The camera station has been abandoned. Similarly, in plate 14, the camera was moved to avoid a building on the original camera station. Several of our recent photographs were taken within a few hundred yards of new houses (plates 35, 42–44, 48, 68, 74, and 81), and one (plate 41) was made beside an interstate highway. With time, such examples will only multiply, and more of the original camera stations will lose their usefulness.

Encroachment of housing developments into

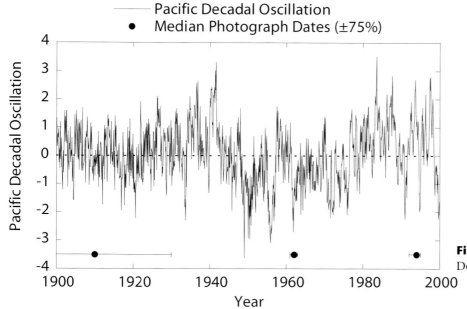

Figure 8.17 The Pacific Decadal Oscillation (PDO).

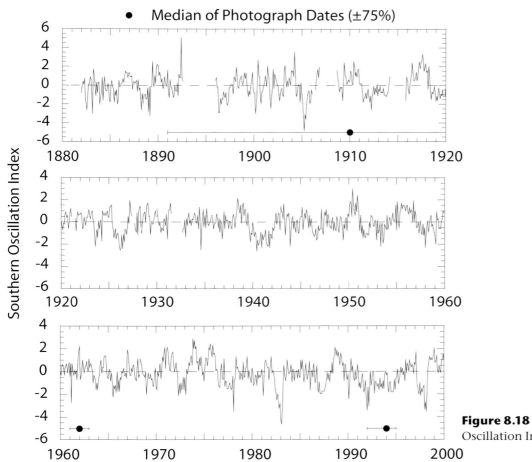

Figure 8.18 The Southern Oscillation Index (SOI).

formerly open land brings numerous forces that hasten the course of change. At elevations where plant communities once experienced frequent burns, for example, fires are now vigorously suppressed in the vicinity of houses. Few homeowners in fire-prone vegetation types provide adequate fire protection for houses and outbuildings. The result is that managers of nearby government lands are unable to restore natural fire regimes even when this is clearly a desirable goal.

Housing developments also serve as epicenters for the spread of exotic plants. Plate 82 is located on the Desert Laboratory grounds, a fenced preserve on the outskirts of Tucson, Arizona. At the time of the earliest photograph, three species of exotic plants grew within the preserve, all on the side nearest town.[129] After nearly a century, the Desert Laboratory has become surrounded by houses and other artifacts of development and is now home to an additional fifty-six exotic plant species.[130] We can safely assume that the scattered houses appearing next to some of our camera stations have promoted the spread of exotic plants to some degree and that the pressure will only increase in the future.

A Tangle of Hypotheses

Of the five general explanations for the vegetational changes documented in our replicated photographs—cattle, woodcutting, rodents, fire suppression, and climate—the rodent and woodcutting hypotheses can be rejected as regional causes. The evidence from sites like MacDougal Crater and the Islas Mellizas indicates that some vegetation changes have occurred where cattle have never been, and are, therefore, probably the result of climatic variation alone.[131] Oaks at the lower edge of the woodland probably succumbed to a combination of aridity and competition from woody invaders from below. The fossil record suggests that climatic stress alone can turn grassland into shrubland,[132] but, historically, it seems likely that a combination of heavy grazing, fire suppression, and climatic stress allowed invasion by woody plants. Unfortunately, because grazing and drought coincided toward the end of the nineteenth century,[133] followed shortly by deliberate fire suppression at the beginning of the twentieth,[134] it is virtually impossible to tease apart the changes that are due to natural causes—a drier and hotter climate—from those due to anthropogenic causes—livestock and fire suppression.

We suspect that certain vegetational changes will prove reversible, especially in cases where population fluctuations are cyclic in nature. For example, at Saguaro National Park East (plate 64), where the grand old cactus forest all but disappeared between the 1930s and the 1980s, a new cactus forest is developing now.[135] In other cases,

irreversible changes have occurred. Where grazing has been protracted and intense, such as around natural water sources, the Semidesert Grassland of historic times most probably is a thing of the past. Given almost complete cover by woody plants and no grass to carry fires, the transition is complete. Where woody plants are scattered and grass cover more or less continuous, a return to frequent fires could probably stop the advance of woody plants and diminish their dominance as well. Although elimination of livestock will not make the shrubs disappear,[136] withholding grazing for short periods to allow fuel buildup, followed by fires, might restore the landscape to a semblance of its presettlement condition.

Desertscrub communities are also experiencing irreversible change. The proliferation of exotic annuals and herbaceous perennials is one example.[137] Another is the large increase in wildfires in recent times. Because there was originally little fuel to carry wildfire in deserts, most desert plants have not evolved to survive fire.[138] Unusually wet years promote heavy growth of herbaceous plants, providing the fuel that is otherwise lacking.[139] Invasive exotics such as buffelgrass, fountain grass, and red brome are particularly likely to form an abundant and continuous source of fine fuel.[140] Lack of adaptation to survive fire, in combination with proliferation of these exotic grasses, means that the species composition of some desertscrub communities will doubtless be permanently altered.[141]

Vegetation in the changing mile of the Sonoran Desert region today reflects a unique combination of climatic and cultural stresses. Periods of low rainfall have reduced oak, saguaro, and paloverde populations at the lower, drier margins of their ranges. Encroachment of woody plants in some areas may have been stimulated by prolonged above-average winter rainfall. At the coldest sites inhabited by mesquites, saguaros, and other plants, catastrophic freezes have kept some populations at bay. Livestock grazing depleted the grass cover, thereby virtually eliminating a natural ingredient, fire. Reduced fire frequency resulted in greater flower and seed production in such plants as the mesquite, in survival of more seeds lying on or near the soil surface, and in greater survival of seedlings. This improvement in reproductive potential, combined with the abundance and ubiquity of livestock, ensured the widespread establishment of those same shrubs and trees. Fire suppression, whether direct or indirect, allowed seedlings to grow into mature plants. With increased age and size, the plants became more resistant to removal, and the pace quickened. For now, this seems to be the best summary consistent with the evidence. It covers a majority of the camera stations viewed in this study.

Climate is implicated in changes involving

more than just vegetation. Recent studies show that the dramatic physical changes along the region's valleys during the period scanned by our photographs is primarily under control of climate. That arroyo cutting commenced at about the same time (around 4,000 years ago) along the San Pedro and Santa Cruz Rivers, and that the subsequent five episodes of arroyo formation were closely timed along these two valleys with watersheds of different sizes, points strongly to a climatic origin for arroyo cutting. Furthermore, the wet periods that have triggered arroyo cutting during the late Holocene have strong ties to changes in the El Niño–Southern Oscillation pattern.[142]

There may be instances where our findings can be used to bolster land-management decisions. However, the complexity of our human-dominated ecosystem makes application of the findings difficult. During the past decade, for example, much emphasis has been placed on "restoring" the region's "natural areas." These calls to restoration rarely specify the ecosystem attributes that are to be restored. Even less frequently discussed is how to maintain the processes that favored the original ecosystem attributes. For example, how could we possibly restore the Semidesert Grassland of our original photographs without periodic fires, and how could periodic fires occur in a landscape now subdivided into small housing tracts? Another example might be restoration of sacaton grasslands in the region's valleys. If this were done, it would result in the replacement of most of the cottonwood-willow forests and adjacent mesquite bosques that have become widely recognized centers of biologic diversity. Even if we knew the causes of changes and had the wherewithal to effect them, we might not be able to return to presettlement conditions because human purpose and "pristine" conditions often conflict.

Our study has documented large changes, many of which could have been detected only through the use of repeat photography. Recognition of these changes may direct research efforts toward explaining their causes. As we find out more about the past climate and vegetation of the desert region and more about the basic behavior of its flora and fauna, better answers to questions of causation will inevitably emerge, and these answers will be based less on speculation and more on fact. For now the changing mile must remain a good subject for debate—but an even better subject for study.

Appendix 1 Climate Stations in Arizona and Sonora

State	Station Name	Date of Record Start	Date of Record End	Precipitation (inches)	Elevation (feet)
Oak Woodland					
Arizona	Bisbee	12/1/1892	2/28/1985	18.63	5,310
Arizona	Bisbee 2	6/1/1961	6/30/1997	15.10	5,020
Arizona	Canelo 1NW	1/1/1910	12/31/1998	18.06	5,010
Arizona	Nogales	7/1/1948	6/30/1983	16.56	3,810
Arizona	Oracle	1/1/1893	3/31/1949	19.40	4,600
Arizona	Oracle 2SE	2/25/1950	12/31/1998	22.66	4,510
Arizona	Ruby 4NW	4/1/1895	12/31/1955	18.10	3,980
Arizona	Santa Rita Exp. Range	7/1/1916	12/31/1998	22.27	4,300
Sonora	Cananea	1/1/1923	10/31/1991	21.21	5,210
Semidesert Grassland					
Arizona	Apache Powder Company	7/1/1923	4/30/1990	13.18	3,690
Arizona	Arivaca	1/1/1956	12/31/1998	17.90	3,590
Arizona	Benson	6/1/1894	5/31/1975	11.33	3,670
Arizona	Elgin	10/1/1912	12/31/1969	15.07	4,850
Arizona	Fort Huachuca	2/1/1900	12/31/1981	15.62	4,670
Arizona	Nogales 6N	10/1/1952	12/31/1999	17.70	3,560
Arizona	Old Nogales	12/1/1892	6/30/1948	15.70	3,900
Arizona	Patagonia	7/1/1921	12/31/1998	17.86	4,000
Arizona	San Rafael Ranch	12/1/1892	3/31/1968	17.27	4,740
Arizona	Sierra Vista	3/1/1982	12/31/1998	14.71	4,600
Arizona	Tombstone	7/1/1893	12/31/1998	13.93	4,610
Sonora	Agua Prieta	2/1/1961	12/31/1986	14.91	3,860
Sonora	Naco	1/1/1923	12/1/1995	14.48	4,560
Arizona Uplands					
Arizona	Ajo	5/11/1914	12/31/1998	8.68	1,800
Arizona	Dudleyville	1/1/1893	2/28/1962	13.89	2,075
Arizona	Red Rock 6SSW	1/1/1893	8/31/1973	9.96	1,880
Arizona	Florence	12/1/1892	12/31/1998	10.14	1,500
Arizona	Organ Pipe National Monument	7/1/1948	12/31/1998	9.76	1,680
Arizona	Silver Bell	2/1/1906	4/30/1974	12.78	2,740
Arizona	Sells	7/1/1948	9/30/1975	12.18	2,380
Arizona	Tucson University of Arizona	1/1/1868	12/31/1998	11.46	2,440

State	Station Name	Date of Record		Precipitation (inches)	Elevation (feet)
		Start	End		
Arizona	Winkleman 6S	4/1/1951	2/28/1995	14.57	2,080
Sonora	Altar	1/1/1922	12/31/1983	11.43	1,290

Lower Colorado Valley

State	Station Name	Start	End	Precipitation	Elevation
Arizona	Dateland	6/1/1972	12/31/1998	4.85	550
Arizona	Casa Grande	6/1/1898	5/31/1999	8.69	1,400
Arizona	Casa Grande National Monument	3/1/1906	5/31/1999	8.94	1,420
Arizona	Gila Bend	1/1/1892	12/31/1998	6.14	730
Arizona	Maricopa 9SSW	6/1/1898	12/31/1958	7.80	1,400
Arizona	Maricopa 4N	3/1/1960	12/31/1998	7.35	1,160
Arizona	Mohawk	7/1/1900	5/31/1951	4.23	540
Arizona	Wellton	3/18/1922	12/31/1980	4.18	260
Arizona	Yuma Citrus Station	9/1/1920	4/30/2000	3.49	190
Arizona	Yuma WB City	1/1/1893	4/31/1974	3.26	240
Arizona	Yuma WSO Airport	9/1/1948	11/30/1995	2.92	210
Sonora	San Luis	1/1/1927	12/31/1983	5.20	90
Sonora	Sonoyta	1/1/1949	12/31/1990	8.89	1,280
Sonora	Puerto Peñasco	1/1/1966	12/31/1979	3.94	0

Central Gulf Coast

State	Station Name	Start	End	Precipitation	Elevation
Sonora	Guaymas	5/1/1921	7/31/1984	9.27	0
Sonora	Empalme	1/1/1923	4/30/2000	7.89	0

Appendix 2 Common and Latin Plant Names

Alphabetical by Common Name (with Latin Equivalent)

agave	*Agave* spp.
alkali sacaton	*Sporobolus airoides* var. *wrightii*
alligator juniper	*Juniperus deppeana*
all thorn	*Koeberlinia spinosa* var. *spinosa*
amole	*Agave schottii*
annual goldeneye	*Viquiera annua*
Arizona carlowrightia	*Carlowrightia arizonica*
Arizona cottontop	*Digitaria californica*
Arizona madrone	*Arbutus arizonica*
Arizona rosewood	*Vauquelinia californica*
Arizona sycamore	*Platanus wrightii*
Arizona white oak	*Quercus arizonica*
banana yucca	*Yucca baccata*
barrel cactus	*Ferocactus* spp.
beardgrass	*Andropogon glomeratus*
beargrass	*Nolina microcarpa*
Bermuda grass	*Cynodon dactylon*
big galleta	*Hilaria rigida*
bisnaga	*Ferocactus wislizeni*
black grama	*Bouteloua eriopoda*
black mangrove	*Avicennia nitida*
blue grama	*Bouteloua gracilis*
blue paloverde	*Cercidium floridum*
blue stem	*Andropogon* spp.
blue yucca	*Yucca baccata* var. *brevifolia*
bristlegrass	*Setaria macrostachya*
brittlebush	*Encelia farinosa*
buffelgrass	*Cenchrus ciliaris*
bullgrass	*Muhlenbergia emersleyi*
bullnettle	*Solanum elaeagnifolium*
burrobrush	*Hymenoclea monogyra*
burroweed	*Haplopappus tenuisectus*
bursage	*Ambrosia deltoidea*
bush muhly	*Muhlenbergia porteri*
cane cholla	*Opuntia spinosior*
canyon grape	*Vitis arizonica*
canyon ragweed	*Ambrosia ambrosioides*
cardón	*Pachycereus pringlei*
catclaw	*Acacia greggii*
century plant	*Agave* spp.

chamiso	*Atriplex canescens*
Chihuahuan whitethorn	*Acacia neovernicosa*
cholla	*Opuntia* spp.
Christmas cactus	*Opuntia leptocaulis*
chuparosa	*Beloperone californica*
cirio	*Fouquieria columnaris*
cliff rose	*Cowania mexicana*
cliff spurge	*Euphorbia misera*
cockroach plant	*Haplophyton crooksii*
coffeeberry	*Rhamnus californica*
coldenia	*Coldenia canescens*
cottonwood	*Populus fremontii*
Coulter hibiscus	*Hibiscus coulteri*
coursetia	*Coursetia glandulosa*
creosote bush	*Larrea tridentata*
crownbeard	*Verbesina encelioides*
curly mesquite	*Hilaria belangeri*
desert broom	*Baccharis sarothroides*
desert cotton	*Gossypium thurberi*
desert hackberry	*Celtis pallida*
desert hibiscus	*Hibiscus coulteri*
desert holly	*Perezia nana*
desert honeysuckle	*Anisacanthus thurberi*
desert lavender	*Hyptis emoryi*
desert mallow	*Sphaeralcea* spp.
desert saltbush	*Atriplex polycarpa*
desert spoon	*Dasylirion wheeleri*
desert willow	*Chilopsis linearis*
desert zinnia	*Zinnia pumila*
devil's club cholla	*Opuntia stanlyi*
Douglas fir	*Pseudotsuga menziesii*
doveweed	*Croton texensis*
elephant tree	*Bursera microphylla*
Emory oak	*Quercus emoryi*
Engelmann prickly pear	*Opuntia engelmannii*
fairyduster	*Calliandra eriophylla*
feather fingergrass	*Chloris virgata*
fluffgrass	*Tridens pulchellus*
foothill paloverde	*Cercidium microphyllum*
fountain grass	*Pennisetum ruppelii*
four-wing saltbush	*Atriplex canescens*
Fremont cottonwood	*Populus fremontii*
giant cactus	*Carnegiea gigantea*
Goodding willow	*Salix gooddingii*
grama grass	*Bouteloua* spp.
gray thorn	*Condalia lycioides*
hairy grama	*Bouteloua hirsuta*
hopbush	*Dodonaea viscosa*

ironwood	*Olneya tesota*
jojoba	*Simmondsia chinensis*
Joshua tree	*Yucca brevifolia*
jumping cholla	*Opuntia fulgida*
juniper	*Juniperus* sp.
kidneywood	*Eysenhardtia polystachya*
Lehmann lovegrass	*Eragrostis lehmanniana*
limber bush	*Jatropha cardiophylla*
little-leaf sumac	*Rhus microphylla*
loosestrife	*Lythrum californicum*
mala mujer	*Cnidoscolus angustidens*
mangrove	*Rhizophora mangle*
mariola	*Parthenium incanum*
mescal	*Agave palmeri*
mesquite	*Prosopis velutina*
Mexican blue oak	*Quercus oblongifolia*
Mexican crucillo	*Condalia spathulata*
Mexican devilweed	*Aster spinosus*
Mexican pinyon pine	*Pinus cembroides*
Mexican tea	*Ephedra trifurca*
mimosa	*Mimosa* spp.
Mormon tea	*Ephedra* spp.
mortonia	*Mortonia scabrella*
mountain mahogany	*Cercocarpus breviflorus*
mountain yucca	*Yucca schottii*
netleaf hackberry	*Celtis reticulata*
netleaf oak	*Quercus rugosa*
oak	*Quercus* spp.
ocotillo	*Fouquieria splendens*
one-seed juniper	*Juniperus coahuilensis*
organpipe cactus	*Lemaireocereus thurberi*
pale lycium	*Lycium pallidum*
Palmer agave	*Agave palmeri*
palmilla	*Yucca elata*
paloverde	*Cercidium* spp.
paperdaisy	*Psilostrophe cooperi*
pencil cholla	*Opuntia arbuscula*
pincushion cactus	*Mammillaria microcarpa*
pine	*Pinus* spp.
plains bristlegrass	*Setaria macrostachya*
plains lovegrass	*Eragrostis intermedia*
pointleaf manzanita	*Arctostaphylos pungens*
poison ivy	*Rhus radicans*
ponderosa pine	*Pinus ponderosa*
prickly pear	*Opuntia* spp.
pygmy cedar	*Peucephyllum schottii*
rabbitbrush	*Chrysothamnus nauseosus* var. *latisquameus*
ragweed	*Ambrosia* spp.

rainbow cactus	*Echinocereus pectinatus*
range ratany	*Krameria parvifolia*
ratany	*Krameria parvifolia*
red brome	*Bromus rubens*
red mangrove	*Rhizophora mangle*
red willow	*Salix laevigata*
Rothrock grama	*Bouteloua rothrockii*
Russian thistle	*Salsola kali*
sacaton	*Sporobolus airoides*
saguaro	*Carnegiea gigantea*
saltbush	*Atriplex* spp.
saltcedar	*Tamarix ramosissima*
Santa-Rita cactus	*Opuntia santa-rita*
seep willow	*Baccharis glutinosa*
senna	*Cassia covesii*
shindagger	*Agave schottii*
shrubby senna	*Cassia wislizenii*
sideoats grama	*Bouteloua curtipendula*
silverleaf oak	*Quercus hypoleucoides*
sinita	*Lophocereus schottii*
skunkbush	*Rhus trilobata*
slender grama	*Bouteloua filiformis*
slender poreleaf	*Porophyllum gracile*
slim tridens	*Tridens muticus*
smoketree	*Dalea spinosa*
snakeweed	*Gutierrezia sarothrae*
soapberry	*Sapindus saponaria*
sotol	*Dasylirion wheeleri*
spidergrass	*Aristida ternipes*
spruce	*Picea* spp.
sprucetop grama	*Boutelous chondrosioides*
staghorn cholla	*Opuntia versicolor*
stickleaf	*Mentzelia* spp.
stickweed	*Stephanomeria pauciflora*
subalpine fir	*Abies lasiocarpa*
tanglehead	*Heteropogon contortus*
tarbush	*Flourensia cernua*
teddybear cholla	*Opuntia bigelovii*
telegraph plant	*Heterotheca* spp.
Texas bluestem	*Schizachyrium cirratum*
Texas mulberry	*Morus microphylla*
threadleaf groundsel	*Senecio longilobus*
three-awn	*Aristida* spp.
tobosa	*Hilaria mutica*
Toumey oak	*Quercus toumeyi*
triangle leaf bursage	*Ambrosia deltoidea*
turpentine bush	*Haplopappus laricifolius*
velvet ash	*Fraxinus pennsylvanica*

velvet mesquite	*Prosopis velutina*
velvet-pod mimosa	*Mimosa dysocarpa*
vine mesquite	*Panicum obtusum*
wait-a-minute	*Mimosa biuncifera*
walnut	*Juglans major*
watercress	*Rorippa nasturtium-aquaticum*
western coral bean	*Erythrina flabelliformis*
western honey mesquite	*Prosopis glandulosa*
white bursage	*Ambrosia dumosa*
white fir	*Abies concolor*
white ratany	*Krameria grayi*
whitethorn	*Acacia constricta*
wild buckwheat	*Eriogonum abertianum*
wolfberry	*Lycium berlandieri*
wolftail	*Lycurus phleoides*
Wright buckwheat	*Eriogonum wrightii*
Wright lippia	*Aloysia wrightii*
Wright's silktassel	*Garrya wrightii*
yellow bird-of-paradise	*Caesalpinia gilliesii*
yerba de pasmo	*Baccharis pteronioides*
yewleaf willow	*Salix taxifolia*
yucca	*Yucca* spp.
zinnia	*Zinnia* spp.

Appendix 3 Details of Photographic Stations and Photo Credits

Stake numbers are the numbers under which photographs are filed at the Desert Laboratory. Archives (Arc) and negative numbers (No.) for the first photographs are designated under Source as follows: A = Arizona Historical Society; B = personal collection of Mr. Albert C. Stewart; C = personal collection of Miss Doris Seibold; D = Desert Laboratory; H = University of Arizona Herbarium; N = National Park Service; U = U.S. Geological Survey Photographic Library, Denver. Abbreviations and full names of photographers are: Has = James R. Hastings; Mac = Daniel T. MacDougal; Nic = E. Tad Nichols; Old = Dominic Oldershaw; Phi = John M. Phillips; Ros = George Roskruge; Sch = Frank C. Schrader; Sha = Homer L. Shantz; Shr = Forrest Shreve; Spa = Volney M. Spalding; Tur = Raymond M. Turner; Unk = Unknown; Wat = Carleton E. Watkins; Web = Robert H. Webb. * = Approximate date.

Pl	Location	Stake	Latitude	Longitude	Elev	Date of photo 1st	2nd	3rd	Source Arc	No.	Photographer 1st	2nd	3rd
1	El Plomo Mine	188	31.6114	110.8594	5050	1891	1962	1995	A	45908	Ros	Has	Old
2	El Plomo Mine	190x	31.6106	110.8660	5805	1891	1962	1995	A	45889	Ros	Has	Old
3	El Plomo Mine	191a	31.6105	110.8642	5435	1891	1962	1995	A	45901	Ros	Has	Old
4	Sonoita Road	164b	31.7950	110.7139	5080	1925	1962	1987	B		Unk	Has	Tur
5	Guajalote Peak	162	31.4056	110.7844	4750	1887	1962	1994	A	45913	Ros	Has	Old
6	Sonoita Creek	216a	31.6511	110.7025	4740	1895	1962	1994	C		Unk	Has	Old
7	Cottonwood Spring	220x	31.6536	110.7136	4575	1895	1962	1998	C		Unk	Has	Tur
8	Fort Crittenden	217a	31.6536	110.7136	4575	1895	1962	1994	C		Unk	Has	Old
9	Monkey Spring	214	31.6308	110.6978	4740	1895	1962	1994	C		Unk	Has	Old
10	Monkey Spring	154	31.6328	110.7047	4630	1889	1962	1994	A	45937	Ros	Has	Old
11	Monkey Spring	157	31.6331	110.7036	4560	1889	1962	1994	A	45933	Ros	Has	Old
12	Alto Gulch	187	31.6083	110.8711	4650	1891	1962	1995	A	45895	Ros	Has	Old
13	Salero Mine	210	31.5753	110.8594	4480	1909	1962	1994	A	17416	Ros	Has	Old
14	Nogales Overpass	305	31.2603	110.9489	4595	1931	1962	1995	H	U-4-31	Sha	Has	Tur
15	San Cayetano Hill	158a	31.4847	110.8289	4200	1890	1962	1994	A	45955	Ros	Has	Old
16	Sanford Butte	158b	31.4847	110.8289	4200	1890	1962	1994	A	46439	Ros	Has	Old
17	Guevavi Canyon	161	31.4861	110.8417	4260	1890	1962	1996	A	45976	Ros	Has	Tur
18	Santa Rita Mts.	205	31.7694	110.8514	4120	1911	1962	1994	D	B1-42	Mac	Has	Old
19	Santa Rita Mts.	204	31.7750	110.8642	3930	1911	1962	1994	D	B1-75	Mac	Has	Old
20	Proto Ridge	166x	31.3703	110.8844	4020	1887	1962	1994	A	45909	Ros	Has	Old
21	Proto Ridge	168	31.3689	110.8889	4030	1887	1962	1994	A	45912	Ros	Has	Old
22	Buena Vista	169	31.3631	110.8475	3970	1890	1962	1994	A	46200	Ros	Has	Old
23	Buena Vista	171	31.3631	110.8472	3920	1890	1962	1994	A	46435	Ros	Has	Old
24	Buena Vista	173	31.3653	110.8425	3780	1890	1962	1996	A	45954	Ros	Has	Tur
25	Near Buena Vista	165	31.5508	110.8481	3770	1887	1962	1994	A	45914	Ros	Has	Old
26	Querobabi	308a	31.1103	110.9072	3500	1931	1962	1996	H	U-7-31	Sha	Has	Tur
27	Research Ranch	744d	31.6083	110.4797	4750	1974	1975	1995	D	744d	Tur	Tur	Tur
28	Red Rock Canyon	213b	31.5519	110.7169	4430	1895	1965	1996	C		Unk	Has	Tur
29	Red Rock Canyon	213a	31.5519	110.7169	4430	1895	1964	1996	C		Unk	Has	Tur
30	Helvetia	47	31.8942	110.8108	4430	1899	1960	1994	A	3502	Ros	Has	Old

Pl	Location	Stake	Latitude	Longitude	Elev	1st	2nd	3rd	Arc	No.	1st	2nd	3rd
							Date of photo			Source		Photographer	
31	Helvetia	46	31.8942	110.8108	4320	1899	1960	1994	A	3501	Ros	Has	Old
32	Helvetia	44	31.8942	110.8108	4320	1899	1962	1994	A	3503	Ros	Has	Old
33	Hacienda Santa Rita	199a	31.5928	110.8992	4150	1891	1962	1994	A	45174	Ros	Has	Old
34	Proto Ridge	167x	31.3692	110.8844	4090	1887	1962	1994	A	45/1915	Ros	Has	Old
35	Davidson Canyon	39a	31.9275	110.6942	4040	1915	1962	1994	D	B1-31	Mac	Has	Old
36	Davidson Canyon	40a	31.9275	110.6942	4040	1915	1962	1994	D	B1-44	Mac	Has	Old
37	Helena Mine	727	31.8725	110.7071	4600	1909	1973	1994	U	1533	Sch	Tur	Old
38	Elephant Head	203a	31.7722	110.8700	3920	1911	1962	1996	D	B1-76	Mac	Has	Web
39	Cerro Colorado	211	31.6531	111.2706	3640	1892	1962	1996	A	45986	Ros	Has	Tur
40	Tumacacori	209	31.5694	111.0550	3350	1891	1962	1994	A	45861	Ros	Has	Old
41	Carmen	283	31.5869	111.0547	3250	1890	1964	1994	A	46203	Ros	Has	Old
42	Monument 98	131	31.3333	110.2392	4330	1891	1962	1994	A	46466	Ros	Has	Old
43	Monument 98	132a	31.3333	110.2392	4330	1891	1962	1995	A	46467	Ros	Has	Old
44	Green Ranch	142	31.3942	110.1275	4250	1891	1962	1998	A	45952	Ros	Has	Tur
45	Spring Creek	147	31.4442	110.0625	4400	1891	1962	1994	A	45950	Ros	Has	Old
46	Lewis Springs	56	31.5775	110.1608	4060	1891	1960	1998	A	19277	Ros	Has	Tur
47	Lewis Springs	55	31.5942	110.1608	4060	1891	1960	1998	A	19276	Ros	Has	Tur
48	Escapule Ranch	134d	31.6275	110.1522	4050	1891	1962	1994	A	45948	Ros	Has	Old
49	Bronco Hill	57	31.6442	110.1942	4000	1880	1960	1994	A	13937	Wat	Has	Old
50	Charleston	58	31.6608	110.1942	4050	1885	1960	1994	A	10237	Unk	Has	Old
51	Charleston	61	31.6608	110.1942	4030	1880	1960	1994	A	14819	Wat	Has	Old
52	Tombstone	52	31.7275	110.0942	4850	1890*	1960	1994	A	51919	Unk	Has	Old
53	Walnut Gulch	49a	31.7442	110.1775	4000	1890*	1960	1994	A	46413	Ros	Has	Old
54	Walnut Gulch	49b	31.7442	110.1775	4000	1890*	1960	1994	A	19324	Ros	Has	Old
55	Walnut Gulch	50	31.7442	110.1775	3990	1890*	1962	1995	A	46223	Ros	Has	Old
56	Walnut Gulch	51	31.7442	110.1942	3980	1890*	1960	1994	A	46414	Ros	Has	Old
57	Fairbank	150	31.7222	110.1964	3830	1890*	1962	1994	A	46404	Ros	Has	Old
58	St. David	153	31.8589	110.2236	3690	1890*	1962	1995	A	46400	Ros	Has	Old
59	St. David	152	31.8681	110.2375	3700	1890*	1962	1994	A	46193	Ros	Has	Old
60	Saguaro National Park	13	32.1975	110.7342	3040	1935*	1960	1998	N		Unk	Has	Tur
61	Saguaro National Park	36	32.2275	110.7608	2940	1935*	1960	2000	N		Unk	Has	Tur
62	Saguaro National Park	23	32.2275	110.7608	2940	1935	1960	2000	H	O-1-1935	Sha	Has	Tur
63	Saguaro National Park	32	32.2275	110.7608	2960	1935	1962	1995	H	T-5-1935	Sha	Has	Tur
64	Saguaro National Park	247b	32.2019	110.7283	2880	1932	1962	1995	H	X-2-1932	Sha	Has	Tur
65	Soldier Canyon	6	32.3033	110.7447	2950	1915	1960	1993	D		Mac	Has	Tur
66	Soldier Canyon	5	32.3033	110.7492	2900	1908	1960	1993	D		Mac	Has	Tur
67	Soldier Canyon	4	32.3022	110.7447	2880	1915	1962	1993	D	B2-85	Mac	Has	Tur
68	Soldier Canyon	7	32.3042	110.7492	2650	1908	1962	1994	D	B2-124	Mac	Has	Tur
69	Sabino Canyon	120	32.3175	110.8094	2840	1911	1962	2000	D		Mac	Has	Tur
70	Sabino Canyon	186f	32.3175	110.8094	2840	1890	1962	1995	A	45838	Ros	Has	Tur

Pl	Location	Stake	Latitude	Longitude	Elev	Date of photo			Source		Photographer		
						1st	2nd	3rd	Arc	No.	1st	2nd	3rd
71	Sabino Canyon	186e	32.3175	110.8094	2840	1890	1962	1995	A	45833	Ros	Has	Tur
72	Cañada del Oro	237a	32.4144	110.9350	2640	1913	1962	1993	D	B1-9	Mac	Has	Tur
73	Cañada del Oro	237b	32.4144	110.9350	2640	1913	1962	1993	D	B4-58	Mac	Has	Tur
74	Robles Pass	175	32.1750	111.0494	2650	1915	1962	1994	D	B1-18	Mac	Has	Old
75	Cat Mountain	180	32.1900	111.0683	2730	1914	1962	1994	D	A3-4	Mac	Has	Old
76	Brown Mountain	182	32.2022	111.1178	2630	1915	1962	1994	D	B2-115	Mac	Has	Old
77	Safford Peak	183	32.3072	111.1467	2900	1916	1962	1988	D	B1-29	Mac	Has	Tur
78	Safford Peak	184a	32.3061	111.1458	2900	1916	1962	1988	D	A2-6	Mac	Has	Tur
79	Picture Rocks	178a	32.3167	111.1100	2410	1908	1962	1996	D	B4-96	Mac	Has	Tur
80	Picture Rocks	178b	32.3167	111.1100	2410	1908	1962	1996	D	B1-8	Mac	Has	Tur
81	Desert Lab	1168	32.2209	111.0139	2430	1928	1985	1998	D	C4-20	Shr	Tur	Tur
82	Desert Lab	377	32.2186	111.0031	2750	1906	1968	1999	D		Spa	Has	Tur
83	MacDougal Pass	63a	31.9876	113.6019	900	1907	1962	1997	D	N-12	Mac	Has	Tur
84	MacDougal Pass	63b	31.9876	113.6019	900	1907	1962	1997	D	N-11	Mac	Has	Tur
85	MacDougal Pass	3408	31.9851	113.6024	900	1907	1958	1997	D		Phi	Nic	Tur
86	Macias Pass	72	32.0275	113.6275	900	1907	1962	1997	D		Mac	Has	Tur
87	MacDougal Crater	64a	32.0108	113.6442	900	1907	1959	1997	D	N-19	Mac	Has	Tur
88	MacDougal Crater	64a	32.0108	113.6442	900	1907	1959	1997	D	N-19	Mac	Has	Tur
89	MacDougal Crater	89	31.9725	113.6263	400	1959	1984	1998	D	89	Has	Tur	Web
90	MacDougal Crater	665	31.9743	113.6264	400	1959	1974	1998	D	665	Has	Tur	Web
91	Puerto Libertad	327	29.9339	112.7058	440	1932	1965	1989	H	B-5-32	Sha	Has	Tur
92	Puerto Libertad	328	29.9339	112.7058	445	1932	1965	1989	H	B-6-32	Sha	Has	Tur
93	Puerto Libertad	329	29.9347	112.6786	200	1932	1965	1989	H	B-10-32	Sha	Has	Tur
94	Punta Cirio	312	29.8353	112.6428	130	1932	1965	1989	H	D-12-32	Sha	Has	Tur
95	Punta Cirio	311	29.8447	112.6472	100	1932	1962	1997	H	D-10-32	Sha	Has	Tur
96	Punta Cirio	321a	29.8322	112.6481	130	1932	1965	1997	H	O-7-32	Sha	Has	Tur
97	Guaymas Bay	128	27.9442	110.9275	5	1903	1961	1996	D	C1-57	Mac	Has	Tur
98	Guaymas Bay	129	27.9442	110.9275	5	1903	1961	1996	D	C1-68	Mac	Has	Tur

Notes

Introduction

1. Cooke and Reeves (1976).
2. Hastings (1959), Brown et al. (1981), Hereford (1993), Hereford and Betancourt (1995).
3. Lumholtz (1912), Cooke and Reeves (1976), Betancourt (1990).
4. Webb et al. (1991).
5. Davis (1982).
6. Griffiths (1904: 29).
7. Griffiths (1910: 22).
8. Ibid., 18.
9. Thornber (1910: 276).
10. Shreve (1910: 240).
11. Benson and Darrow (1954).
12. Parker and Martin (1952).
13. Alcorn and May (1962), Niering et al. (1963).
14. Pierson and Turner (1998).
15. Steenbergh and Lowe (1977, 1983), Turner (1990, 1992).
16. Bahre (1991), Bahre and Shelton (1993).
17. For example, Bahre (1991).
18. For example, Hastings and Turner (1965b).
19. Hattersley-Smith (1966).
20. Rogers et al. (1984).
21. Shantz and Turner (1958).
22. Phillips (1963), Hastings and Turner (1965b).
23. Turner and Karpiscak (1980), Gruell (1980), Rogers (1982), Klett et al. (1984), Humphrey (1987), Johnson (1987), Stephens and Shoemaker (1987), Webb et al. (1991), McGinnies et al. (1991), Anonymous (1993), Wright and Bunting (1994), Melis et al. (1996), Webb (1996), Meagher and Houston (1998), Gruell (2001).
24. Rogers et al. (1984).
25. Webb (1996).
26. Masse (1979), Fish et al. (1985).
27. Perevolotsky (1999).
28. Bahre (1991).
29. Ibid.
30. McPherson (1997).

Chapter 1

1. Tundra is an example of vegetation that falls within the precipitation requirements of deserts, yet it is clearly not desert because of low temperatures. McDonald (1959) gives a short discussion of the world's deserts and of the meteorology that dictates their existence.
2. One approach might define "desert" in terms of the ratio of potential evapotranspiration to rainfall (Thornthwaite 1948), or it might, as with the widely known Köppen system or one of its modifications, use yearly mean temperature, yearly rainfall, and seasonal distribution of rainfall as a combined index (Trewartha 1954: 382).
3. Sellers (1965).
4. McDonald (1959: 7).

5. Axelrod (1950, 1986), Van Devender and Spaulding (1986).
6. Turner (1994b). Grayson (1993) provides a thorough, readable account of the natural history and prehistory of both the Great Basin Desert and the Mojave Desert.
7. Turner (1994b), McLaughlin and Bowers (1999).
8. Shreve (1942a), Brown (1994b).
9. Shreve (1942a), Benson and Darrow (1954), Brown and Lowe (1994), Brown et al. (1998). Small, isolated patches of this desert also occur in northern Sonora near the headwaters of the Bavispe and Yaqui Rivers.
10. McDonald (1962: 4) has estimated the daily influx of water vapor into Arizona from the south during July to be about 2,000,000 acre-feet.
11. Shreve (1964), Brown and Lowe (1994).
12. See Turner and Brown (1994: 182) for a comparison of continentality among Sonoran Desert subdivisions and the Mojave Desert.
13. Pyke (1972).
14. Pyke (1972), Eidemiller (1978), W. Smith (1986).
15. Eychaner and Rehmann (1989).
16. Hastings and Turner (1965a).
17. Bryson and Lowry (1955).
18. Randerson (1986).
19. Hansen and Shwarz (1981).
20. Durrenberger and Ingram (1978).
21. Dzerdzeevskii (1970).
22. See discussion in Webb and Betancourt (1992).
23. Several important review articles have been written on ENSO phenomena, including Philander (1990), Rasmusson (1984), Enfield (1989), and Diaz and Markgraf (1992).
24. Webb and Betancourt (1992).
25. Cayan and Peterson (1989), Cayan and Webb (1992).
26. Andrade and Sellers (1988), Woolhiser et al. (1993).
27. Ropelewski and Halpert (1986).
28. Hirschboeck (1985).
29. For a discussion of Tropical Storm Octave and its effects on southern Arizona, see Saarinen et al. (1984); for discussions of large winter storms, see Larson and Slosson (1997).
30. Webb and Betancourt (1992).
31. Shreve (1964: 26)
32. Ibid.
33. Ibid., 31.
34. Ibid., 9.
35. Turner and Brown (1994). Several summaries of Arizona climate have been published, including Green and Sellers (1964) and Sellers et al. (1985). Monthly climate summaries can be downloaded from the Internet at http://www.wrcc.dri.edu/climsum.html.
36. The precipitation values are derived by interpolation from Turner and Brown (1994: 113).
37. Shreve (1944).

38. The CV is defined as the standard deviation divided by the mean for the precipitation record. More properly, since precipitation distributions are positively skewed, the CV is the root mean-square-deviation divided by the mean. Ignoring the question of whether the data are distributed normally or not, the coefficient may be roughly defined as the percentage of the mean precipitation by which the actual precipitation may be expected to depart during about one-third of the time. For data that are not normally distributed, see Hastings (1965) and Hastings and Turner (1965a). McDonald (1956: 6) calculated CVs for some Arizona stations, Wallén (1955: 67) for some Sonora stations, and Hastings and Humphrey (1969a) for some Baja California stations.

39. The regression equation is $CV = 0.61 - 0.020 \, P$, where CV = the coefficient of variation of annual precipitation and P = annual precipitation in inches and R = 0.83.

40. Hastings and Turner (1965a).

41. Shreve (1934b: 373).

42. Shreve (1944: 108).

43. Bryson (1957: 4).

44. See discussions of these species in Turner et al. (1995).

45. Martin et al. (1961).

46. Shreve (1934a).

47. The initial stages of the experiment were described by Hastings (1961b). The unpublished findings are from dendrograph data on file at the Desert Laboratory.

48. MacDougal (1924).

49. Nobel and Sanderson (1984), Nobel (1988).

50. The absence of succulents from both the Mojave and the Chihuahuan Deserts might lie largely with the fact that these deserts have substantially colder winters than the Sonoran Desert.

51. Sims et al. (1978).

52. Neilson (1986) discusses vegetation changes in the northern Chihuahuan Desert in the context of carbon pathway physiology and suggests that the Chihuahuan Desert vegetation of 100 years ago was established under the compulsion of 300 years of "little ice age" climate and that the grassland-shrubland mosaic of today is only marginally supported by the present climate.

53. Went (1949), Shreve (1964: 121).

54. Seed germination studies have been conducted by numerous investigators. For germination requirements of mesquite, see Hull (1956); for saguaro, see Alcorn and Kurtz (1959); for foothill paloverde and ocotillo, see Bowers (1994); for creosote bush, see Went and Westergaard (1949), Sheps (1973), Beatley (1974a, 1974b), Rivera and Freeman (1979), Zedler (1981), and Boyd and Brum (1983); and for triangle-leaf bursage and brittlebush, see Turner et al. (1995) and Bowers (1994).

55. Cannon (1916).

56. Nobel (1984), Nobel et al. (1986).

57. Steenbergh and Lowe (1977), Nobel (1980a, 1980b), Webb and Bowers (1993), Webb (1996).

58. Cottam (1937), Fosberg (1938), Daubenmire (1957), Glinski and Brown (1982), Jones (1979), Bowers (1980–81).

59. The difference between a "secular trend" and an "unusual occurrence" is not obvious, and may be only a matter of how frequently an extreme recurs. In the case of large perennials, the microclimates in which the seedlings and the adults reside are substantially different. Close to the ground it is hotter by day and colder by night, and a mature plant that has survived these rigors as a seedling may find that even the intensity of once-in-100-years events in the more elevated, adult environment does not exceed what it has already experienced. The same event, more intense near the ground, may extinguish all seedlings, but since the seed source has survived, this is a matter of small consequence and can be remedied during the following year. Thus, only a continuous series of rare events can ultimately diminish the adult population by depriving it, season after season, of replacements.

60. Although as Shreve (1964: 9) points out, the climate is nowhere very maritime.

61. Green and Sellers (1964), Hastings (1964a, 1964b), Hastings and Humphrey (1969a, 1969b).

62. Temperatures for the Pinacates are estimated from the Sonoita record (Hastings 1964b: 118). Hastings (1964b: 53) examined the Guaymas temperature record.

63. Sinclair (1922).

64. Bryson (1957: 4), Page (1930: 1).

65. Turnage and Hinckley (1938: 547), Shreve (1934b: 379).

66. Turnage and Hinckley (1938: 544).

67. Shreve (1934b: 379–80), Turnage and Hinckley (1938: 547). Gentry, the observer at Cedros for Turnage and Hinckley, and their authority for the statement about the southern limit of the frost zone, states elsewhere that "in the Thorn Forest frost occasionally occurs, and during the excessive cold wave of January 1937 frost struck into the Short-tree Forest region" (Gentry 1942: 15).

68. Green and Sellers (1964: 25–34).

69. Turnage and Hinckley (1938).

70. Ibid., 541.

71. These factors, plus calm nights, are the classical requirements for the formation of an inversion. However, Dickson (1958: 39) points out that "inversions probably are more common in regions of diversified relief than has before been realized and . . . they frequently occur when a moderate cloud cover and light winds are present." Shreve (1912) provides some data dealing with the variation of inversions from season to season.

72. Conduction and convection are the processes usually given (e.g. Sutton 1953: 132). However, observations that minimum air temperatures under certain conditions are found not at the soil surface, but from 1 inch to 1 foot or more higher, have led to hypotheses that direct radiation losses by the air may also be involved (Lake 1956).

73. The phrase is borrowed from Geiger (1957: 195). Some useful temperature profiles through an inversion layer are presented by Young (1921).

74. Bisbee, which is too warm for its elevation, can probably be explained by another topographic anomaly. The town occupies a narrow strip along the bottom of a steep canyon flanked on either side by mountains that rise sharply to a height of about 500 feet. The canyon falls too rapidly to permit the accumulation of cold air; radiation and counterradiation from the slopes on either side tend at night to warm the town or, more accu-

rately, to depress its cooling rate. The same factor may operate in the case of Fort Grant, nestled at the foot of the massive Pinaleño range.

75. In cases this extreme, however, moisture considerations apart from temperature are probably involved too. The example is from Shreve (1922).

76. See Yang and Lowe (1956), McAuliffe (1994).

77. At Death Valley, Went and Westergaard found that creosote bush, given adequate rainfall, germinated at temperatures intermediate between those optimum for, on the one hand, summer annuals and, on the other, winter annuals. They found abundant *Larrea* seedlings after an October rain when minimum temperatures were 59.0°–60.8°F, but not after rains toward the end of August (78.8°F) or November (46.4°–50.0°F). A similar seasonal pattern of germination was shown for a Chihuahuan Desert population (Rivera and Freeman 1979). The distribution of *Larrea* on the lower, cooler part of a bajada may be influenced by these germination requirements (Went and Westergaard 1949).

78. A case in point is the January 1937 cold wave already referred to. The duration of freezing at both the hill and the garden stations was 19 hours; at Summerhaven, 130 hours (Turnage and Hinckley 1938: 538–42).

79. Ibid., 538.

80. Shreve (1911b).

81. Turnage and Hinckley (1938: 542–47).

82. Green and Sellers (1964).

83. Turner and Brown (1994).

84. Turner et al. (1995). For what it is worth in the largely philosophical debate among ecologists over the relative importance of means and extremes in controlling plant distributions: Shreve's Sonoran Desert boundary also coincides precisely with a mean monthly isotherm of 64.2°F. If extremes, in the form of record lows, are plotted, *no* isotherm will fit. That for a record low of 8.5°F comes closest; however, it throws Bagdad, Miami, Globe, and Santa Rita Experimental Range headquarters into the Sonoran Desert, and excludes Tucson, Maricopa, and Casa Grande National Monument. The mean minimum temperature for January—more or less a hybrid between mean and extreme—gives somewhat better results; the isotherm for 47.5°F excludes only Wickenburg, Reno Ranger Station, and Aguila. This is not to say, of course, that the physiological processes of a saguaro respond in any magical fashion to a mean temperature of 64.2°F. It merely states a truism that is frequently overlooked by "extremists" and that, therefore, needs to be emphasized: a mean is more likely than an extreme to correlate well with a large variety of other temperature measures, at least one of which—like the duration of freezing—may have real physiological significance. A mean is therefore more likely than an extreme to be useful in defining a plant's distribution, even though the actual limiting temperature factor may be unknown and may be neither a mean nor an extreme.

85. See Lowe (1964) for a thorough description of the region's biotic communities and Whittaker and Niering (1964) for a description of the vegetation on one of the region's mountain ranges.

86. See Brown (1994c) for a definition of Semidesert Grassland. Oak Woodland as used here is equivalent to Madrean Evergreen Woodland in Brown (1994a).

87. Dick-Peddie (1993).

88. Shreve (1922: 270–73), Marshall (1957).

Chapter 2

1. Sprugel (1991) advances the discussion of what is "natural" beyond a consideration of man's role alone in changing environments. Perevolotsky (1999) proposes the concept of "pastoral ecosystems" to describe lands permanently molded by the actions of human domesticates.

2. Sauer (1935: 32) says, referring to the area between the Gila and the Rio Grande de Santiago, "aboriginal populations and present ones are much the same."

3. Hack (1939) and Betancourt and Van Devender (1981) describe how fuelwood cutting by native peoples has decimated pinyon-juniper woodlands at their lower, drier margins during the past 1,000 years.

4. See Spicer (1962: 8–15).

5. Figure 2.1 has been redrawn from Sauer (1934). For a discussion of the Sumas and their relation to the Apaches who later occupied the same area, see Forbes (1957), who disagrees with Sauer's interpretation.

6. McGee (1898: pt. 1, 206ff.).

7. Felger and Moser (1985), Sheldon (1993).

8. Sauer (1935: 5).

9. Spicer (1962: 12, 99–100) estimates that the ranchería groups numbered about 150,000. These include, however, 30,000 Mayo who fell outside the boundaries of the Sonoran Desert, and perhaps 15,000 of the 25,000 Opata. To the remaining 105,000 must be added Seri, Yuman, and a few Apache to arrive at the number who inhabited the area within.

10. See Reff (1991) for a description of the precipitous population decline in our region. Studies that pertain to other parts of New Spain but nevertheless illuminate the decline of the aboriginal population under the impact of Spanish culture are Cook (1940) and Cook and Simpson (1948).

11. See Sauer (1935: 32).

12. See Martin and Szuter (1999) for discussion of this concept in an area to the north of our region.

13. From Aschmann (1959: 78–93).

14. From various sources (Pfefferkorn 1949: 46–78; Russell 1908: 68ff.; Del Barco 1973; Felger and Moser 1985; Rea 1997), the list of useful plants can be greatly expanded.

15. From Carter (1950).

16. McGee (1898: 207).

17. Aschmann (1959: 79).

18. McGee (1898: 214).

19. McGregor et al. (1962: 266). These authors estimate that an average saguaro bears "about 4 flowers per day over a 30-day period each year," and that "half of these flowers normally set."

20. We have observed a density of 116 plants per acre in a saguaro stand near Redington, Arizona. The stand shown in plate 60 has about 33 plants per acre. A stand with 6 plants per acre, the figure used here, would be very open, but the density has been deliberately assumed low in order to arrive at a maximum impact for food gathering.

21. With the saguaro, 12,000 acres would suffice to feed 100,000 people if the stand were as dense as that at Redington, Arizona.

22. Computed on the basis of 2,000 seeds per fruit (McGregor et al. 1962: 266).

23. This discussion must be slightly qualified by the

observation that other American Indian groups may have made thriftier use of cactus fruits than did the Seri, on whose habits these calculations are based. Pfefferkorn (1949: 76, 200) observes that the Pima made cakes of the fruits, which "keep for two or three months without spoiling at all." Russell (1908: 72) describes both syrup and dried balls made from the fruit. "The supply [of fresh fruit] is a large one and only industry is required to make it available throughout the entire year." In any event, one can inject a factor of two or even four into the calculations without affecting the conclusion.

24. Alcorn (1961: 24).

25. Masse (1979), Fish et al. (1985, 1992), Fish (2000). A view of such a once heavily populated area is shown in plate 81.

26. The approximate site of the southernmost ranchería fields, Santa Cruz de Gaybanipitea, appears in plate 57.

27. Mange (1926), Bolton (1948).

28. Bolton (1948: I, 123; 1960: 364).

29. Bolton (1948: I, 233; 1960: 385–86).

30. Decorme (1941: 11, 410).

31. Nentvig (1951: 79).

32. See Seymour (1989) for a detailed description of the seventeenth- and eighteenth-century movements of the Sobaipuri along the San Pedro River Valley.

33. Elton (1958: 147), Anderson (1956).

34. For examples of the timing of mesquite encroachment, see plates 6, 7, 10, 42–60. For an example of post-1880 mesquite encroachment on Cienega Creek, tributary to the Santa Cruz, see photographs filed under number 3411 at the Desert Laboratory Archive, Tucson, Arizona.

35. Sauer (1944), Stewart (1956).

36. The list of those who have expounded the fire hypothesis is a long and distinguished one and dates at least back to 1819. Humphrey (1958) and Wright and Bailey (1982) provide excellent reviews of this topic.

37. Thornber (1910), Griffiths (1910), Branscomb (1956), Humphrey (1958).

38. Bahre (1985, 1991).

39. Wright and Bailey (1982).

40. Matson and Schroeder (1957), Pfefferkorn (1949), Dobyns (1981), Pyne (1982).

41. Pfefferkorn (1949).

42. Bahre (1985, 1991). Reap (1986) shows for much of the *Changing Mile* region in the United States a cloud-to-ground strike frequency of twenty-five to fifty strikes per square kilometer per season. Minnich et al. (1993) note that in the mountains of northern Baja California as few as 2–5 percent of the strikes would account for all of the fires started there in recent years.

43. Humphrey (1958), Cable (1967), Bock et al. (1976), Caprio and Zwolinski (1992).

44. Baisan (1988), Baisan and Swetnam (1990), Kaib et al. (1996), Swetnam et al. (1992).

45. It is obviously impossible to treat here the involved questions of pre-Spanish ecology, and no attempt has been made even to describe those cultures next preceding the ones found by the conquistadors. An extensive literature exists for the Hohokam of southern Arizona; Sauer and Brand (1931) have surveyed what they consider to be the counterpart of the Hohokam in northern Sonora.

46. Szuter (1991) shows that the lowland Hohokam, with large commitments to tending cultivated fields, ate a larger proportion of rodents, such as jackrabbits, than of artiodactyls, such as javelina and deer. The upland inhabitants ate a larger proportion of artiodactyls than of rodents. Part of this dietal difference resulted from the prey's preference for disturbed versus undisturbed land.

47. Bolton and Marshall (1920), Bolton (1949). The route of Fray Marcos and Coronado is in dispute. Undreiner (1947) and Schroeder (1956) disagree with Bolton and believe that Coronado went all the way down the San Pedro, crossed the Gila, thence to the Salt, and from it to Zuñi. See also Sauer (1932), Reff (1991), and Weber (1992).

48. Pérez de Ribas (1645).

49. The Jesuit advance has been well studied and much written about; a score of studies deal with special phases of it. Decorme (1941) presents a general survey.

50. Bolton (1917).

51. Mange (1926), Alegre (1956–60: III, 11–21).

52. Dunne (1957: 87–88).

53. Ewing (1934).

54. There is general agreement that the Apache appeared on the northern frontier about 1680–90. Whether their migration was linked to the Pueblo Revolt, to pressures from the Comanche (Sauer 1934: 59, 74, 81; Schroeder 1956: 28–29; Spicer 1962: 230–33; Forbes 1958; Worchester 1941), or was the result of passive movement along the Rocky Mountains from the north (Perry 1991) is a matter of dispute. See Wilcox (1981) for a recent summary.

55. Chapman (1916: 65–66), Thomas (1941: 10–12), Goodwin and Basso (1971).

56. Nentvig (1951: 136–37).

57. Pfefferkorn (1949: 43–45).

58. Tamarón (1937: 237–51, 279–86, 303–11). Tamarón's figures agree substantially with those given by Nicolas de Lafora (1939). They are considerably under Teodoro de Croix's figures for 1781 (Thomas 1941: 133).

59. For a thorough historical study of the colonization of Sonora, see Radding (1997). Hastings (1961a) has surveyed the more obvious aspects.

60. Priestley (1916).

61. See Radding (1997).

62. One crop, winter wheat (introduced by missionaries such as Kino) had a marked effect upon those American Indians practicing agriculture. Prior to the arrival of the Europeans, the major crops raised by these people were summer-growing crops such as beans, corn, and squash. The gift of winter wheat allowed the natives to farm year round, planting in December and harvesting in June (Sheridan 1995: 23).

63. Although West (1949: 39–46) writes that the area around Parral is no longer Oak Woodland, Reichenbacher et al. (1998) map this area as Oak Woodland today.

64. West (1949: 116, n. 166).

65. Simpson (1952: frontispiece).

66. Ewing (1934: 18).

67. Bannon (1955: 137–39).

68. Bolton (1948: map, I, 331).

69. Spanish speakers often use the phrase *"cien mil"* to express the idea of large numbers in a nonliteral sense. These uncritical references to livestock numbers often become exact values when translated into English.

70. Bolton (1948: I, 58, 357).

71. Pfefferkorn (1949: 43, 45, 91, 94–95, 98, 285).

72. Ibid., 45, 98.

73. Bancroft (1883: I, 490).

74. Thomas (1941: 31).

75. Bancroft (1889: II, 628–48). Hardy (1829), Velasco (1850), and Zuñiga (1835) give good contemporary descriptions of Sonora during this period. Berber (1958) and Almada (1990) present general histories of the state.

76. Bancroft (1889: II, 633–52).

77. Stevens (1964).

78. Velasco (1850: 238) says the Indian attacks recommenced in 1832.

79. Bancroft (1962: 399–407), Bancroft (1889: II, 653ff.).

80. Officer (1987) provides an excellent summary of early Arizona history under the Spanish and Mexican flags.

81. U.S. Congress (1880).

82. Sheridan (1995: 128).

83. Mattison (1946: 300).

84. Officer (1987).

85. Bieber (1938: 132–33). For other descriptions of San Bernardino, see Bartlett (1854: II, 255–56), Powell (1931: 127), Clarke (1852: 77).

86. Bartlett (1854: I, 386). Bartlett mistakenly calls Sonoita "Calabazas." It is clear from his description, however, that he is just west of Patagonia; later, when he visits the true Calabasas, he corrects his error (ibid., II, 307–8).

87. Ibid., II, 307.

88. Ibid., II, 311–12.

89. Powell (1931: 136), Evans (1945: 148).

90. In 1848, Couts (1961: 58–59) found Guebabi, Tubac, and Tumacácori inhabited and, in between Santa Cruz and Guebabi, he passed "several deserted as well as inhabited ranches." In 1852, Bartlett (1854: II, 304ff.) found Tubac inhabited but notes that the preceding year it was deserted. Tumacácori and the establishments along the river all the way to Santa Cruz were abandoned. Two years later, James Bell found Tubac inhabited and a thriving establishment owned by Manuel Gandara at Calabasas as well: "I learn that the owner—the governor of Sonora—intends abandoning it. . . . [I]t has only been a few weeks since, that the Indians killed fifty head of sheep, and are continually driving off the stock" (Haley 1932: 311). Another account of Gandara's operation at Calabasas may be found in Senate Executive Document 207 (U.S. Congress 1880), where the involved history of the Tumacácori-Calabasas grants is traced in some detail.

91. Bartlett (1854: I, 396–97).

92. Bieber (1938: 132).

93. Ibid., 143.

94. Durivage (1937: 205), Cox (1925: 140–42), Evans (1945: 144–45), Clarke (1852: 78), Powell (1931: 126–28).

95. Sheridan (1995).

96. Christiansen (1988).

97. McAuliffe (1998) describes the effect of water development on Arizona range resources during the twentieth century.

98. Sheridan (1995).

99. A hint of the changes produced by livestock during the Mexican period can be gleaned from a series of matched photographs first taken of the monuments along the U.S.-Mexico boundary in the early 1890s (Humphrey 1987). Out of roughly 70 matched photograph pairs, the only stations in our region that show heavy mesquite growth at the time of the first photographs (1892–93) were those near permanent water sources (Monument 46 near Ojo de Los Moscos [Mosquito Springs], Chihuahua; Monument 77 near San Bernardino Ranch). Thus, it appears that by the early 1890s, at a time when few artificial water sources had been developed, the decades-long cumulative effect of livestock was highly localized around natural waters. By the time of Humphrey's second photographs (1982–83), following decades of artificial rangeland water development and greater stocking rates, numerous sites along the boundary in our region supported heavy mesquite growth. Many of Humphrey's photographic matches, as well as additional unpublished views of the monuments, have been repeated (1994–2001) and corroborate his earlier findings (R. M. Turner, unpublished photographs, Desert Laboratory Archive).

100. Hadley and Sheridan (1995) provide an excellent, detailed presentation of the history of human impact on the San Rafael Valley, Arizona, an area bordering our region of interest.

Chapter 3

1. Carson (1926), Pattie (1905). Secondary accounts of fur-trapping in the region may be found in Lavender (1954), Cleland (1950), and Lockwood (1929).

2. Davis (1982) has gleaned all wildlife references from travelers' journals for the period 1824–65. Ricketts (1996) documents the day-by-day travel of the Mormon Battalion as it passed through our region in the mid-1840s.

3. Eaton (1933).

4. Hastings (1959).

5. Etz (1939). For other references to beaver in the San Pedro River in the late nineteenth century, see the following manuscripts in the Arizona Historical Society, Tucson: Pool (1935); Ohnesorgen (1929); and Boedecker (1930). Davis (1982) provides a valuable review of Arizona wildlife occurrences during the 1824–65 period.

6. Bieber (1938: 42). The fish referred to here as salmon trout is probably *Ptychocheilus lucius*. Also known as Colorado River pikeminnow, it was formerly found along the San Pedro as far south as Fairbank (Minckley 1973). References to fish in the Santa Cruz River, the San Pedro River, and Sonoita Creek are abundant. For some of the later ones, see *Arizona Daily Star*, August 20, 1886, and July 17, 1887; *Arizona Weekly Star*, July 26, 1877; Spring (1902); *The Tombstone*, August 27, 1885; and *Weekly Arizonan*, May 12, 1859. See also Davis (1982) for a summary of wildlife sightings in Arizona between 1824 and 1865.

7. Tevis (1954: 55).

8. Itinerary (1858: 33). The "trout" referred to is probably *Gila intermedia*, a species of chub found along the Gila River system.

9. An early picture of the junction of the San Pedro River and Babocomari Creek appears as plate 57 and illustrates the ill-defined nature of the region's rivers.

The creek seeped through a rank growth of marsh grass and can hardly be said to have had a course.

10. Powell (1931: 145).

11. Bartlett (1854: I, 379–81). For another account by Lt. Col. Graham, see U.S. Congress (1852: 35–36).

12. Bandelier (1892: pt. 2, 478). In plate 51, a picture of Charleston at about the time of Bandelier's visit, a clear, small trench is visible.

13. Emory (1848: 75).

14. Johnston (1848: 592).

15. Golder (1928: 193). The estimate of mesquite forest acreage is hugely inflated.

16. Itinerary (1858: 32).

17. U.S. Congress (1859: 87).

18. Parke (1857: 25).

19. Eccleston (1950: 192–93).

20. Allison (n.d.).

21. Cochise County District Court (1889).

22. Powell (1931: 130–31), Bartlett (1854: II, 324).

23. *Report on Barracks* (1870: 465–66).

24. Bourke (1950: 8), Spring (1903), Rothrock (1875). See also newspaper accounts in *Arizona Daily Star*, July 21, 1880; *Arizona Weekly Star*, September 25, 1879; and *Arizona Citizen*, August 29, 1874.

25. Schumm and Hadley (1957).

26. Compare Bryan (1928a), Cottam and Stewart (1940). See Shmida and Burgess (1988) and Burgess (1995) for discussions of the effect of waterlogging in shaping the vegetation of seasonally flooded valleys in our region.

27. Hereford (1993: 14) points out how misinterpretation of "inset" and "superimposed" stratigraphic relations has led to incorrectly assuming that all steep embankments encountered along the San Pedro River in the late nineteenth century were products of the same cutting and filling cycle, when actually they were formed during different unrelated cycles.

28. Powell (1931: 130).

29. Bartlett (1854: II, 325).

30. Parke (1857: 24).

31. U.S. Congress (1859: 99).

32. Itinerary (1858: 30).

33. The notion that mesquite was brought into this country from Mexico a few centuries ago is a persistent misconception portrayed by a letter from James C. Malin, inquiring about an article in *Life* (August 18, 1952) dealing with the mesquite problem. Malin wrote that "a century and a half ago, there was hardly any [mesquite] in the U.S., but during the next fifty years it was brought into this country from Mexico by Spanish ponies and by wandering herds of wild buffalo . . . so that, in 1850 . . . scattered stands of mesquite were growing along the creeks and river beds of the southwest. This first generation mesquite, however, was exceedingly sparse" (Malin 1956: 449–50). Studies of fossil packrat middens show beyond a doubt that mesquite arrived in our area thousands of years before the date set by Malin (Van Devender 1990).

34. Parke (1857: 24).

35. The term "invasion" or "invader" is used broadly to refer to three classes of plants that have increased at sites where they were missing or at low density before: (1) exotic plants recently introduced by humans from distant localities; (2) plants that were primarily asso-ciated with adjacent life zones but have recently moved upward or downward; and (3) plants that were local residents but have recently increased in density in place. We do not employ the term "invade" in any pejorative sense. We do not regard the new plants on the scene as interlopers, impostors, transgressors, etc.; they are merely plants with greater presence now than before.

36. Bahre (1985), Kaib (1998).

37. Swetnam and Baisan (1996a, 1996b), Kaib (1998), McPherson et al. (1993), Villanueva (1996).

38. Bahre (1991).

39. Hastings and Turner (1965b).

40. See Dobyns (1981: 27–43), Pyne (1982), and Bahre (1985) for a summary of American Indian fire use in our region.

41. Eaton (1933), Pumpelly (1870).

42. Keleher (1952).

43. Bell (1869: II, 92), *Report on Barracks* (1870: 464).

44. Among the best sources for recreating the Arizona scene in the 1870s are the lively, but frequently inaccurate, contemporary handbooks: Hinton (1954), Hamilton (1881), *History of Arizona Territory* (1884).

45. Greever (1957).

46. U.S. Bur. Census (1895: XV).

47. Several studies recount the history of ranching in the Gadsden Purchase: Haskett (1935), Morrisey (1950), Wagoner (1951, 1952). The regional context is treated by Love (1916). The national picture is presented in such works as Osgood (1929), Dale (1930), and Pelzer (1936).

48. Although figures from the Census Bureau (U.S. Bur. Census 1872: III, 75) show only 5,132 cattle in Arizona Territory, more recent evaluation of all records show the value in 1871 to be about 38,000 (Sheridan 1995: 132, 140).

49. Wagoner (1952: 36).

50. Ibid., 120; U.S. Bur. Census (1883: III, 141–42).

51. Osgood (1929: 85–86).

52. Ibid., 92–93.

53. *Report of Governor* (1896: 21).

54. Ibid. (1885: 8).

55. *Southwestern Stockman*, January 10, March 14, and April 11, 1885.

56. Ibid., February 7, 1885.

57. Ibid., July 25, 1885.

58. *Tombstone Daily Epitaph*, April 4, 1886.

59. *Report of Governor* (1896: 22).

60. U.S. Bur. Census (1895: I, 29).

61. *Southwestern Stockman*, January 3, 1891.

62. *Report of Governor* (1896: 22). Sheridan (1995: 140) agrees with the larger figure, citing other observers of the time.

63. *Report of Governor* (1896: 22).

64. Land (1934).

65. Wagoner (1952: 120–21). Sayre (1999) describes the "political ecology" that drove the 1873-93 cattle boom and that continues to drive the ranching-to-urbanization trend today.

66. Herbel et al. (1972), Schmutz et al. (1985).

67. Fleischner (1994 and references therein).

68. Graf (1988).

69. Pfefferkorn (1949: 37–38, 41–42, 282), Thomas (1933), Wyllys (1931: 118).

70. Waters and Haynes (2001).

71. See *Arizona Quarterly* (1880: 18), Hinton (1890:

112), Murphy (1928), and *Arizona Weekly Star*, September 12, 1878, for reports of flooding. See Conkling and Conkling (1947: 11, 49), Ohnesorgen (1929), and *Arizona Weekly Star*, September 25, 1879, for descriptions of bridge construction projects.

72. *Arizona Daily Star*, August 30, 1886.

73. *Arizona Weekly Enterprise*, August 14, 1886.

74. *Arizona Daily Star*, August 3, 1886.

75. Ibid., August 9, 1886.

76. Ibid., August 16, 1886.

77. *Arizona Weekly Enterprise*, September 17, 1887.

78. Ibid., September 3, 1887.

79. Ibid., July 16, 1887; *Arizona Daily Star*, July 17, 1887.

80. *Arizona Daily Star*, July 8, 1887.

81. Ibid., July 15, 1887.

82. *Arizona Weekly Enterprise*, July 28, 1888; *Arizona Citizen*, September 9, 1889.

83. *Arizona Daily Star*, August 5, 6, 7, 8, 9, and 13, 1890.

84. Ibid., August 8, 1890.

85. Ibid., August 14, 1890.

86. *Arizona Weekly Enterprise*, September 6, 1890.

87. *Arizona Daily Star*, October 2, 1890.

88. Ibid., August 13, 1890.

89. Olmstead (1919: 79), Schwennesen (1917: 6), Swift (1926: 70–71).

90. The Sonoyta River, which drains toward the Gulf of California, should not be confused with Sonoita Creek, a tributary of the Santa Cruz River.

91. Lumholtz (1912: 178–79).

92. U.S. Congress (1898: 23). In addition to the foregoing historical accounts of flooding and downcutting, see the following recent studies of southern Arizona's rivers. Hereford (1993) and Hereford and Betancourt (1995) review the historical evidence along the San Pedro River for these phenomena and bring their treatment up-to-date by concluding that entrenchment has stabilized since about 1955. For discussions of these phenomena along just the Santa Cruz River, see Betancourt (1990) and Webb and Betancourt (1992). For more general regional treatments, see Cooke and Reeves (1976), Dobyns (1981), and Bahre (1991).

93. Several authors consider the overall question of the cause of arroyos. See Cooke and Reeves (1976), Graf (1983), and Webb (1985).

94. Rich (1911), Thornber (1910: 336–38), Cottam and Stewart (1940: 626), Winn (1926), Thornthwaite et al. (1942), Dittmer (1951).

95. Lusby et al. (1971), Lusby (1979).

96. Duce (1918), Brady (1936), Cooke and Reeves (1976).

97. Bryan (1940).

98. Waters (1985, 1988).

99. Leopold (1951a).

100. Schumm and Hadley (1957).

101. Webb (1985), Webb and Baker (1987), Ely et al. (1992).

102. Bryan (1940).

103. Bryan (1928a: 477).

104. Hack (1939); Leopold (1951b).

105. Reagan (1924) asserted from little evidence that past periods of alluviation can be explained by aboriginal farming and dam building. According to his ar-

gument, past channeling was caused by the coming of great herds of herbivores, although no evidence for these herds has ever surfaced. Dellenbaugh (1912) stated that erosion occurred "in places where there were no cattle and never had been any." He viewed channeling as continuous, not cyclic, occurring at irregular and independent intervals along individual streams on the basis of local conditions; his arguments ultimately led to the idea of intrinsic geomorphic processes as first espoused by Schumm and Hadley (1957). Huntington (1914) argued that arroyo cutting was merely the inexorable shaping of land forms by the omnipotent hand of climate.

106. Antevs (1952).

107. Thornthwaite et al. (1942: 127).

108. Ibid., 123.

109. Ibid., Hack (1939), Leopold and Snyder (1951: 17). See also Leopold (1951a: 305).

110. Leopold (1951a). Elsewhere (Leopold et al. 1963), Leopold reiterates this explanation but suggests, like Martin (1963), that the arroyo cutting of the thirteenth century may not have been caused by aridity but by a different balance of summer and winter precipitation.

111. Leopold and Miller (1954: 85).

112. Judson (1952).

113. Waters (1985, 1988, 1992), Waters and Haynes (2001).

114. Waters and Haynes (2001).

115. Gregory (1917: 132).

116. Peterson (1950: 421).

117. Cooke and Reeves (1976).

118. Betancourt (1990).

119. Notably Euler et al. (1979), who took the concept of "cycles of erosion" a bit too literally.

120. Reviews of the various positions may be found in Schumm and Hadley (1957: 161) and Antevs (1952: 375–76).

121. Bryan (1925).

122. Graf (1983), Hereford (1984), Webb (1985), Elliott et al. (1999).

123. Cottam and Stewart (1940: 614).

124. Gregory and Moore (1931: 30ff.).

125. Webb et al. (1991).

126. Webb (1985), Webb and Baker (1987).

127. Bryan (1927: 17–19; 1928b).

128. Sauer and Brand (1931: 122–24).

129. Schumm and Hadley (1957).

130. Cottam and Stewart (1940: 614) note the relationship between settlement (1861) and arroyo cutting (1884) in Mountain Meadows in southwestern Utah. On the basis of the dates given by Gregory and Moore, Bailey (1935: 350) predicates a similar relationship between settlement (1870) and channeling (1883) in the Kanab and Long Valleys of Utah. Colton (1937: 18–19) states that settlement (1878 or 1879) and overgrazing (1880s) initiated floods (early 1890s) on the Little Colorado (but, see Hereford 1985 for a different opinion).

131. Hereford (1993: 15).

132. Bahre and Hutchinson (1985).

133. Bahre (1991).

134. Bahre and Hutchinson (1985). The Tombstone "woodshed" did not include the Tombstone Hills. These hills, in which the town of Tombstone is located, are low

and mostly of limestone and provide a habitat too dry to support many trees.

135. Bahre (1991).

136. Bahre and Hutchinson (1985: 185).

Chapter 4

1. Shreve (1915: 24–29).

2. Marshall (1957).

3. Ibid.

4. Brown (1994a).

5. Castetter (1956), Dick-Peddie (1993).

6. White (1948), Marshall (1957), Brown (1994a).

7. See plates 42–60. In this particular case, soil differences may also be involved; limestone soils support more xerophytic communities than soils derived from granite (Shreve 1922: 272, 1942b: 192), and calcareous substrates are common in the San Pedro Valley.

8. See, for example, plates 70–72.

9. Although there has been little quantitative ecology performed in the Oak Woodland, a number of descriptive treatments exist. In addition to those already referred to are Darrow (1944), Wallmo (1955), and McPherson (1992).

10. Schrader (1915: 197–98).

11. Turner et al. (1995).

12. Phillips (1912: 12–14), Swain (1893).

13. *Arizona Weekly Star*, July 28, 1892.

14. Brandes (1960: 26–27).

15. *Arizona Citizen*, August 10, 1872.

16. Glinski and Brown (1982).

17. Granger (1960).

18. *Visita* is a mission station, usually a small village but without a resident priest. A traveling priest periodically visits the village, saying mass, baptizing, etc.

19. See plates 15, 16, 20, 22, 23, 42, and 43 for similar views of sparsely vegetated landscape.

20. Phillips (1912: 12).

21. Sanchini (1981).

22. Bowers et al. (1995).

23. Bahre (1991).

24. At one Oak Woodland station (plate 1), mesquite is actually absent on a north slope at mesquite's upper elevational limit but is seen on a nearby south-facing slope.

25. At one mine site (plate 37 in chapter 5), the presence of undamaged oak carcasses in 1994 argues against woodcutting as the cause of decline. See also plates 38 and 49 in chapter 5 for other examples of oak decline.

26. If one could be sure that the Charleston patch of 1883 was, in fact, a relict colony, one could use its attrition during recent years to infer something about the magnitude of recent stresses. The fact that, having survived the vicissitudes of several thousand years of isolation, the colony now is on the verge of extinction would argue that recent events have been severe, indeed. Although Martin (1963: 68) finds "no reliable pollen evidence that postglacial droughts, if they occurred, were sufficient to shift biotic zones above their present level," this seeming contradiction probably stems from the different level of resolution permitted by photographic analysis versus palynological analysis.

27. Hastings and Turner (1965b).

28. Bahre (1991).

29. Herbel et al. (1972), Martin and Turner (1977), Meko et al. (1980), Turner (1990), Betancourt et al. (1993).

30. Nyandiga and McPherson (1992).

31. Pase (1969).

32. If one uses estimates of plant numbers per area covered by the photographs (density) instead of plant volume or biomass, different conclusions will be made about the direction of changes shown by our matched photographs. For example, using density as the datum, several species showed little or no change in the Oak Woodland during the period from 1962 to the 1990s. Among these are ocotillo, one-seed juniper, and mesquite, all of which increased significantly using biomass as the datum for judging change. We use plant biomass in our semiquantitative, photogrammetric analysis of vegetation change rather than plant density because the former is a better measure of the role a given plant species plays in an ecosystem.

Chapter 5

1. Brown (1994b, 1994c).

2. Nelson (1934).

3. White (1948), Shreve (1942b), Dick-Peddie (1993).

4. McLaughlin et al. (2001).

5. Bowers and McLaughlin (1984, 1996).

6. The interested reader may refer to several general discussions of arid region grasslands: Clements (1920), Brand (1936), White (1948), Shantz (1924), Shreve (1917b, 1942b, 1942c), Humphrey (1958), McClaran and Van Devender (1995), and Burgess (1995), or to more detailed studies of the grasslands of the desert region: Wallmo (1955), Darrow (1944), and Johnson (1961).

7. Schrader (1915).

8. The minimum elevation for oaks was found to be roughly 1,000 feet lower at our other camera stations showing the lower edge of the Oak Woodland. Shreve (1922) describes the influence that substrates such as rhyolite and limestone have in elevating plant ranges. See also Whittaker and Niering (1968).

9. Browne (1951: 265–66).

10. Granger (1960).

11. Applegate (1981).

12. Ibid.

13. The flood of 1977 was greatly exceeded by a flood in October 1983 that probably was responsible for establishment of some of the trees seen in this view.

14. Emory (1857: I, 99ff.).

15. Anable et al. (1992).

16. Emory (1857: II, 18).

17. Bahre and Shelton (1993).

18. Hereford and Betancourt (1995).

19. Ibid.

20. Brown et al. (1981).

21. Hereford and Betancourt (1995).

22. Bieber (1938).

23. For references to Gird Dam destruction and reconstruction, see *Tombstone Daily Epitaph*, June 27, 1881, and July 16, 1887.

24. Careful scrutiny of the vicinity in 1962 and again in 1996 disclosed one living oak tree—an Emory oak—among many carcasses belonging to the same genus. The USDA Forest Products Laboratory at Madison, Wisconsin, has shown the wood from one of these carcasses to

belong to a member of the red oak group, but identification to species is not possible (personal communication, February 12, 1963). Of the two red oaks in this part of Arizona (*Quercus hypoleucoides* being the other), only Emory oak can be expected at this elevation.

25. Sheridan (1995).

26. This photograph was taken by Carleton E. Watkins. See Palmquist (1983) for details of Watkins's trip to the region.

27. Bahre (1991).

28. Hereford (1993).

29. Glinski and Brown (1982).

30. Hastings and Turner (1965b, plate 60a).

31. See McClaran and Van Devender (1995) for thorough discussions and reviews of this topic.

32. Shreve (1942a: 235) recognized the changes from grassland to Chihuahuan Desert that have recently altered the appearance of the landscape in the San Pedro Valley. He assigned the area to grassland, an interpretation more recently followed by Martin (1963). Benson and Darrow (1954) described the same vegetation as Chihuahuan Desert without raising the question of recent change; this area is shown on their map, however, as Sonoran Desert. Nichol (1952) described the tarbush–creosote bush desert of this area but did not point out its resemblance to the Chihuahuan Desert. Brown and Lowe (1994) place much of the San Pedro Valley between Cascabel and the Mexican border in the Chihuahuan Desert.

Chapter 6

1. Shreve (1964).

2. Shreve (1951, 1964) recognizes seven provinces. We follow Brown and Lowe (1994) and Turner and Brown (1994), who place one of these, Shreve's Foothills of Sonora province, outside the Sonoran Desert.

3. McAuliffe (1994) has provided a detailed description of the basis for shifts in vegetation along the piedmonts descending from mountain fronts in the Sonoran Desert. The complex nature of the bajada vegetation was acknowledged earlier by Shreve (1951, 1964) and partially explained by Yang and Lowe (1956), but a thorough understanding of the complex interaction of geomorphology, soils, vegetation, climate, and time awaited McAuliffe's lucid analysis of the phenomenon.

4. Shreve (1964: 21).

5. Yang and Lowe (1956).

6. See Shantz and Piemeisel (1924) for an example of soil salt emphasis; Shreve (1964) emphasizes soil texture; McAuliffe (1994) places emphasis on soil age.

7. Green (1959).

8. Turner and Brown (1994).

9. The entire complex has been referred to as the paloverde-saguaro association by Yang and Lowe (1956), as the paloverde, bursage, and cacti desert by Nichol (1952), as the paloverde–triangle bursage range by Humphrey (1960), and as the paloverde-cacti–mixed scrub series by Turner and Brown (1994).

10. Darrow (1943).

11. Hastings (1961b) provides water absorption and growth data for saguaro during the summer. Unpublished data extending the time covered by the original report to include winter are held at the Desert Laboratory, Tucson, Arizona.

12. Desert plant flowering season is described in Shreve (1964), Turner (1963), and McGinnies (n.d.), but especially see Bowers and Dimmitt (1994) and Bowers (1996). In saguaro populations at the dry margin of the plant's range, plants growing on the slightly elevated and relatively dry interfluves between adjacent runnels do not flower in dry years, whereas the plants in the adjacent minor waterways do flower (Brum 1973).

13. Turner and Brown (1994).

14. Shreve (1964), Allred et al. (1963), Turner (1994b).

15. Green and Sellers (1964), Hastings (1964a, 1964b), Hastings and Humphrey (1969a, 1969b).

16. Turner and Brown (1994).

17. Benson and Darrow (1954: 208).

18. Ibid., 313, Turner et al. (1995).

19. Turner (1994b).

20. Shantz and Piemeisel (1924), Turner and Brown (1994).

21. See Bowers (1982, 1998).

22. Shreve (1964).

23. Hastings (1964a, 1964b), Turner and Brown (1994).

24. Hastings and Humphrey (1969b).

25. Humphrey and Humphrey (1990).

26. Escoto Rodríguez (1999).

27. Aschmann (1959).

28. Steenbergh and Lowe (1976).

29. Alcorn and May (1962: 157).

30. Turner (1992).

31. Several photographs taken in 1964 about 0.3 mile from this camera station and on a line toward Agua Caliente Hill were repeated in 1968 and 1996 (Stake nos. 542f, g, h, and i) and corroborate our conclusions regarding the changes in plate 60. Another photograph pair (Stake 1670) near the cottonwood trees in the right background shows that cottonwoods have increased along Tanque Verde Creek in recent years.

32. Berry and Steelink (1961).

33. The approximate heights of these five saguaros were four at 3.3 feet and one at 1.6 feet.

34. This fraction has, of course, no meaning without reference to the maximum life span, which is about 200 years (Bowers and Turner, 2002).

35. The appearance of the saguaro at far right results from relocating the camera.

36. Shreve (1911a: 296).

37. Pierson and Turner (1998).

38. Shreve (1915: 24).

39. Bowers (1980–81).

40. Steenbergh and Lowe (1976).

41. Shreve (1911a: 293).

42. Saguaro ages were determined by measuring the saguaro heights and using a height-to-age model developed for a nearby population (Pierson and Turner, 1998).

43. Masse (1979).

44. Vasek (1980).

45. Blydenstein et al. (1957), Goldberg and Turner (1986).

46. MacDougal and Cannon (1910: 12) state that, of the several plant species found growing with range ratany, the parasite is most frequently found growing with creosote bush.

47. Tolman (1909).

48. Pierson and Turner (1998).

49. Bowers and Turner (2001).

50. Hornaday (1908).

51. Turner (1990).

52. See chapter 7 for a discussion of saguaro age composition at this and adjacent sites in the 1990s.

53. Hornaday (1908).

54. Ibid.

55. This modest recovery is similar to findings from a permanent plot established in the area with densest saguaros at left (Turner 1990).

56. Turner (1992).

57. Meko et al. (1980).

58. Turner (1990).

59. This is the Permanent Playa Plot described in Turner (1990: Table 5).

60. Shreve (1917a: 216).

61. Several individuals of this valuable horticultural species have been removed from adjacent areas in recent years, however.

62. Indirect corroboration of the recent increase can be had from Shreve (1964: 164), who almost certainly was familiar with the Islas Mellizas yet stated that the only heavy stand of the cardón he had encountered in Sonora twenty or thirty years earlier was located above the salt flats near Empalme. An inspection in 1961 showed the Empalme stand to be not only thinner than this one but less dense than many of the insular colonies that one must pass when driving along the coast toward Empalme.

63. See a detailed description of this island population in chapter 7.

Chapter 7

1. Multiple views were photographed from the same point at several of our camera stations.

2. The mesquites we have seen in our photographs actually represent two and perhaps three different species. Mesquite has been the subject of several taxonomic studies (Benson 1941, Johnston 1962, Burkart 1976), all showing the close relationship between the taxonomic units we have encountered. These are western honey mesquite (*Prosopis glandulosa* Torrey var. *torreyana* [L. Benson] M. C. Johnston) and velvet mesquite (*P. velutina* Wooton). The taxa commonly intergrade, "forming perplexing series of intermediate forms and of plants with recombinations of the characters" of the parent types (Benson 1941). For our purposes, we assume that the taxonomic groups encountered are very similar in their environmental requirements and that the time and effort required to sort out the confusing intergradations and recombinations of physical traits would serve little purpose, since we would still be left in some doubt as to the ecological and physiological relevance of these physical differences. The difficulty of assigning mesquites in our region to a particular species has become even more complex since *The Changing Mile* first appeared in 1965. Individuals of a third taxon, Texas honey mesquite (*Prosopis glandulosa* Torrey), are now commonly seen along the median and shoulder strips of the west-bound lane of the interstate highway east of Tucson, where seeds have probably fallen from westbound cattle trucks. How far beyond the highway shoulders their influence has spread is not known.

3. Glinski and Brown (1982).

4. Glendening and Paulsen (1955), Cox et al. (1993).

5. Kramp et al. (1998).

6. Martin (1948, 1970).

7. Cox et al. (1993).

8. Glendening and Paulsen (1955) found that near optimum germination for velvet mesquite occurred at 86°F with germination percentages falling at temperatures above and below that value. Siegel and Brock (1990) found 100 percent germination occurring through a broader range of high temperatures. Hull (1958) found that this plant exhibited maximum growth at a combination of high night and day temperatures and noted that "velvet mesquite is one of a very few species of plants which has exhibited maximum growth at these high temperatures."

9. See Wilson et al. (2001) for the role of winter rain in mesquite encroachment.

10. Griffiths (1904, 1910), Thornber (1910), Grover and Musick (1990).

11. Glendening (1952), Hastings and Turner (1965b: plate 8b), and plate 8b in this volume.

12. Turner et al. (1995).

13. Bahre and Shelton (1993).

14. Bowers (1980–81), Glinski and Brown (1982). See also Webb (1996) and Bowers et al. (1995) for evidence of freezing in the Grand Canyon.

15. The critical physical character here is not diameter, of course, but the relationship between stem surface area and stem volume among stem segments with different diameters. The intensity of the freezing event determines whether a particular surface-to-volume ratio proves frost resistant or not.

16. The changes in this plot through 1982 are described in Turner (1990: table 5). The plot was resurveyed in March 2000.

17. The decline with time in saguaro biomass at the camera stations featured in chapter 6 is less pronounced than that seen in the larger set of photographs noted here.

18. Pierson and Turner (1998).

19. Parker (1993), Turner (1990).

20. Steenbergh and Lowe (1969, 1977).

21. Steenbergh and Lowe (1977).

22. The protection provided may be from heat (Turner et al. 1966, Despain 1974, Franco and Nobel 1989), cold (Steenbergh and Lowe 1976, Nobel 1980a), or herbivores (Niering et al. 1963, Steenbergh and Lowe 1977). Several reports by Joseph McAuliffe and co-workers have greatly expanded our knowledge of the phenomenon of saguaro–nurse plant interactions. See McAuliffe (1984, 1990), Hutto et al. (1986), and McAuliffe and Janzen (1986).

23. Creosote bush served less frequently as a nurse plant than mesquite and foothill paloverde (Hutto et al. 1986) at Organ Pipe Cactus National Monument.

24. Several studies of saguaro longevity have been made (Shreve 1910, Hastings and Alcorn 1961, Steenbergh and Lowe 1983, Turner 1990, Pierson and Turner 1998). Because habitat differences can influence lifespan, a value of 175 years is a conservative average.

25. Hastings and Turner (1965b: 246).

26. For a description of the model used to estimate saguaro age, see Turner (1990) and Pierson and Turner (1998). Besides the areas defined in plates 87–

90, the MacDougal Pass sample includes the population shown in negative number 1693 in the Desert Laboratory Archives.

27. The low number of plants falling in the last decade results from the difficulty of finding small, newly established plants and because the time period actually included only seven instead of ten years.

28. Turner et al. (1995).

29. Mortality and recruitment for cardón are apparently closely related to soil development. Mortality exceeded establishment on ancient geomorphic surfaces in Baja California. In contrast to those shrinking populations, high recruitment and low mortality were found on young geomorphic surfaces (McAuliffe 1991). Recruitment exceeded mortality at 74 percent of fifty-three sites examined in Baja California (S. B. Bullock and N. E. Martijena, written communication, 1999).

30. The years listed are median values for the groups of years defining the beginnings and ends of the two observation periods. See table 1.

31. Greater biomass probably reflects growth of individual plants and the tendency of foothill paloverde to produce root sprouts within several feet of the trunk. In some locations, increased coverage might have occurred at the expense of seedling establishment. On Tumamoc Hill, a site where biomass in 1999 was much higher than in 1906 (plate 82), mortality has generally exceeded establishment since the 1940s despite many years that should have been favorable for seedling recruitment (Bowers and Turner 2001).

32. Turner (1963), McGinnies (n.d.).

33. Bowers (1994).

34. Shreve (1911b, 1917a).

35. Bowers and Turner (2001).

36. Bowers and Turner (2002).

37. Shreve concluded that foothill paloverdes could attain an age of 400 years (Shreve 1911a). Turner (1963), in a short study involving five paloverde trees near Tucson, Arizona, showed that annual radial growth in this tree was only about 0.01 inch to 0.05 inch. Using these growth values, large trees with radii of 6.0 inches would be only 120 to 175 years of age. (See also Bowers and Turner 2002.) Some of the dead trees in the crater exceeded 6.0 inches in diameter.

38. Many of these trees apparently only suffered dieback to the base and then regrew with the return of favorable conditions.

39. Hutto et al. (1986).

40. McAuliffe (1984).

41. The absence of grazing influence has been noted by others; see Bowers et al. (1995) and Webb (1996).

42. More specifically, December to February rain enhances the proportion of flowers that set fruit (Bowers 1996). Fruit set is poor in dry years, as is pad production (Bowers 1997). Moreover, because pads do not produce flowers until they are a year in age, the size of the fruit crop in any one year depends to some extent on how many new pads were produced in the previous year (Bowers 1996). The seeds probably do not germinate upon dispersal but must first undergo a year or two of leaching in the soil. Because prickly pear seeds require warmth and prolonged, continuous moisture for germination, it seems likely that germination occurs only during unusually wet summers (Potter et al. 1984).

Webb and Bowers (1993) suggest that a recent increase in prickly pear in the Grand Canyon is the result of less frequent freezing events.

43. Zedler (1981).

44. Goldberg and Turner (1986), Bowers et al. (1995). Based upon an old photograph, one individual next to the parking lot at the Desert Laboratory in Tucson attained the age of ninety years before dying recently.

45. White (1969).

46. Whitfield and Anderson (1938).

47. Rogers and Steele (1980).

48. Turner et al. (1995).

49. For creosote bush germination requirements, see Went and Westergaard (1949), Sheps (1973), Beatley (1974a, 1974b), Rivera and Freeman (1979), Zedler (1981), and Boyd and Brum (1983).

50. Beatley (1974a, 1974b), Dalton (1961).

51. Humphrey (1974).

52. Although the largest clone of creosote bush in Lucerne Valley, California, is estimated to have an age of 12,000 years (Vasek 1980), studies of fossil packrat middens throughout the Mojave Desert suggest that this plant first appeared in the area closer to 8,500 years before present (Spaulding 1990). The average longevity of creosote bush at a study site in Dateland, California, was estimated to be 1,250 years; at a San Luis site, the estimated age is 625 years (McAuliffe 1988).

53. See Grover and Musick (1990).

54. McAuliffe (1988), Vasek (1979/80), Webb et al. (1988).

55. Chew and Chew (1965).

56. Net photosynthesis is maintained with leaf water potentials as low as −80 bars (Smith and Nobel 1986).

57. MacDougal and Cannon (1910), Cannon (1911).

58. The *Juniperus* that we refer to as one-seed juniper in our photographs was formerly referred to as *Juniperus monosperma* (Engelm.) Sarg. but is now regarded as *J. coahuilensis* (M. Martinez) Gausson ex. R. P. Adams. See Adams and Zanoni (1979) and Adams (1994) for a description of the changes in nomenclature.

59. Shaw (1998).

60. See Tschirley (1963) and Benson (1982).

61. See Tschirley and Wagle (1964) and Shreve (1935).

62. Bahre and Hutchinson (1985) give a thorough discussion of fuelwood cutting in southeastern Arizona.

63. Fowells (1965).

64. Fenner et al. (1984, 1985).

65. Asplund and Gooch (1988), Fenner et al. (1985), Glinski (1977).

66. Glinski (1977).

67. Asplund and Gooch (1988).

68. Brinkman (1974), Burns and Honkala (1990).

69. Burns and Honkala (1990), Pope et al. (1990).

70. Howell and Roth (1981) propose a close tie between mescal seed set and numbers of nectar-feeding bats. Slauson (2000) has shown that, although bats are important pollinators, adequate pollination occurs when these nocturnal pollinators are excluded. See also Cockrum and Petryszyn (1991).

71. Howell and Roth (1981).

72. Liz Slauson, personal communication (2001).

73. Ibid.

74. See plate 52c, this volume.

75. Bowers and Dimmitt (1994).

76. Bowers (1994).

77. Goldberg and Turner (1986).

78. Webb and Bowers (1993).

79. Bowers (1994), Turner et al. (1995).

Chapter 8

1. Roeske et al. (1978), Bull (1974).

2. Brown et al. (1997).

3. Bahre (1998).

4. Bahre and Hutchinson (1985).

5. Ibid.; Bahre (1991: chap. 7).

6. The presence of wood-burning mills within several miles of some of our camera stations has been advanced as proof that the disappearance of oaks resulted from wood harvesting rather than from some other force (Bahre and Hutchinson 1985, Bahre 1991). The photograph in plate 13a and photographs of the Helena Mine (plate 37a and Desert Laboratory Stake No. 726) all show oaks and mesquites surviving in close proximity to wood-burning mills. The wood stacked at these mills was probably hauled in from distant sources.

7. Haul roads are necessary where intensive woodcutting occurs. If these were present, they are not visible on our photographs.

8. Weltzin and McPherson (1995: 181) assert that the oak deaths documented in Hastings and Turner (1965b) are the result of top kill among mature oaks, not the result of death of entire plants. Our field investigations in the 1960s and 1990s often revealed that oak carcasses remain where living plants once grew, an observation hard to reconcile with the top-kill hypothesis. The theory that the oak deaths occurred during the 1950s drought has been strengthened with the passage of time.

9. Bahre and Hutchinson (1985).

10. For the effect of small herbivores on grass cover, see Taylor et al. (1935), Taylor (1936), Reynolds and Glendening (1949), Norris (1950), and Cox et al. (1993). For their seed predation, see Brown and Heske (1990). Reynolds and Glendening (1949) and Reynolds (1950) describe positive effects these small herbivores have on shrub establishment. Because black-tailed jackrabbits (*Lepus californicus*) have been shown to browse more heavily on mesquite than on grass (Reynolds and Martin 1968, Vorhies and Taylor 1933), they might influence shrub establishment as well.

11. Taylor et al. (1935), Reynolds and Glendening (1949).

12. Brown and Heske (1990). Following several years of above-average winter rain, one of the keystone kangaroo rat species became extinct throughout the area at the same time that the general landscape experienced an increase in woody species. Changes that would usually be expected to occur in response to increased aridity or anthropogenic desertification actually occurred in response to an increase in seasonal precipitation (Brown et al. 1997).

13. Reynolds (1958) describes the importance of kangaroo rats and white-throated woodrats as agents for increasing shrub populations.

14. Brown and Archer (1989: 25).

15. Cox et al. (1993). Other investigators, in studies not involving rodents, compared mesquite seedling emergence on sites with and without cattle, and found ten to thirteen seedlings per square yard in the former compared to no seedlings in the latter (Brown and Archer 1989).

16. Turner et al. (1969), Steenbergh and Lowe (1977: 84).

17. In addition to the white-throated woodrat, several other species of *Neotoma* occur in our region; they are often referred to as packrats.

18. R. M. Turner, unpublished data, 2002. The observation of packrat damage on saguaros has been made during repeated examinations of nine long-term permanent saguaro monitoring plots. At only one of these nine plots has packrat damage been severe, and this plot stands out among all of the plots by the virtual absence of prickly pear cactus there. This relationship deserves further study.

19. Hoffmeister (1986).

20. Dobyns (1981).

21. Davis (1982).

22. See plates 36 and 37.

23. "Beavers improve San Pedro habitat," *Arizona Daily Star*, May 29, 2000.

24. Oakes (2000).

25. Mearns (1907: 341).

26. The black-tailed prairie dog still occurs in three colonies along the San Pedro drainage between Cananea, Sonora, and the U.S.-Mexican border (*Fide* William VanPelt, Arizona Game and Fish Dept. 1997).

27. Archer et al. (1987).

28. Oakes (2000) shows black-tailed prairie dogs occurring as far west as the vicinity of Elgin.

29. Balda (1987).

30. Reynolds and Glendening (1949), Glendening and Paulsen (1955), Morello and Saravia Toledo (1959), Cox et al. (1993), Kramp et al. (1998).

31. See Cox et al. (1993) and Kramp et al. (1998) for examples of nonbovine vectors of shrub seeds.

32. Glendening and Paulsen (1955).

33. Morello and Saravia Toledo (1959), Reynolds and Glendening (1949), Kramp et al. (1998).

34. Glendening and Paulsen (1955), Went (1957), Morello and Saravia Toledo (1959).

35. See Kurtén and Anderson (1980) for a description of extinct North American Pleistocene Artiodactylids.

36. Evidence that small granivorous mammals occurred in grassland in Pleistocene and early Holocene is by inference from packrat middens in nearby nongrassland sites (Mead et al. 1983, Van Devender et al. 1991, Van Devender 1995).

37. Shantz (1905), Glendening (1952), Paulsen (1950), Glinski (1977).

38. Bock and Bock (1997), Curtin and Brown (2001).

39. Whitfield and Beutner (1938), Whitfield and Anderson (1938).

40. Glendening and Paulsen (1955) and Martin (1975) support the view that seedling establishment is inhibited in dense grass cover. Counter argument to this hypothesis is provided by Kramp et al. (1998) and Brown and Archer (1989), albeit with a different species of mesquite.

41. Kramp et al. (1998).

42. Bock and Bock (1997), Curtin and Brown (2001).

43. Brown et al. (1997).

44. Griffiths (1904, 1910).

45. Hastings and Turner (1965b).

46. Ibid., 285.

47. Sheridan (1995: 129).

48. In addition, any increase in woody plants initiated during the brief incursion of livestock in the 1820s–30s could have been erased by grassland fires before widespread fire suppression, and before our earliest photographs.

49. Swetnam and Betancourt (1990, 1992, 1998), Minnich et al. (1993).

50. Rogers and Steele (1980), McLaughlin and Bowers (1982).

51. Bahre (1985), McLaughlin and Bowers (1982), Humphrey (1962, 1974), McPherson (1995).

52. Humphrey (1974), Martin (1983), McPherson (1995), Wilson et al. (2001).

53. Humphrey (1949), Reynolds and Bohning (1956).

54. Humphrey (1949), Glendening and Paulsen (1955), Reynolds and Bohning (1956).

55. This observation is based on examination of recently burned sites in southeastern Arizona and needs to be more rigorously tested.

56. Cox et al. (1993).

57. Dobyns (1981), Bahre (1985), Swetnam (1990), Kaib et al. (1996).

58. Hastings and Turner (1965b).

59. Caprio and Zwolinski (1992, 1995).

60. Pase and Lindenmuth (1971).

61. Sanchini (1981).

62. Ibid. At a shorter time scale, establishment is probably an inconstant process regulated by drought, which creates openings, and wet periods, which promote germination and establishment.

63. Wright and Bailey (1982), Baisan and Swetnam (1990), Swetnam and Baisan (1996a).

64. Sanchini (1981).

65. Caprio and Zwolinski (1992).

66. Bahre (1985), Swetnam and Baisan (1996b).

67. Curtin and Brown (2001).

68. Wilson (2001), Wilson et al. (2001). See also McPherson (1995).

69. Wilson et al. (2001).

70. See Cottam and Evans (1945), Denevan (1967), and Orodho et al. (1990).

71. Simanton and Emmerich (1994).

72. Cooke and Reeves (1976: chap. 2).

73. Waters and Haynes (2001).

74. This view is also held by Graf (1988) but would be rejected by Cooke and Reeves (1976).

75. Cooke and Reeves (1976), Betancourt (1990), Betancourt and Turner (1988).

76. See plates 6, 7, 11, 26, 42, 43, 46, 47, 48, 49, 50, 51, 56, 57, 59, 70, and 71. From the Desert Laboratory Archives, see Stakes 3409 and 3411.

77. Stromberg (1993), Carothers (1977).

78. Steinhart (1994).

79. McNamee (1994).

80. See Collier et al. (1996) for a basic review of downstream effects of dams throughout the United States. See Minckley and Deacon (1991) for native fish losses in the American West.

81. Shmida and Burgess (1988).

82. Hereford and Betancourt (1995).

83. Hereford (1993).

84. Phillips et al. (1964: 194, 200).

85. Glinski (1977), Rucks (1984).

86. Archer et al. (1995) provide an excellent review of this subject. See also Bazzaz (1990).

87. Mayeux et al. (1991), Polley et al. (1992, 1994), Idso (1992), Johnson et al. (1993).

88. Polley et al. (1994). The CO_2 levels in this study were much higher than current values as seen in figure 8.1. Therefore, the relevance to present conditions is questionable.

89. For example, Thompson et al. (1998) address changes in the western United States; Mitchell and Williams (1996) address changes in New Zealand.

90. Thompson et al. (1998).

91. General atmospheric circulation models cannot possibly address all of the intricacies of biotic communities as they undergo change. Adjustments through interspecific competition may occur in the plant assemblages, offsetting the competitive advantage of C_3 species over C_4 species; for example, if winter rainfall should decline under a continued increase of CO_2.

92. The data on recent increases in atmospheric CO_2 come from Keeling et al. (1989). The background CO_2 concentration of 275 ppm comes from Neftel et al. (1985).

93. Archer (1994), Archer et al. (1995).

94. Merideth (2001).

95. Ahlmann (1949).

96. As documented in *The Changing Mile*, scientists reported on the "recent climatic fluctuation" between 1920 and 1960. Willett (1950: 205–6) concluded that the increase varies somewhat with latitude; in general, warming of about 2.2°F in winter mean temperatures and 1.0°F in annual mean temperatures began in 1885. Mitchell (1963: 161–62) reported "rather uniform rates of warming from the 1880s to the early 1940s and a marked tendency for cooling since the early 1940s" of about 1.6°F for winter and 0.9°F for annual mean temperature. Callendar (1961) concurred, except for the southern hemisphere (Callendar 1961, Landsberg and Mitchell 1961). Butzer (1957) associated a tendency toward aridity with the temperature changes in North Africa. The fluctuation in the United States was reported by several workers, including Kincer (1933, 1946) and Page (1937).

97. Fleming (1998: 133).

98. Roberts (1989).

99. Gutzler (2000).

100. The high average temperatures around 1900 in southern Arizona may be an anomaly resulting from few stations recording data. Some of the increase in recent decades may be the result of urban heat-island effects (Karl et al. 1988).

101. Balling and Brazel (1987). Roden (1966) found increases in urban areas and no change in rural areas between 1821 and 1964 in the western United States.

102. Webb and Bowers (1993), Webb (1996).

103. As of spring 2001, the last severe frost in our region had occurred on December 9, 1978, establishing a frequency approaching two decades.

104. Brown and Davis (1995).

105. Compare the conclusions of Hastings and Turner (1965b) and Bahre (1991). Wilson et al. (2001) recently showed seasonal changes in precipitation in the

Semidesert Grassland of southern Arizona and south-western New Mexico.

106. Schulman (1956: 67).

107. Thornthwaite et al. (1942).

108. Leopold (1951b: 351–52).

109. Tuan (1966).

110. Cooke and Reeves (1976: 74, 78–79).

111. Sellers (1960: 85).

112. Von Eschen (1958) noted comparable decreases in winter precipitation for several New Mexico stations; Hubbs (1957) postulated a long-term trend toward aridity throughout the desert region.

113. For each climate station, we calculated the standardized seasonal precipitation, Ps, by

$$P_s = \Sigma \{\Sigma [(x_{i,j} - \mu_i) / \sigma_i] / k\} / n$$

where $x_{i,j}$ = monthly precipitation for climate station i in month j (mm); μ_i = the mean and σ_i = the standard deviation of monthly precipitation for climate station i (mm); k = the number of months in the season; and n = the number of climate stations with data. See Hereford and Webb (1992).

114. Hereford and Webb (1992), Webb and Betancourt (1992).

115. Curtin and Brown (2001). The changes noted in this study occurred within a livestock-free exclosure.

116. Neilson (1986: 31).

117. Ibid.

118. As noted in chapter 7, seeds of velvet mesquite require relatively high temperatures to germinate, and seedlings require equally high temperature for subsequent growth. Creosote bush seeds germinate in the summer or fall. Thus, establishment of two of the most important plants in our photographs is not likely to respond to winter precipitation.

119. Wilson et al. (2001) support the view that mesquite has increased as a result of increased winter rainfall.

120. Although not addressing competition with oaks directly, the competitive effect of mesquites on perennial grasses was seen up to 49 feet from the base of mesquites and was shown to be much more severe in dry years than in wet years (Cable 1977).

121. Zhang et al. (1997), Mantua et al. (1997).

122. Nigam et al. (1999).

123. Andrade and Sellers (1988).

124. McCabe and Dettinger (1999).

125. Ibid.

126. Webb and Betancourt (1992).

127. McCabe and Dettinger (1999).

128. Schmidt and Webb (2001).

129. Spalding (1909).

130. Burgess et al. (1991). Although, by 1991, most of these new species had merely entered the preserve along its periphery from nearby yards, twelve exotic species had become generally distributed about the Desert Laboratory grounds.

131. Turner (1990), Martin and Turner (1977).

132. Van Devender (1995).

133. Wagoner (1952).

134. Bahre (1991).

135. Turner (1992).

136. Glendening (1952), Branscomb (1958), Brown and Archer (1989), Laycock (1991).

137. Burgess et al. (1991).

138. McLaughlin and Bowers (1982), Rogers (1986).

139. McLaughlin and Bowers (1982), Rogers and Vint (1987), Burgess et al. (1991).

140. Rogers and Steele (1980).

141. Burgess et al. (1991).

142. Waters and Haynes (2001).

Literature Cited

Adams, R. P. 1994. Geographic variation and systematics of monospermous *Juniperus* (Cupressaceae) from the Chihuahua Desert based on RAPDS and terpenes. *Biochemical Systematics and Ecology* 22(7):699–710.

Adams, R. P., and T. A. Zanoni. 1979. The distribution, synonymy, and taxonomy of three junipers of the southwestern United States and northern Mexico. *Southwestern Naturalist* 24(2):323–29.

Ahlmann, H. W. 1949. The present climatic fluctuation. *Geographical Journal* 112:165–95.

Alcorn, S. M. 1961. Natural history of the saguaro. *University of Arizona, Arid Lands Colloquia* 1959–60/1960–61:23–29.

Alcorn, S. M., and E. B. Kurtz. 1959. Some factors affecting the germination of seeds of the saguaro cactus (*Carnegiea gigantea*). *American Journal of Botany* 46:526–29.

Alcorn, S. M., and C. May. 1962. Attrition of a saguaro forest. *Plant Disease Reporter* 46:156–58.

Aldous, A. E., and H. L. Shantz. 1924. Types of vegetation in the semiarid portion of the United States and their economic significance. *Journal of Agricultural Research* 28:99–127.

Aldrich, L. D. 1950. *A Journal of the Overland Route to California and the Gold Mines*. Los Angeles: Dawson's Book Shop.

Alegre, F. J. 1956–60. *Historia de la Provincia de la Compañía de Jesús de Nueva España*. 4 vols. Rome: Institutum Historicum, S.J.

Allison, W. n.d. Arizona, the last frontier. Unpublished ms., Arizona Historical Society, Tucson.

Allred, D. M., D. E. Beck, and C. D. Jorgensen. 1963. *Biotic Communities of the Nevada Test Site*. Brigham Young University Science Bulletin, Biological Series no. 2.

Almada, F. R. 1990. *Diccionario de Historia, Geografía y Biografía Sonorenses*. Instituto Sonorense de Cultura, Hermosillo.

Anable, M. E. 1989. Alien plant invasion in relation to site characteristics and disturbance: *Eragrostis lehmanniana* on the Santa Rita Experimental Range, Arizona, 1937–89. Master's thesis, University of Arizona, Tucson.

Anable, M. E., M. P. McClaran, and G. B. Ruyle. 1992. Spread of introduced Lehmann lovegrass (*Eragrostis lehmanniana* Nees.) in southern Arizona, USA. *Biological Conservation* 61:181–88.

Anderson, E. 1956. Man as a maker of new plants and new plant communities. Pages 763–77 in W. L. Thomas et al. (eds.), *Man's Role in Changing the Face of the Earth*. Chicago: University of Chicago Press.

Andrade, E. R., and W. D. Sellers. 1988. El Niño and its effect on precipitation in Arizona. *Journal of Climatology* 8:403–10.

Anonymous. 1993. *Vegetation Changes on the Manti-LaSal National Forest: A Photographic Study Using Comparative Photographs from 1902–1992*. USDA Forest Service, Manti-LaSal National Forest, Price, Utah.

Antevs, E. 1952. Arroyo-cutting and filling. *Journal of Geology* 60:375–85.

Applegate, L. H. 1981. Hydraulic effects of vegetation changes along the Santa Cruz River near Tumacacori, Arizona. Master's thesis, University of Arizona, Tucson.

Archer, S. R. 1994. Woody plant encroachment into southwestern grasslands and savannas: Rate, patterns and proximate causes. Pages 13–68 in M. Vavra, W. Laycock, and R. Pieper (eds.), *Ecological Implications of Livestock Herbivory in the West*. Denver, Colo.: Society for Range Management.

Archer, S. R., M. G. Garrett, and J. K. Detling. 1987. Rates of vegetation change associated with prairie dog (*Cynomys ludovicianus*) grazing in North American mixed-grass prairie. *Vegetatio* 72:159–66.

Archer, S. R., D. S. Schimel, and E. A. Holland. 1995. Mechanisms of shrubland expansion: Land use, climate or CO_2? *Climate Change* 29(1):91–99.

ARIS. 1975. Arizona Precipitation, Arizona Resoures Information System Cooperative Publication no. 5.

Arizona Citizen, passim. Tucson and Florence.

Arizona Daily Star, passim. Tucson.

Arizona Quarterly Illustrated, July 1880. Tucson.

Arizona Weekly Enterprise, passim. Florence.

Arizona Weekly Star, passim. Tucson.

Aschmann, H. 1959. *The Central Desert of Baja California: Demography and Ecology*. Berkeley: University of California Press.

Asplund, K. K., and M. T. Gooch. 1988. Geomorphology and the distributional ecology of Fremont cottonwood (*Populus fremontii*) in a desert riparian canyon. *Desert Plants* 9(1):17–27.

Axelrod, D. I. 1950. *Studies in Late Tertiary Paleobotany*. Carnegie Institution of Washington Publication no. 590. Washington, D.C.

———. 1986. Paleobotanical history of the western deserts. Pages 113–29 in S. G. Wells and D. R. Haragan (eds.), *Origin and Evolution of Deserts*. Albuquerque: University of New Mexico Press.

Baegert, J. J. 1952. *Observations in Lower California*. Berkeley: University of California Press.

Bahre, C. J. 1985. Wildfire in southeastern Arizona between 1859 and 1890. *Desert Plants* 7(4):190–94.

———. 1991. *A Legacy of Change: Historic Human Impact on Vegetation in the Arizona Borderlands*. Tucson: University of Arizona Press.

———. 1998. Late 19th century human impacts on the woodlands and forests of southeastern Arizona's sky islands. *Desert Plants* 14(1):8–21.

Bahre, C. J., and C. F. Hutchinson. 1985. The impact of historic fuelwood cutting on the semidesert woodlands of southern Arizona. *Journal of Forest History* 29:175–86.

Bahre, C. J., and M. L. Shelton. 1993. Historic vegetation change, mesquite increases, and climate in southeastern Arizona. *Journal of Biogeography* 20:489–504.

Bailey, R. W. 1935. Epicycles of erosion in the valleys of the Colorado Plateau province. *Journal of Geology* 43: 337–55.

Baisan, C. H. 1988. Fire history of the Rincon Mountain Wilderness. Unpublished ms., Saguaro National Park, Tucson, Arizona.

Baisan, C. H., and T. W. Swetnam. 1990. Fire history on a desert mountain range: Rincon Mountain Wilderness, USA. *Canadian Journal of Forest Research* 20: 1559–69.

Balda, R. P. 1987. Avian impacts on pinyon-juniper woodlands. Pages 525–33 in R. L. Everett (compiler), *Proceedings: Pinyon-Juniper Conference*. USDA Forest Service General Technical Report INT-215. Intermountain Research Station, Ogden, Utah.

Balling, R. C., Jr., and S. W. Brazel. 1987. Time and space characteristics of the Phoenix urban heat island. *Journal of the Arizona-Nevada Academy of Science* 21:75–81.

Bancroft, H. H. 1883. *The Native Races. Vol. 1: Wild Tribes.* San Francisco: A. L. Bancroft.

———. 1889. *History of the North Mexican States and Texas.* 2 vols. San Francisco: History Company.

———. 1962. *History of Arizona and New Mexico, 1530–1888.* Facsimile of 1889 edition. Albuquerque: Horn and Wallace.

Bandelier, A. F. 1892. *Final Report of Investigations among the Indians of the Southwestern United States.* Papers of the Archaeological Institute of America no. 4. Cambridge, Mass.: Harvard University Press.

——— (ed.). 1905. *The Journey of Alvar Nuñez Cabeza de Vaca.* New York: A. S. Barnes.

Bannon, J. F. 1955. *The Mission Frontier in Sonora, 1620–1687.* New York: U.S. Catholic Historical Society.

Barry, K. G., G. Kiladis, and R. S. Bradley. 1981. Synoptic climatology of the western United States in relation to climatic fluctuations during the twentieth century. *Journal of Climatology* 1:97–113.

Bartlett, J. R. 1854. *Personal Narrative.* 2 vols. New York: D. Appleton.

Bazzaz, R. A. 1990. The response of natural ecosystems to the rising global CO_2 levels. *Annual Review of Ecology and Systematics* 21:167–96.

Beatley, J. E. 1974a. Effects of rainfall and temperature on the distribution and behavior of *Larrea tridentata* (creosote bush) in the Mojave Desert of Nevada. *Ecology* 55:245–61.

———. 1974b. Phenologic events and their environmental triggers in Mohave Desert ecosystems. *Ecology* 55: 856–63.

Becker, C. L. 1932. *The Heavenly City of the Eighteenth Century Philosophers.* New Haven, Conn.: Yale University Press.

Bell, W. A. 1869. *New Tracks in North America.* 2 vols. London: Chapman and Hall.

Benson, L. 1941. The mesquites and screwbeans of the United States. *American Journal of Botany* 28:748–54.

———. 1950. *The Cacti of Arizona.* Tucson: University of Arizona Press.

———. 1962. *Plant Taxonomy: Methods and Principles.* New York: Ronald Press.

———. 1982. Cacti of the United States and Canada. Stanford, Calif.: Stanford University Press.

Benson, L., and R. A. Darrow. 1954. *Trees and Shrubs of the Southwestern Deserts.* Tucson: University of Arizona Press.

Berry, J. W., and C. Steelink. 1961. Chemical constituents of the saguaro. *University of Arizona, Arid Lands Colloquia* 1959–60/1960–61:39–45.

Betancourt, J. L. 1990. Tucson's Santa Cruz River and the arroyo legacy. Ph.D. diss., University of Arizona, Tucson.

Betancourt, J. L., E. A. Pierson, K. A. Rylander, J. A. Fairchild-Parks, and J. S. Dean. 1993. Influence of history and climate on New Mexico piñon-juniper woodlands. Pages 42–62 in E. F. Aldon and D. W. Shaw (eds.), *Managing Piñon-Juniper Ecosystems for Sustainability and Social Needs.* USDA Forest Service General Technical Report RM-236. Rocky Mountain Forest and Range Experiment Station, Fort Collins, Colo.

Betancourt, J. L., and R. M. Turner. 1988. Historic arroyo-cutting and subsequent channel changes at the Congress Street crossing, Santa Cruz River, Tucson, Arizona. Pages 1353–71 in E. E. Whitehead, C. F. Hutchinson, B. N. Timmerman, and R. G. Varady (eds.), *Arid Lands: Today and Tomorrow, Proceedings of an International Research and Development Conference.* Boulder, Colo.: Westview Press.

Betancourt, J. L., and T. R. Van Devender. 1981. Holocene vegetation in Chaco Canyon, New Mexico. *Science* 214:656–58.

Bieber, R. P. 1938. Cooke's journal of the march of the Mormon Battalion, 1846–1847. Pages 65–240 in *Exploring in Southwest Trails, 1846–1854.* Southwest Historical Series no. 3. Glendale, Calif.: Arthur H. Clark.

Blydenstein, J., C. R. Hungerford, G. I. Day, and R. R. Humphrey. 1957. Effect of domestic livestock on vegetation in the Sonoran Desert. *Ecology* 38:522–26.

Bock, C. E., and J. H. Bock. 1997. Shrub densities in relation to fire, livestock grazing, and precipitation in an Arizona desert grassland. *Southwestern Naturalist* 42(2):188–93.

Bock, J. H., C. E. Bock, and J. R. McKnight. 1976. A study of the effects of grassland fires at the Research Ranch in southeastern Arizona. *Journal of the Arizona Academy of Science* 11:49–57.

Boedecker, L. A. 1930. Letter to Frank Lockwood, December 9, 1930. Arizona Historical Society, Tucson.

Bolton, H. E. 1917. The mission as a frontier institution in the Spanish-American colonies. *American Historical Review* 23:42–61.

———. 1948. *Kino's Historical Memoir of Pimería Alta.* 2 vols. Berkeley: University of California Press.

———. 1949. *Coronado, Knight of Pueblos and Plains.* Albuquerque: University of New Mexico Press.

———. 1960. *Rim of Christendom.* New York: Russell and Russell.

Bolton, H. E., and T. M. Marshall. 1920. *The Colonization of North America, 1492–1783.* New York: Macmillan.

Borchert, J. R. 1950. The climate of the central North American grassland. *Annals of the Association of American Geographers* 40:1–39.

Bourke, J. G. 1950. *On the Border with Crook.* Columbus, Ohio: Long's College.

Bowers, J. E. 1980–81. Catastrophic freezes in the Sonoran Desert. *Desert Plants* 2(4):232–36.

———. 1982. Plant ecology of inland dunes in western North America. *Journal of Arid Environments* 5:199–220.

———. 1994. Natural conditions for seedling emergence

of three woody species in the northern Sonoran Desert. *Madroño* 41(2):73–84.

———. 1996. More flowers or new cladodes? Environmental correlates and biological consequences of sexual reproduction in a Sonoran Desert prickly pear cactus, *Opuntia engelmannii*. *Bulletin of the Torrey Botanical Club* 123(1):34–40.

———. 1997. The effect of drought on Engelmann prickly pear (Cactaceae: *Opuntia engelmannii*) fruit and seed production. *Southwestern Naturalist* 42:240–42.

———. 1998. *Dune Country*. Tucson: University of Arizona Press.

Bowers, J. E., and M. A. Dimmitt. 1994. Flowering phenology of six woody plants in the northern Sonoran Desert. *Bulletin of the Torrey Botanical Club* 121(3): 215–29.

Bowers, J. E., and S. P. McLaughlin. 1984. Flora and vegetation of the Rincon Mountains, Pima County, Arizona. *Desert Plants* 8:51–94.

———. 1996. Flora of the Huachuca Mountains, a botanically rich and historically significant sky island in Cochise County, Arizona. *Journal of the Arizona-Nevada Academy of Science* 29:66–107.

Bowers, J. E., and R. M. Turner. 2001. Dieback and episodic mortality of *Cercidium microphyllum* (foothill paloverde), a dominant Sonoran Desert tree. *Bulletin of the Torrey Botanical Club* 128:128–40.

———. 2002. The influence of climatic variability on local population dynamics of foothill paloverde (*Cercidium microphyllum*). *Oecologia* 130:105–13.

Bowers, J. E., R. H. Webb, and R. J. Rondeau. 1995. Longevity, recruitment and mortality of desert plants in Grand Canyon, Arizona, USA. *Journal of Vegetation Science* 6:551–64.

Boyd, R. S., and G. D. Brum. 1983. Postdispersal reproductive biology of a Mojave Desert population of *Larrea tridentata* (Zygophyllaceae). *American Midland Naturalist* 110:25–36.

Brady, L. F. 1936. The arroyo of the Rio de Flag. *Museum Notes, Museum of Northern Arizona* 9 (Dec.):33–37.

Brand, D. D. 1936. *Notes to Accompany a Vegetation Map of Northwest Mexico*. University of New Mexico Bulletin no. 280, Biological Series IV, no. 4.

Brandes, R. 1960. *Frontier Military Posts of Arizona*. Globe, Ariz.: Dale Stuart King.

Branscomb, B. L. 1956. Shrub invasion of a southern New Mexico desert grassland range. Master's thesis, University of Arizona, Tucson.

———. 1958. Shrub invasion of a southern New Mexico desert grassland range. *Journal of Range Management* 11:129–32.

Brinkman, K. A. 1974. *Salix* L. willow. Pages 746–50 in C. S. Schopmeyer (technical coordinator), *Seeds of Woody Plants in the United States*. U.S. Department of Agriculture Handbook 450. Washington, D.C.

Brown, D. E. 1994a. Madrean evergreen woodland. Pages 59–65 in D. E. Brown (ed.), *Biotic Communities: Southwestern United States and Northwestern Mexico*. Salt Lake City: University of Utah Press.

———. 1994b. Chihuahuan desertscrub. Pages 169–79 in D. E. Brown (ed.), *Biotic Communities: Southwestern United States and Northwestern Mexico*. Salt Lake City: University of Utah Press.

———. 1994c. Semidesert grassland. Pages 123–31 in D. E. Brown (ed.), *Biotic Communities: Southwestern United States and Northwestern Mexico*. Salt Lake City: University of Utah Press.

Brown, D. E., N. B. Carmony, and R. M. Turner. 1981. *Drainage Map of Arizona Showing Perennial Streams and Some Important Wetlands*. 2-page map. Arizona Game and Fish Department, Phoenix.

Brown, D. E., and R. Davis. 1995. One hundred years of vicissitude: Terrestrial bird and mammal distribution changes in the American Southwest, 1890–1990. Pages 231–44 in L. F. DeBano et al. (technical coordinators), *Biodiversity and Management of the Madrean Archipelago: The Sky Islands of Southwestern United States and Northwestern Mexico*. USDA Forest Service General Technical Report RM-GTR-264. Rocky Mountain Forest and Range Experiment Station, Fort Collins, Colo.

———. 1996. Terrestrial bird and mammal distribution changes in the American Southwest, 1890–1990. Pages 47–64 in B. Tellman, D. M. Finch, C. Edminster, and R. Hamre (eds.), *The Future of Arid Grasslands: Identifying Issues, Seeking Solutions*. USDA Forest Service Proceedings RMRS-P-3. Rocky Mountain Forest and Range Experiment Station, Fort Collins, Colo.

Brown, D. E., and C. H. Lowe. 1994. *Biotic Communities of the Southwest*. 1-page map. Salt Lake City: University of Utah Press.

Brown, D. E., F. Reichenbacher, and S. E. Franson. 1998. *A Classification of North American Biotic Communities*. Salt Lake City: University of Utah Press.

Brown, J. H., and E. J. Heske. 1990. Control of a desert-grassland transition by a keystone rodent guild. *Science* 250:1705–7.

Brown, J. H., T. J. Valone, and C. G. Curtin. 1997. Reorganization of an arid ecosystem in response to recent climate change. *Proceedings of the National Academy of Science* 94:9729–33.

Brown, J. R., and S. Archer. 1989. Woody plant invasion of grasslands: Establishment of honey mesquite (*Prosopis glandulosa* var. *glandulosa*) on sites differing in herbaceous biomass and grazing history. *Oecologia* 80:19–26.

Browne, J. R. 1951. *A Tour through Arizona, 1864*. Tucson: Arizona Silhouettes.

Brum, G. D. 1973. Ecology of the saguaro (*Carnegiea gigantea*): Phenology and establishment in marginal populations. *Madroño* 22:195–204.

Bryan, K. 1925. Date of channel trenching (arroyo cutting) in the arid Southwest. *Science* 62:338–44.

———. 1927. *Channel Erosion of the Rio Salado, Socorro County, New Mexico*. USGS Bulletin no. 790. Washington, D.C.

———. 1928a. Change in plant associations by change in ground water level. *Ecology* 9:474–78.

———. 1928b. Historic evidence on changes in the channel of Rio Puerco. *Journal of Geology* 36:265–82.

———. 1940. Erosion in the valleys of the Southwest. *New Mexico Quarterly* 10:227–32.

Bryson, R. A. 1957. *The Annual March of Precipitation in Arizona, New Mexico, and Northwestern Mexico*. Technical Reports on the Meteorology and Climatology of Arid Regions no. 6. University of Arizona, Institute of Atmospheric Physics, Tucson.

Bryson, R. A., and W. P. Lowry. 1955. Synoptic climatology of the Arizona summer precipitation singularity. *Bulletin of the American Meteorological Society* 36:329–39.

Bull, W. B. 1974. Playa processes in the volcanic craters of the Sierra Pinacate, Sonora, Mexico. *Zeitschrift für Geomorphologie, neue Folge, Supplementum Band* 20:117–29.

Burgess, T. L. 1995. The dilemma of coexisting growth forms. Pages 31–67 in M. P. McClaran and T. R. Van Devender (eds.), *The Desert Grassland*. Tucson: University of Arizona Press.

Burgess, T. L., J. E. Bowers, and R. M. Turner. 1991. Exotic plants at the Desert Laboratory, Tucson, Arizona. *Madroño* 38(2):96–114.

Burkart, A. 1976. A monograph of the Genus *Prosopis* (Leguminosae subfam. Mimosoideae). *Journal of the Arnold Arboretum* 57:217–49, 450–85.

Burns, R. M., and B. H. Honkala (technical coordinators). 1990. *Silvics of North America. Vol. 2. Hardwoods.* USDA Forest Service Agricultural Handbook 654. Washington, D.C.

Butzer, K. W. 1957. The recent climatic fluctuation in lower latitudes and the general circulation of the Pleistocene. *Geografiska Annaler* 39:105–13.

Cable, D. R. 1967. Fire effects on semidesert grasses and shrubs. *Journal of Range Management* 20:170–76.

———. 1977. Seasonal use of soil water by mature velvet mesquite. *Journal of Range Management* 30(1):4–11.

Callendar, G. S. 1961. Temperature fluctuations and trends over the Earth. *Quarterly Journal of the Royal Meteorological Society* 87:1–12.

Calvo B. L. 1958. *Nociones de Historia de Sonora*. Mexico, D.F.: Librería de Manuel Porrua.

Cannon, W. A. 1911. *The Root Habits of Desert Plants*. Carnegie Institution of Washington Publication no. 131.

———. 1916. Distribution of the cacti with especial reference to the role played by the root response to soil temperature and soil moisture. *American Naturalist* 50:435–42.

Caprio, A. C., and M. J. Zwolinski. 1992. Fire effects on Emory and Mexican Blue oaks in southeastern Arizona. Pages 150–54 in P. F. Pfolliott, G. J. Gottfried, D. A. Bennett, V. M. Hernandez C., A. Ortega Rubio, and R. H. Hamre (technical coordinators), *Ecology and Management of Oak and Associated Woodlands: Perspectives in the Southwestern United States and Northern Mexico, a Symposium*. USDA Forest Service General Technical Report RM GTR-218. Rocky Mountain Forest and Range Experiment Station, Fort Collins, Colo.

———. 1995. Fire and vegetation in a Madrean oak woodland, Santa Catalina Mountains, southeastern Arizona. Pages 389–98 in L. F. DeBano et al. (technical coordinators), *Biodiversity and Management of the Madrean Archipelago: The Sky Islands of Southwestern United States and Northwestern Mexico*. USDA Forest Service General Technical Report RM GTR-264. Rocky Mountain Forest and Range Experiment Station, Fort Collins, Colo.

Carothers, S. W. 1977. Importance, preservation, and management of riparian habitats: An overview. Pages 2–4 in R. R. Johnson and D. A. Jones (technical coordinators), *Importance, Preservation and Management of Riparian Habitat: A Symposium*. USDA Forest Service General Technical Report RM-43. Rocky Mountain Forest and Range Experiment Station, Fort Collins, Colo.

Carson, C. 1926. *Kit Carson's Own Story of His Life*. Taos, N.M.: n.p.

Carter, G. F. 1950. Ecology—geography—ethnobotany. *Scientific Monthly* 70:73–80.

Castetter, E. F. 1956. The vegetation of New Mexico. *New Mexico Quarterly* 26:257–88.

Cayan, D. R., and D. H. Peterson. 1989. The influence of North Pacific atmospheric circulation on streamflow in the west. Pages 325–98 in D. H. Peterson (ed.), *Aspects of Climate Variability in the Pacific and Western Americas*. Geophysical Monograph 55. Washington, D.C.: American Geophysical Union.

Cayan, D. R., and R. H. Webb. 1992. El Niño/Southern Oscillation and streamflow in the western United States. Pages 29–68 in H. F. Diaz and V. Markgraf (eds.), *El Niño: Historical and Paleoclimatic Aspects of the Southern Oscillation*. Cambridge, Engl.: Cambridge University Press.

Chamberlain, W. H. 1945. From Lewisburg to California in 1849. *New Mexico Historical Review* 20:144–80, 239–68.

Chapman, C. E. 1916. *The Founding of Spanish California*. New York: Macmillan.

Chew, R. M., and A. E. Chew. 1965. The primary productivity of a desert shrub (*Larrea tridentata*) community. *Ecological Monographs* 35:355–75.

Christiansen, L. D. 1988. The extinction of wild cattle in southern Arizona. *The Journal of Arizona History* 29:89–100.

Clarke, A. B. 1852. *Travels in Mexico and California*. Boston: Wright and Hasty's Steam Press.

Cleland, R. G. 1950. *This Reckless Breed of Men*. New York: Alfred A. Knopf.

Clements, F. E. 1920. *Plant Indicators: The Relation of Plant Communities to Process and Practice*. Carnegie Institution of Washington Publication no. 290. Washington, D.C.

Cochise County District Court. 1889. Antonio Crijalba *et al.* vs. Thos. Dunbar *et al.* Unpublished ms., Arizona Historical Society, Tucson.

Cockrum, E. L., and Y. Petryszyn. 1991. The long-nosed bat, *Leptonycteris*: An endangered species in the Southwest? *Occasional Papers of the Texas Tech University* 142:1–32.

Cole, L. C. 1957. Biological clock in the unicorn. *Science* 125:874–76.

Collier, M., R. H. Webb, and J. C. Schmidt. 1996. *Dams and Rivers: Primer on the Downstream Effects of Dams*. U.S. Geological Survey Circular 1126. Reston, Va.

Colton, H. S. 1937. Some notes on the original condition of the Little Colorado River: A side light on the problems of erosion. *Museum Notes, Museum of Northern Arizona* 10 (Dec.):17–20.

Conkling, R. P., and M. B. Conkling. 1947. *The Butterfield Overland Mail, 1857–1869*. 3 vols. Glendale, Calif.: Arthur H. Clark.

Contreras Arias, A. 1942. *Mapas de las provincias climatológicas de la República Mexicana*. Secretaría de Agricultura y Fomento, Dirección de Geografía, Meteorología e Hidrología, Mexico.

Cook, S. F. 1940. *Population Trends among the California Mission Indians*. Berkeley: University of California Press.

———. 1942. The population of Mexico in 1793. *Human Biology* 14:499–515.

Cook, S. F., and L. B. Simpson. 1948. *The Population of Central Mexico in the Sixteenth Century*. Berkeley: University of California Press.

Cooke, R. U., and R. W. Reeves. 1976. *Arroyos and Environmental Change in the American South-West*. Oxford: Clarendon Press.

Cottam, W. P. 1937. Has Utah lost claim to the Lower Sonoran Zone? *Science* 85:563–64.

Cottam, W. P., and F. R. Evans. 1945. A comparative study of the vegetation of grazed and ungrazed canyons of the Wasatch Range, Utah. *Southwest Quarterly* 29:128–46.

Cottam, W. P., and G. Stewart. 1940. Plant succession as a result of grazing and of meadow desiccation by erosion since settlement in 1862. *Journal of Forestry* 38:613–26.

Coupland, R. T. 1959. Effects of changes in weather conditions upon grasslands in the northern Great Plains. In *Grasslands*. Washington, D.C.: American Association for the Advancement of Science.

Couts, C. J. 1961. *Hepah, California*. Tucson: Arizona Historical Society.

Cox, C. C. 1925. From Texas to California in 1849. *Southwestern Historical Quarterly* 29:128–46.

Cox, J. R., A. de Alba-Avila, R. W. Rice, and J. N. Cox. 1993. Biological and physical factors influencing *Acacia constricta* and *Prosopis velutina* establishment in the Sonoran Desert. *Journal of Range Management* 46:43–48.

Curtin, C. G., and J. H. Brown. 2001. Climate and herbivory in structuring the vegetation of the Malpai Borderlands. Pages 84–94 in G. L. Webster and C. J. Bahre (eds.), *Changing Plant Life of La Frontera: Observations of Vegetation in the United States/Mexico Borderlands*. Albuquerque: University of New Mexico Press.

Daily Alta Californian, passim. San Francisco.

Dale, E. E. 1930. *The Range Cattle Industry*. Norman: University of Oklahoma Press.

Dalton, P. D., Jr. 1961. Ecology of the creosotebush (*Larrea tridentata* [DC.] Cov.). Ph.D. diss., University of Arizona, Tucson.

Darrow, R. A. 1943. Vegetative and floral growth of *Fouquieria splendens. Ecology* 24:310–22.

———. 1944. *Arizona Range Resources and Their Utilization: I. Cochise County*. University of Arizona Agricultural Experiment Station Bulletin no. 103.

Daubenmire, R. F. 1956. C1imate as a determinant of vegetation distribution in eastern Washington and northern Idaho. *Ecological Monographs* 26:131–54.

———. 1957. Injury to plants from rapidly dropping temperature in Washington and northern Idaho. *Journal of Forestry* 55:581–85.

———. 1959. *Plants and Environment*. New York: John Wiley.

Davis, G. P., Jr. 1982. *Man and Wildlife in Arizona: The American Exploration Period 1824–1865*. Phoenix: Arizona Game and Fish Department.

Decorme, G. 1941. *La Obra de los Jesuitas Mexicanos Durante la Época Colonial, 1572–1767*. 2 vols. Mexico: José Porrua.

Del Barco, M. 1973. *Historia Natural y Crónica de la Antigua California*. Instituto de Investigaciones Histó-ricas, Universidad Nacional Autónoma de México, Mexico.

Dellenbaugh, F. S. 1912. Cross cutting and retrograding of stream-beds. *Science* 35:656–58.

Denevan, W. M. 1967. Livestock numbers in nineteenth-century New Mexico, and the problem of gullying in the Southwest. *Annals of the Association of American Geographers* 57:691–703.

Despain, D. G. 1974. The survival of saguaro (*Carnegiea gigantea*) seedlings on soils of differing albedo and cover. *Journal of the Arizona Academy of Science* 9:102–7.

Diaz, H. F., and V. Markgraf (eds.). 1992. *El Niño: Historical and Paleoclimatic Aspects of the Southern Oscillation*. Cambridge, Engl.: Cambridge University Press.

Dick-Peddie, W. A. 1993. *New Mexico Vegetation: Past, Present, and Future*. Albuquerque: University of New Mexico Press.

Dickson, C. R. 1958. *Ground Layer Temperature Inversions in an Interior Valley and Canyon*. Department of Meteorology, University of Utah.

Di Peso, C. C. 1953. *The Sobaipuri Indians of the Upper San Pedro River Valley, Southeastern Arizona*. Dragoon, Ariz.: Amerind Foundation.

Dittmer, H. J. 1951. Vegetation of the Southwest—past and present. *Texas Journal of Science* 3: 350–55.

Dobyns, H. F. 1981. *From Fire to Flood: Historic Human Destruction of Sonoran Desert Riverine Oases*. Ballena Press Anthropological Papers no. 20, Socorro, New Mexico.

Dorroh, J. H., Jr. 1946. *Certain Hydrologic and Climatic Characteristics of the Southwest*. Albuquerque: University of New Mexico Press.

Duce, J. T. 1918. The effect of cattle on the erosion of canon bottoms. *Science* 47:450–52.

Dunne, P. M. 1957. *Juan Antonio Balthasar: Padre Visitador to the Sonora Frontier, 1744–1745*. Tucson: Arizona Historical Society.

Durivage, J. E. 1937. Journal. Pages 159–255 in R. P. Bieber (ed.), *Southern Trails to California in 1849*. Glendale, Calif.: Arthur H. Clark.

Durrenberger, R. W., and R. S. Ingram. 1978. *Major storms and floods in Arizona 1862–1977*. Office of the State Climatologist, Climatological Publications, Precipitation Series no. 4. Phoenix, Ariz.

Dzerdzeevskii, B. L. 1970. *Circulation Mechanisms in the Atmosphere of the Northern Hemisphere in the 20th Century*. Institute of Geography, Soviet Academy of Sciences, Moscow.

Eaton, W. C. 1933. Frontier life in southern Arizona, 1858–61. *Southwestern Historical Quarterly* 36:173–92.

Eccleston, R. 1950. *Overland to California on the Southwestern Trail: 1849*. Berkeley: University of California Press.

Eidemiller, D. I. 1978. *The Frequency of Tropical Cyclones in the Southwestern United States and Northwestern Mexico*. The State Climatologist for Arizona, Climatological Publications, Scientific Papers no. 1. Phoenix, Ariz.

Elliott, J. G., A. C. Gellis, and S. B. Aby. 1999. Evolution of arroyos: Incised channels of the Southwestern United States. In S. E. Darby and A. Simon (eds.), *Incised River Channels*. New York: John Wiley and Sons.

Elton, C. S. 1958. *The Ecology of Invasions by Animals and Plants*. London: Methuen.

Ely, L. L., R. H. Webb, and Y. Enzel. 1992. Accuracy of post-bomb ^{137}Cs and ^{14}C in dating fluvial deposits. *Quaternary Research* 38:196–204.

Emory, W. H. 1848. Notes of a military reconnoissance. U.S. Congress, House, Executive Document 41, 30th Cong., 1st Sess., 15–126. Washington, D.C.

———. 1857. *Report on the United States and Mexican Boundary Survey.* 2 vols. U.S. Congress, Senate, Executive Document 108, 34th Cong., 1st Sess. Washington, D.C.

Enfield, D. B. 1989. El Niño, past and present. *Reviews in Geophysics* 27:159–87.

Escoto Rodríguez, M. 1999. Variaciones temporales y espaciales en la tasa de crecimiento del cirio (*Fouquieria columnaris*). Master's thesis, Universidad de Baja California, Ensenada.

Etz, D. B. 1939. Reminiscences. Unpublished ms., Arizona Historical Society, Tucson.

Euler, R. C., G. J. Gumerman, T. N. V. Karlstrom, J. S. Dean, and R. H. Hevly. 1979. The Colorado Plateaus: cultural dynamics and paleoenvironment. *Science* 205:1089–1101.

Evans, G. W. B. 1945. *Mexican Gold Trail: The Journal of a Forty-Niner.* San Marino, Calif.: Huntington Library.

Ewing, R. C. 1934. The Pima uprising, 1751–52. Ph.D. diss., University of California, Berkeley.

Extracts from the journal of Henry W. Bigler. 1932. *Utah Historical Quarterly* 5:35–64, 87–112, 134–60.

Eychaner, J. H., and M. R. Rehmann. 1989. Arizona floods and droughts. Pages 181–88 in R. W. Paulson, E. B. Chase, R. S. Roberts, and D. W. Moody (compilers), *National Water Summary 1988–89: Hydrologic Events and Floods and Droughts.* U.S. Geological Survey Water-Supply Paper 2374, Washington, D.C.

Felger, R. S., and M. B. Moser. 1985. *People of the Desert and Sea: Ethnobotany of the Seri Indians.* Tucson: University of Arizona Press.

Fenner, P., W. W. Brady, and D. R. Patton. 1984. Observations on seeds and seedlings of Fremont cottonwood. *Desert Plants* 6(1):55–58.

———. 1985. Effects of regulated water flows on regeneration of Fremont cottonwood. *Journal of Range Management* 38(2):135–38.

Fish, S. K. 2000. Hohokam impacts on Sonoran Desert Environment. Pages 251–80 in D. L. Lentz (ed.), *Imperfect Balance: Landscape Transformations in the Precolumbian Americas.* New York: Columbia University Press.

Fish, S. K., P. R. Fish, C. Miksicek, and J. H. Madsen. 1985. Prehistoric agave cultivation in Southern Arizona. *Desert Plants* 7(2):107–12.

———. 1992. *The Marana Community in the Hohokam World.* Anthropological Papers of the University of Arizona, Number 56. Tucson: University of Arizona Press.

Fleishner, T. L. 1994. Ecological costs of livestock grazing in western North America. *Conservation Biology* 8:629–44.

Fleming, J. R. 1998. *Historical Perspectives on Climate Change.* New York: Oxford University Press.

Forbes, J. D. 1957. The Janos, Jocomes Mansos and Sumas Indians. *New Mexico Historical Review* 32:319–34.

Forbes, R. H. 1958. *The Expanding Sahara.* University of Arizona Physical Science Bulletin no. 3. Tucson, Ariz.

Fosberg, F. R. 1938. The Lower Sonoran in Utah. *Science* 87:39–40.

Fowells, H. A. 1965. *Silvics of Forest Trees of the United States.* U.S. Department of Agriculture Handbook 271. Washington, D.C.

Franco, A. C., and P. S. Nobel. 1989. Effect of nurse plants on the microhabitat and growth of cacti. *Journal of Ecology* 77:870–86.

Furness, F. N. 1961. Solar variations, climatic change, and related geophysical problems. *Annals of the New York Academy of Sciences* 95:1–740.

Galbraith, F. W. 1959. Craters of the Pinacates. Pages 161–64 in L. A. Heindl (ed.), *Southern Arizona Guidebook II.* Tucson: Arizona Geological Society.

Geiger, R. 1957. *The Climate Near the Ground.* Cambridge, Mass.: Harvard University Press.

Gentry, H. S. 1942. *Rio Mayo Plants.* Carnegie Institution of Washington Publication no. 527. Washington, D.C.

Glendening, G. E. 1952. Some quantitative data on the increase of mesquite and cactus on a desert grassland range in southern Arizona. *Ecology* 33:319–28.

Glendening, G. E., and H. A. Paulsen, Jr. 1955. *Reproduction and Establishment of Velvet Mesquite as Related to Invasion of Semidesert Grasslands.* USDA Forest Service Technical Bulletin 1127. Washington, D.C.

Glinski, R. L. 1977. Regeneration and distribution of sycamore and cottonwood trees along Sonoita Creek, Santa Cruz County, Arizona. Pages 116–23 in R. R. Johnson and D. A. Jones (technical coordinators), *Importance, Preservation and Management of Riparian Habitat: A Symposium.* USDA Forest Service General Technical Report RM-43. Rocky Mountain Forest and Range Experiment Station, Fort Collins, Colo.

Glinski, R. L., and D. E. Brown. 1982. Mesquite (*Prosopis juliflora*) response to severe freezing in southeastern Arizona. *Journal of Arizona Academy of Science* 17:15–18.

Goldberg, D. E., and R. M. Turner. 1986. Vegetation change and woody plant demography in permanent plots in the Sonoran Desert. *Ecology* 67(3):695–712.

Golder, F. A. (ed.). 1928. *The March of the Mormon Battalion from Council Bluffs to California Taken from the Journal of Henry Standage.* New York: Century.

Goodwin, G., and K. Basso (ed.). 1971. *Western Apache Raiding and Warfare.* Tucson: University of Arizona Press.

Gould, F. W. 1951. *Grasses of Southwestern United States.* University of Arizona, Biological Science Bulletin no. 7. Tucson, Ariz.

Graf, W. L. 1983. The arroyo problem—paleohydrology and paleohydraulics in the short term. Pages 279–302 in K. G. Gregory (ed.), *Background to Paleohydrology.* New York: John Wiley and Sons.

———. 1988. *Fluvial Processes in Dryland Rivers.* New York: Springer-Verlag.

Granger, B. H. 1960. *Arizona Place Names.* Tucson: University of Arizona Press.

Grayson, D. K. 1993. *The Desert's Past: A Natural Prehistory of the Great Basin.* Washington, D.C.: Smithsonian Institution Press.

Green, C. R. 1959. *Arizona Statewide Rainfall.* Technical Reports on the Meteorology and Climatology of Arid Regions no. 7. University of Arizona, Institute of Atmospheric Physics, Tucson.

———. 1962. *Heating and Cooling Degree-Day Characteristics in Arizona*. Technical Reports on the Meteorology and Climatology of Arid Regions no. 10. University of Arizona, Institute of Atmospheric Physics, Tucson.

Green, C. R., and W. D. Sellers. 1964. *Arizona Climate*. Tucson: University of Arizona Press.

Greever, W. S. 1957. Railway development in the Southwest. *New Mexico Historical Review* 32:151–203.

Gregory, H. E. 1917. *Geology of the Navajo Country*. USGS Professional Paper 93. Washington, D.C.

Gregory, H. E., and R. C. Moore. 1931. *The Kaiparowits Region*. USGS Professional Paper 164. Washington, D.C.

Griffin, J. S. 1943. *A Doctor Comes to California*. California Historical Society, San Francisco.

Griffiths, D. 1904. *Range Investigations in Arizona*. USDA, Bureau of Plant Industry Bulletin no. 67. Washington, D.C.

———. 1910. *A Protected Stock Range in Arizona*. USDA, Bureau of Plant Industry Bulletin no. 177. Washington, D.C.

Grover, H. D., and H. B. Musick. 1990. Shrubland encroachment in southern New Mexico, U.S.A.: An analysis of desertification processes in the American Southwest. *Climatic Change* 17:305–30.

Gruell, G. E. 1980. Fire's influence on wildlife habitat on the Bridger-Teton National Forest, Wyoming. *Volume I: Photographic Record and Analysis*. USDA Forest Service Research Paper INT-235. Intermountain Research Station, Ogden, Utah.

———. 2001. *Fire in Sierra Nevada Forests: A Photographic Interpretation of Ecological Change Since 1849*. Missoula, Mont.: Mountain Press Publishing.

Gutzler, D. S. 2000. Evaluating global warming: A post-1990s perspective. *GSA Today* 10:1–7.

Hack, J. T. 1939. The Late Quarternary history of several valleys of northern Arizona: A preliminary announcement. *Museum Notes, Museum of Northern Arizona* 11 (May):67–73.

———. 1942. *The Changing Physical Environment of the Hopi Indians of Arizona*. Papers of the Peabody Museum of American Archaeology and Ethnology no. 35. Harvard University, Cambridge, Mass.

Hadley, D., and T. E. Sheridan. 1995. *Land Use History of the San Rafael Valley, Arizona (1540–1960)*. USDA Forest Service General Technical Report RM-GTR-269. Rocky Mountain Forest and Range Experiment Station, Fort Collins, Colo.

Haley, J. E. (ed.). 1932. A log of the Texas-California cattle trail, 1854. *Southwestern Historical Quarterly* 35:290–316.

Hallenbeck, C. 1940. *The Journey and Route of Alvar Nuñez, Cabeza de Vaca*. Glendale, Calif.: Arthur H. Clark.

Hamilton, P. 1881. *The Resources of Arizona*. Prescott, Ariz.: n.p.

Hansen, E. M., and F. K. Shwarz. 1981. *Meteorology of Important Rainstorms in the Colorado River and Great Basin Drainages*. Hydrometeorological Report no. 50. U.S. Department of Commerce, National Ocean and Atmospheric Administration, Washington, D.C.

Hardy, R. W. H. 1829. *Travels in the Interior of Mexico in 1825, 1826, 1827, and 1828*. London: H. Colburn and R. Bentley.

Haskett, B. 1935. Early history of the cattle industry in Arizona. *Arizona Historical Review* 6:3–42.

Hastings, J. R. 1959. Vegetation change and arroyo cutting in southeastern Arizona. *Journal of the Arizona Academy of Science* 1:60–67.

———. 1961a. People of reason and others: The colonization of Sonora to 1767. *Arizona and the West* 111:321–40.

———. 1961b. Precipitation and saguaro growth. *University of Arizona, Arid Lands Colloquia* 1959–60/1960–61:30–38.

———. 1964a. *Climatological Data for Baja California*. Technical Reports on the Meteorology and Climatology of Arid Regions no. 14. University of Arizona, Institute of Atmospheric Physics, Tucson.

———. 1964b. *Climatological Data for Sonora and Northern Sinaloa*. Technical Reports on the Meteorology and Climatology of Arid Regions no. 15. University of Arizona, Institute of Atmospheric Physics, Tucson.

———. 1965. On some uses of nonnormal coefficients of variation. *Journal of Applied Meterology* 4(4):475–78.

Hastings, J. R., and S. M. Alcorn. 1961. Physical determinations of growth and age in the giant cactus. *Journal of the Arizona Academy of Science* 2:32–39.

Hastings, J. R., and R. R. Humphrey. 1969a. *Climatological Data and Statistics for Baja California*. Technical Reports on the Meteorology and Climatology of Arid Regions no. 18. University of Arizona, Institute of Atmospheric Physics, Tucson.

———. 1969b. *Climatological Data and Statistics for Sonora and Northern Sinaloa*. Technical Reports on the Meteorology and Climatology of Arid Regions no. 19. University of Arizona, Institute of Atmospheric Physics, Tucson.

Hastings, J. R., and R. M. Turner. 1965a. Seasonal precipitation regimes in Baja California. *Geografiska Annaler, Ser. A* 47:204–23.

———. 1965b. *The Changing Mile: An Ecological Study of Vegetation Change with Time in the Lower Mile of an Arid and Semiarid Region*. Tucson: University of Arizona Press.

Hattersley-Smith, G. 1966. The symposium on glacier mapping. *Canadian Journal of Earth Sciences* 3(6):737–43.

Herbel, C. H., F. N. Ares, and R. A. Wright. 1972. Drought effects on a semidesert grassland range. *Ecology* 53:1084–93.

Hereford, R. 1984. Climate and ephemeral-stream processes: Twentieth-century geomorphology and alluvial stratigraphy of the Little Colorado River, Arizona. *Geological Society of America Bulletin* 95:654–68.

———. 1986. Modern alluvial history of the Paria River drainage basin, southern Utah. *Quaternary Research* 25:293–311.

———. 1993. *Entrenchment and Widening of the San Pedro River, Arizona*. Geological Society of America Special Paper 282. Boulder, Colo.

Hereford, R., and J. L. Betancourt. 1995. Historic geomorphology of the San Pedro River: Archival and physical evidence. Unpublished ms. Desert Laboratory, Tucson.

Hereford, R., and R. H. Webb. 1992. Historic variation in warm-season rainfall on the Colorado Plateau, U.S.A. *Climate Change* 22:239–56.

Hinton, R. J. 1890. *Irrigation in the United States*. Vol. IV of Report of the Special Committee of the United States Senate on the Irrigation and Reclamation of

Arid Lands. Senate Report 928, 51st Cong., 1st Sess. Washington, D.C.

Hinton, R. J. 1954. *The Handbook to Arizona.* Tucson: Arizona Silhouettes.

Hirschboeck, K. K. 1985. Hydroclimatology of flow events in the Gila River Basin, central and southern Arizona. Ph.D. diss., University of Arizona, Tucson.

History of Arizona Territory. 1884. San Francisco: Wallace W. Elliott.

Hitchcock, A. S. 1950. *Manual of the Grasses of the United States.* 2nd ed. USDA Miscellaneous Publication 200. Washington, D.C.

Hoffmeister, D. F. 1986. *Mammals of Arizona.* Tucson: University of Arizona Press.

Hornaday, W. T. 1908. *Camp-Fires on Desert and Lava.* New York: Charles Scribner's Sons.

Howell, D. J., and V. S. Roth. 1981. Sexual reproduction in agaves: The benefits of bats; the cost of semelparous advertising. *Ecology* 62:1-7.

Hubbs, C. L. 1957. Recent climatic history in California and adjacent areas. Pages 10-22 in *Proceedings of the Conference of Recent Research in Climatology.* La Jolla, Calif.: Scripps Institute of Oceanography.

Hull, H. M. 1956. Studies on herbicidal absorption and translocation in velvet mesquite seedlings. *Weeds* 4:22-42.

———. 1958. The effect of day and night temperature on growth, foliar wax content, and cuticle development of velvet mesquite. *Weeds* 6:133.

Humphrey, R. R. 1949. Fire as a means of controlling velvet mesquite, burroweed, and cholla on southern Arizona ranges. *Journal of Range Management* 2:175-82.

———. 1953. The desert grassland, past and present. *Journal of Range Management* 6:159-64.

———. 1958. The desert grassland: A history of vegetational change and an analysis of causes. *Botanical Review* 24:193-252.

———. 1960. *Forage Production on Arizona Ranges V: Pima, Pinal and Santa Cruz Counties.* University of Arizona Agricultural Experiment Station Bulletin 302.

———. 1962. *Range Ecology.* New York: Ronald Press.

———. 1974. Fire in the deserts and desert grassland of North America. Pages 365-400 in T. T. Kozlowski and C. E. Ahlgren (eds.), *Fire and Ecosystems.* New York: Academic Press.

———. 1987. *90 Years and 535 Miles: Vegetation Changes Along the Mexican Border.* Albuquerque: University of New Mexico Press.

Humphrey, R. R., and A. B. Humphrey. 1990. *Idria columnaris*: Age as determined by growth rate. *Desert Plants* 10(2):51-54.

Huntington, E. 1914. *The Climatic Factor.* Carnegie Institution of Washington Publication no. 192. Washington, D.C.

Hutto, R. L., J. R. McAuliffe, and L. Hogan. 1986. Distributional associates of the saguaro (*Carnegiea gigantea*). *Southwestern Naturalist* 31:469-76.

Huxley, J. 1961. *Of Wild Life and Natural Habitats in Central and East Africa.* Paris: UNESCO.

Huzayyin, S. 1956. Changes in climate, vegetation, and human adjustment in the Saharo-Arabian belt with special reference to Africa. Pages 304-23 in W. L. Thomas et al. (eds.), *Man's Role in Changing the Face of the Earth.* Chicago: University of Chicago Press.

Idso, S. B. 1992. Shrubland expansion in the American Southwest. *Climate Change* 22:205-46.

Itinerary of the El Paso and Fort Yuma Wagon Road Expedition under the Superintendence of James B. Leach. 1858. Records of the Office of the Secretary of the Interior relating to wagon roads, 1857-1881. National Archives, Microfilm no. 95, roll 3.

Ives, R. L. 1949. Climate of the Sonoran Desert region. *Annals of the Association of American Geographers* 39:143-87.

Jahns, R. H. 1959. Collapse depressions of the Pinacate Volcanic Field, Sonora, Mexico. Pages 165-83 in L. A. Heindl (ed.), *Southern Arizona Guidebook II.* Tucson: Arizona Geological Society.

Johnson, D. E. 1961. Edaphic factors affecting the distribution of creosotebush, *Larrea tridentata* (DC) Cov., in desert grassland sites of southeastern Arizona. Master's thesis, University of Arizona, Tucson.

Johnson, H. B., H. W. Polley, and H. S. Mayheux. 1993. Increasing CO_2 and plant-plant interactions: Effects on natural vegetation. *Vegetatio* 104-5:157-70.

Johnson, K. L. 1987. *Rangeland Through Time.* University of Wyoming Agricultural Experiment Station Miscellaneous Publication 50.

Johnston, A. R. 1848. Journal. U.S. Congress, House, Executive Document 41, 30th Cong., 1st Sess., 565-614. Washington, D.C.

Johnston, M. C. 1962. The North American mesquites *Prosopis* sect. *Algarobia* (Leguminosae). *Brittonia* 14:72-90.

Jones, W. D. 1979. Effects of 1978 freeze on native plants of Sonora. *Desert Plants* 1:33-36.

Journal of Nathaniel V. Jones, with the Mormon Battalion. 1931. *Utah Historical Quarterly* 4:6-23.

Journal of Robert S. Bliss, with the Mormon Battalion. 1931. *Utah Historical Quarterly* 4:67-96, 110-28.

Judson, S. 1952. Arroyos. *Scientific American* 187:71-76.

Kaib, J. M. 1998. Fire history in riparian canyon pine-oak forests and intervening desert grasslands of the Southwest borderlands: A dendrochronological, historical, and cultural inquiry. Master's thesis, University of Arizona, Tucson.

Kaib, J. M., C. H. Baisan, H. D. Grissino-Mayer, and T. W. Swetnam. 1996. Fire history in the gallery pine-oak forests and adjacent grasslands of the Chiricahua Mountains of Arizona. Pages 253-64 in P. Ffolliott et al. (technical coordinators), *Effects of Fire on Madrean Province Ecosystems: A Symposium Proceedings.* USDA Forest Service General Technical Report RM-GTR-289. Rocky Mountain Forest and Range Experiment Station, Fort Collins, Colo.

Karl, T. R., H. F. Diaz, and G. Kukla. 1988. Urbanization: Its detection and effect in the United States climate record. *Journal of Climate* 1:1099-123.

Kearney, T. H., R. H. Peebles, et al. 1960. *Arizona Flora.* 2nd ed. Berkeley: University of California Press.

Keeling, C. D. 1978. The influence of Mauna Loa Observatory on the development of atmospheric CO_2 research. Pages 36-54 in J. Miller (ed.), *Mauna Loa Observatory: A 20th Anniversary Report.* National Oceanic and Atmospheric Administration Special Report. Washington, D.C.

Keeling, C. D., R. B. Bacastow, A. F. Carter, S. C. Piper, T. P. Whorf, M. Heimann, W. G. Mook, and H. Roeloffzen. 1989. A three-dimensional model of atmospheric

CO_2 transport based on observed winds: 1. Analysis of observational data. Pages 165–235 in D. H. Peterson (ed.), *Aspects of Climate Variability in the Pacific and the Western Americas*. Washington, D.C.: American Geophysical Union.

Keleher, W. A. 1952. *Turmoil in New Mexico*. Santa Fe: Rydal Press.

Kincer, J. B. 1933. Is our climate changing? *Monthly Weather Review* 61:251–59.

———. 1946. Our changing climate. *Transactions of the American Geophysical Union* 27:342–47.

Klett, M., E. Manchester, J. Verburg, G. Bushaw, R. Dingus, and P. Berger. 1984. *Second View: The Rephotographic Survey Project*. Albuquerque: University of New Mexico Press.

Kramp, B. A., R. J. Ansley, and T. R. Tunnell. 1998. Survival of mesquite seedlings emerging from cattle and wildlife feces in a semi-arid grassland. *Southwestern Naturalist* 43(3):300–312.

Kurtén, B., and E. Anderson. 1980. *Pleistocene Mammals of North America*. New York: Columbia University Press.

Lafora, N. de. 1939. *Relación del Viaje que Hizo a los Presidios Internos Situados en la Frontera de la América Septentrional*. Mexico: Editorial Pedro Robredo.

Lake, J. V. 1956. The temperature profile above bare soil on clear nights. *Quarterly Journal of the Royal Meteorological Society* 82:187–97.

Land, E. 1934. Reminiscences. Unpublished ms., Arizona Historical Society, Tucson.

Landsberg, H. E., and J. M. Mitchell, Jr. 1961. Temperature fluctuations and trends over the Earth. *Quarterly Journal of the Royal Meteorological Society* 87:435–36.

Larson, R. A., and J. E. Slosson. 1997. *Storm-induced Geologic Hazards: Case Histories from the 1992–1993 Winter in Southern California and Arizona*. Geological Society of America, Reviews in Engineering Geology 11. Boulder, Colo.

Lavender, D. 1954. *Bent's Fort*. Garden City, N.J.: Doubleday.

Laycock, W. A. 1991. Stable states and thresholds of range condition on North American rangelands: A viewpoint. *Journal of Range Management* 44:427–33.

Leopold, L. B. 1951a. Vegetation of Southwestern watersheds in the nineteenth century. *Geographical Review* 41:295–316.

———. 1951b. Rainfall frequency: An aspect of climatic variation. *Transactions of the American Geophysical Union* 32:347–57.

Leopold, L. B., E. B. Leopold, and F. Wendorf. 1963. Some climatic indicators in the period A.D. 1200–1400 in New Mexico. Pages 265–70 in *Changes of Climate*. Paris: UNESCO.

Leopold, L. B., and J. P. Miller. 1954. *A Postglacial Chronology for Some Alluvial Valleys in Wyoming*. USGS Water-Supply Paper 1261. Washington, D.C.

Leopold, L. B., and C. T. Snyder. 1951. *Alluvial Fills Near Gallup, New Mexico*. USGS Water-Supply Paper 1110-A. Washington, D.C.

Life, August 18, 1952.

Lockwood, F. C. 1929. American hunters and trappers in Arizona. *Arizona Historical Review* 2:70–85.

Love, C. M. 1916. History of the cattle industry in the Southwest. *Southwestern Historical Quarterly* 19:370–99, 20:1–18.

Lowe, C. H. 1964. *Arizona's Natural Environment*. Tucson: University of Arizona Press.

Lumholtz, C. 1912. *New Trails in Mexico*. New York: Charles Scribner's Sons.

Lusby, G. C. 1979. *Effects of Grazing on Runoff and Sediment Yield from Desert Rangeland at Badger Wash in Western Colorado, 1953–1973*. U.S. Geological Survey Water Supply Paper 1532-I. Reston, Va.

Lusby, G. C., V. H. Reid, and O. D. Knipe. 1971. *Effects of Grazing on the Hydrology and Biology of the Badger Wash Basin in Western Colorado, 1953–66*. U.S. Geological Survey Water-Supply Paper 1532-D. Washington, D.C.

MacDougal, D. T. 1908. Across Papagueria. *American Geographical Society Bulletin* 40:705–25.

———. 1924. *Growth in Trees and Massive Organs of Plants—Dendrographic Measurements*. Carnegie Institution of Washington Publication no. 350. Washington, D.C.

MacDougal, D. T., and W. A. Cannon. 1910. *Conditions of Parasitism in Plants*. Carnegie Institution of Washington Publication no. 129. Washington, D.C.

MacPhee, R. D. E. (ed.). 1999. *Extinctions in Near Time: Causes, Contexts, and Consequences*. New York: Kluwer Academic/Plenum Publishers.

Mahall, B. E., and R. M. Callaway. 1991. Root communication among desert shrubs. *Proceedings of the National Academy of Sciences (U.S.A.)* 88:874–76.

Malin, J. C. 1956. *The Grassland of North America*. Lawrence, Kan.: James C. Malin.

Mange, J. M. 1926. *Luz De Tierra Incógnita En La América Septentrional Y Diario De Las Exploraciones En Sonora*. Ed. F. del Castillo. Mexico: Talleres Gráficos de la Nación.

Mann, D. E. 1963a. *The Politics of Water in Arizona*. Tucson: University of Arizona Press.

———. 1963b. Political and social institutions in arid regions. Pages 397–428 in C. Hodge and P. C. Duisberg (eds.), *Aridity and Man*. Washington, D.C.: American Association for the Advancement of Science.

Mantua, N. J., S. R. Hare, Y. Zhang, J. M. Wallace, and R. C. Francis. 1997. A Pacific interdecadal climate oscillation with impacts on salmon production. *Bulletin of the American Meteorological Society* 78:1069–79.

Marks, J. B. 1950. Vegetation and soil relations in the lower Colorado Desert. *Ecology* 31:176–93.

Marshall, J. T., Jr. 1957. *Birds of Pine-Oak Woodland in Southern Arizona and Adjacent Mexico*. Berkeley, Calif.: Cooper Ornithological Society.

Martin, P. S. 1963. *The Last 10,000 Years*. Tucson: University of Arizona Press.

———. 1990. 40,000 years of extinctions on the "planet of doom." *Palaeogeography, Palaeoclimatology, Palaeoecology (Global and Planetary Change Section)* 82:187–201.

Martin, P. S., and R. Klein (eds.). 1984. *Quaternary Extinctions: A Prehistoric Revolution*. Tucson: University of Arizona Press.

Martin, P. S., J. Schoenwetter, and B. C. Arms. 1961. *The Last 10,000 Years*. University of Arizona Geochronology Laboratories, Tucson.

Martin, P. S., and C. R. Szuter. 1999. War zones and game sinks in Lewis and Clark's West. *Conservation Biology* 13:36–45.

Martin, P. S., and H. E. Wright (eds.). 1967. *Pleistocene Extinctions: The Search for a Cause*. New Haven, Conn.: Yale University Press.

Martin, S. C. 1948. Mesquite seeds remain viable after 44 years. *Ecology* 3:393.

———. 1970. Longevity of velvet mesquite in the soil. *Journal of Range Management* 23:69-70.

———. 1975. *Ecology and Management of Southwestern Semi-desert Grass-shrub Ranges: The Status of Our Knowledge*. USDA Forest Service Research Papers RM-156. Rocky Mountain Forest and Range Experiment Station, Fort Collins, Colo.

———. 1983. Responses of semi-desert grasses and shrubs to fall burning. *Journal of Range Management* 36:604-10.

Martin, S. C., and R. M. Turner. 1977. Vegetation change in the Sonoran Desert region, Arizona and Sonora. *Journal of the Arizona Academy of Science* 12:59-69.

Masse, W. B. 1979. An intensive survey of prehistoric dry farming systems near Tumamoc Hill in Tucson, Arizona. *The Kiva* 45:141-86.

Matson, D. S., and A. H. Schroeder (eds.). 1957. Cordero's description of the Apache—1796. *New Mexico Historical Review* 32:335-56.

Mattison, R. H. 1946. Early Spanish and Mexican settlements in Arizona. *New Mexico Historical Review* 21:273-327.

Mayeux, H. S., H. B. Johnson, and H. W. Polley. 1991. Global change and vegetation dynamics. Pages 62-74 in L. F. James, J. O. Evans, M. H. Ralphs, and B. J. Sigler (eds.), *Noxious Range Weeds*. Boulder, Colo.: Westview Press.

McAuliffe, J. R. 1984. Sahuaro-nurse tree associations in the Sonoran Desert: Competitive effects of sahuaros. *Oecologia* 64:319-21.

———. 1988. Markovian dynamics of simple and complex desert plant communities. *American Naturalist* 131(4):459-90.

———. 1990. Paloverdes, pocket mice, and bruchid beetles: Interrelationships of seeds, dispersers, and seed predators. *Southwestern Naturalist* 35:329-37.

———. 1991. Demographic shifts and plant succession along a late Holocene soil chronosequence in the Sonoran Desert of Baja California. *Journal of Arid Environments* 20:165-78.

———. 1994. Landscape evolution, soil formation, and ecological patterns and processes in Sonoran Desert bajadas. *Ecological Monographs* 64(2):111-48.

———. 1995. Landscape evolution, soil formation, and Arizona's desert grasslands. Pages 100-129 in M. P. McClaran and T. R. Van Devender (eds.), *The Desert Grassland*. Tucson: University of Arizona Press.

———. 1998. Rangeland water developments: Conservation solution or illusion? Pages 310-35 in *Proceedings of a Symposium on Environment, Economics, and Legal Issues Related to Rangeland Water Developments, Nov. 13-18, 1997*. The Center for the Study of Law, Science and Technology, Arizona State University, Tempe.

———. 1999. The Sonoran Desert: Landscape complexity and ecological diversity. Pages 68-114 in R. H. Robichaux (ed.), *Ecology of Sonoran Desert Plants and Plant Communities*. Tucson: University of Arizona Press.

McAuliffe, J. R., and F. J. Janzen. 1986. Effects of intraspe-

cific crowding on water uptake, water storage, apical growth, and reproductive potential in the sahuaro cactus, *Carnegiea gigantea*. *Botanical Gazette* 147:334-41.

McCabe, G. J., and M. D. Dettinger. 1999. Decadal variations in the strength of ENSO teleconnections with precipitation in the western United States. *International Journal of Climatology* 19:1399-410.

McClaran, M. P., and T. R. Van Devender. 1995. *The Desert Grassland*. Tucson: University of Arizona Press.

McDonald, J. E. 1956. *Variability of Precipitation in an Arid Region: A Survey of Characteristics for Arizona*. Technical Reports on the Meteorology and Climatology of Arid Regions, no. 1. University of Arizona, Institute of Atmospheric Physics, Tucson.

———. 1959. Climatology of arid lands. *Arid Lands Colloquia* (1958-59):3-13.

———. 1962. The evaporation-precipitation fallacy. *Weather* 17:1-9.

McGee, W. J. 1898. *The Seri Indians*. Annual Report of the Bureau of American Ethnology no. 17, 1895-96. Washington, D.C.: GPO.

McGinnies, W. G. n.d. Flowering periods for common desert plants of southwestern Arizona. Office of Arid Lands Studies, College of Agriculture, University of Arizona, Tucson.

McGinnies, W. G., B. J. Goldman, and P. Paylore (eds.). 1968. *Deserts of the World: An Appraisal of Research into Their Physical and Biological Environments*. Tucson: Unversity of Arizona Press.

McGinnies, W. J., H. L. Shantz, and W. G. McGinnies. 1991. *Changes in Vegetation and Land Use in Eastern Colorado: A Photographic Study, 1904 to 1986*. U.S.D.A. Agricultural Research Service, ARS-85. Washington, D.C.

McGregor, S. E., S. M. Alcorn, and G. Olin. 1962. Pollination and pollinating agents of the saguaro. *Ecology* 43:259-67.

McLaughlin, S. P., and J. E. Bowers. 1982. Effects of fire on a Sonoran Desert plant community. *Ecology* 63:246-48.

———. 1999. Diversity and affinities of the flora of the Sonoran Floristic Province. Pages 12-35 in R. H. Robichaux (ed.), *Ecology of Sonoran Desert Plants and Plant Communities*. Tucson: University of Arizona Press.

McLaughlin, S. P., E. L. Geiger, and J. E. Bowers. 2001. A flora of the Appleton-Whittell Research Ranch, northeastern Santa Cruz County, Arizona. *Journal of the Arizona-Nevada Academy of Science* 33(2):113-31.

McNamee, G. 1994. *Gila: The Life and Death of an American River*. New York: Orion Books.

McPherson, G. R. 1992. Ecology of oak woodlands in Arizona. Pages 24-33 in P. F. Ffolliott, G. J. Gottfried, D. A. Bennett, V. M. Hernandez C., A. Ortega-Rubio, and R. H. Hamre (technical coordinators), *Proceedings of the Symposium on Ecology and Management of Oak and Associated Woodlands: Perspectives in the Southwestern United States and Northern Mexico*. USDA Forest Service General Technical Report RM 218. Rocky Mountain Forest and Range Experiment Station, Fort Collins, Colo.

———. 1995. The role of fire in the Desert Grassland. Pages 130-51 in M. P. McClaran and T. R. Van Devender (eds.), *The Desert Grassland*. Tucson: University of Arizona Press.

———. 1997. *Ecology and Management of North American Savannas*. Tucson: University of Arizona Press.

McPherson, G. R., T. W. Boutton, and A. J. Midwood. 1993. Stable carbon isotope analysis of soil organic matter illustrates vegetation change at the grassland/woodland boundary in southeastern Arizona, USA. *Oecologia* 93:95–101.

Mead, J. I., T. R. Van Devender, and K. L. Cole. 1983. Late Quaternary small mammals from Sonoran Desert packrat middens, Arizona and California. *Journal of Mammalogy* 64:173–80.

Meagher, M., and D. B. Houston. 1998. *Yellowstone and the Biology of Time: Photographs Across a Century*. Norman: University of Oklahoma Press.

Mearns, E. A. 1907. *Mammals of the Mexican Boundary of the United States*. Smithsonian Institution Bulletin 56, Part I. Washington, D.C.

Mehrhoff, L. A., Jr. 1955. Vegetation changes on a southern Arizona grassland range—An analysis of causes. Master's thesis, University of Arizona, Tucson.

Meko, D. M., C. W. Stockton, and W. R. Bogess. 1980. A tree-ring reconstruction of drought in southern California. *Water Resources Bulletin* 16:594–600.

Melis, T. S., W. M. Phillips, R. H. Webb, and D. J. Bills. 1996. *When the Blue-Green Waters Turn Red: Historical Flooding in Havasu Creek, Arizona*. U.S. Geological Survey Water Resources Investigations Report 96-4059. Washington, D.C.

Merideth, R. 2001. *A Primer on Climatic Variability and Change in the Southwest*. Udall Center for Studies in Public Policy, Tucson, Arizona.

Meteorological Abstracts and Bibliography. 1950. 1:473–75.

Minckley, W. L. 1973. *Fishes of Arizona*. Phoenix: Arizona Game and Fish Department.

Minckley, W. L., and J. E. Deacon (eds.). 1991. *Battle Against Extinction: Native Fish Management in the American West*. Tucson: University of Arizona Press.

Minnich, R. A., E. F. Vizcaino, J. Sosaramirez, and Y. H. Chou. 1993. Lightning detection rates and wildland fire in the mountains of Baja California, Mexico. *Atmosfera* 6(4):235–53.

Mitchell, J. M., Jr. 1953. On the causes of instrumentally observed secular temperature trends. *Journal of Meteorology* 10:244–61.

———. 1961. *Bibliographic List of Recent Studies of Climatic Change by United States Citizens*. USDC, Weather Bureau. Washington, D.C.

———. 1963. On the world-wide pattern of secular temperature change. Pages 161–81 in *Changes of Climate*. Paris: UNESCO.

Mitchell, N. D., and J. E. Williams. 1996. The consequences for native biota of anthropogenic-induced climate change. Pages 308–24 in W. J. Bouma, G. I. Pearman, and M. R. Manning (eds.), *Greenhouse: Coping with Climate Change*. Collingwood, Victoria: CSIRO.

Morello, J. H., and C. Saravia Toledo. 1959. El bosque chaqueño. I. Paisaje primitivo, paisaje natural y paisaje cultural en el oriente de Salta. *Revista Agronómica del Noroeste Argentino* 3:5–81.

Morgan, W. D., and H. M. Lester. 1954. *Graphic Graflex Photography*. 10th ed. New York: Morgan and Lester.

Morris, R. C. 1926. The notion of a Great American Desert east of the Rockies. *Mississippi Valley Historical Review* 13:190–200.

Morrisey, R. J. 1950. The early range cattle industry in Arizona. *Agricultural History* 24:151–56.

———. 1951. The northward expansion of cattle ranching in New Spain, 1550–1600. *Agricultural History* 25:115–21.

Muller, C. H. 1947. Vegetation and climate of Coahuila, Mexico. *Madroño* 9:33–57.

Murphy, F. 1928. Reminiscences. Unpublished ms., Arizona Historical Society, Tucson.

Neftel, A., E. Moore, H. Oeschger, and B. Stauffer. 1985. Evidence from polar ice cores for the increase in atmospheric CO_2 in the past two centuries. *Nature* 315:45–47.

Neilson, R. P. 1986. High-resolution climatic analysis and southwest biogeography. *Science* 232:27–34.

Nelson, E. W. 1934. *The Influence of Precipitation and Grazing upon Black Grama Grass Range*. USDA Technical Bulletin 409. Washington, D.C.

Nentvig, J. 1951. *Rudo Ensayo*. Trans. Eusebio Guiteras. Tucson: Arizona Silhouettes.

Nichol, A. A. 1952. *The Natural Vegetation of Arizona*. The University of Arizona Agricultural Experiment Station Technical Bulletin 127. Tucson.

Niering, W. A., and C. H. Lowe. 1985. Vegetation of the Santa Catalina Mountains: Community types and dynamics. Pages 159–84 in R. K. Peet (ed.), *Plant Community Ecology: Papers in Honor of Robert H. Whittaker*. Boston: Dr. W. Junk, Publishers.

Niering, W. A., R. H. Whittaker, and C. H. Lowe. 1963. The saguaro: A population in relation to environment. *Science* 142:15–23.

Nigam, S., M. Barlow, and E. H. Berbery. 1999. Analysis links Pacific decadal variability to drought and streamflow in United States. *EOS* 80:621–24.

Nobel, P. S. 1980a. Morphology, nurse plants, and minimum apical temperatures for young *Carnegiea gigantea*. *Botanical Gazette* 141:188–91.

———. 1980b. Influences of minimum stem temperatures on ranges of cacti in the southwestern United States and central Chile. *Oecologia* 47:10–15.

———. 1984. Extreme temperatures and the thermal tolerances for seedlings of desert succulents. *Oecologia* 62:310–17.

———. 1988. *Environmental Biology of Agaves and Cacti*. New York: Cambridge University Press.

Nobel, P. S., G. N. Geller, S. C. Kee, and A. D. Zimmerman. 1986. Temperatures and thermal tolerances for cacti exposed to high temperatures near the soil surface. *Plant, Cell and Environment* 9:279–87.

Nobel, P. S., and J. Sanderson. 1984. Rectifier-like activities of roots of two desert succulents. *Journal of Experimental Botany* 35:727–37.

Norris, J. J. 1950. *Effect of Rodents, Rabbits, and Cattle on Two Vegetation Types in Semidesert Range Land*. New Mexico Agricultural Experiment Station Bulletin 353. Las Cruces.

Nyandiga, C. O., and G. R. McPherson. 1992. Germination of two warm-temperate oaks, *Quercus emoryi* and *Q. arizonica*. *Canadian Journal of Forest Research* 22:1395–401.

Oakes, C. L. 2000. History and consequences of keystone mammal eradication in the desert grassland: The Arizona black-tailed prairie dog (*Cynomys ludovicianus arizonensis*). Ph.D. diss., University of Texas, Austin.

Officer, J. 1987. *Hispanic Arizona, 1536–1856.* Tucson: University of Arizona Press.

Ohnesorgen, W. 1929. Reminiscences. Unpublished ms., Arizona Historical Society, Tucson.

Olmstead, F. H. 1919. *A Report on Flood Control of the Gila River in Graham County, Arizona.* U.S. Congress, Senate, Document 436, 65th Cong., 3d Sess. Washington, D.C.

Orodho, A. B., M. J. Trlica, and C. D. Bonham. 1990. Long-term heavy grazing effects on soil and vegetation in the Four Corners region. *Southwestern Naturalist* 35:9–14.

Osgood, E. S. 1929. *The Day of the Cattleman.* Minneapolis: University of Minnesota Press.

Page, J. L. 1930. *Climate of Mexico.* Monthly Weather Review Supplement no. 33. Washington, D.C.: GPO.

Page, L. F. 1937. Temperature and rainfall changes in the United States during the past forty years. *Monthly Weather Review* 65:46–55.

Palmquist, P. E. 1983. *Carleton E. Watkins: Photographer of the American West.* Albuquerque: University of New Mexico Press.

Parke, J. G. 1857. *Report of Explorations for Railroad Routes.* Vol. 7 of *Reports of Explorations and Surveys.* U.S. Congress, House, Executive Document 91, 33d Cong., 2d Sess. Washington, D.C.

Parker, K. C. 1993. Climatic effects on regeneration trends for two columnar cacti in the northern Sonoran Desert. *Annals of the Association of American Geographers* 83:452–74.

Parker, K. W., and S. C. Martin. 1952. *The Mesquite Problem on Southern Arizona Ranges.* USDA Circular 908. Washington, D.C.

Pase, C. P. 1969. Survival of *Quercus turbinella* and *Q. emoryi* seedlings in an Arizona chaparral community. *Southwestern Naturalist* 14:149–56.

Pase, C. P., and A. W. Lindenmuth, Jr. 1971. Effects of prescribed fire on vegetation and sediment in oak-mountain mahogany chaparral. *Journal of Forestry* 69:800–805.

Pattie, J. O. 1905. *The Personal Narrative of James O. Pattie, of Kentucky.* Ed. R. Goldthwaites. Cleveland: Arthur H. Clark.

Paulsen, H. A., Jr. 1950. Mortality of velvet mesquite seedlings. *Journal of Range Management* 3:281–86.

Pelzer, L. 1936. *The Cattlemen's Frontier.* Glendale, Calif.: Arthur H. Clark.

Perevolotsky, A. 1999. Natural conservation, reclamation, and livestock grazing in the northern Negev: Contradictory or complementary concepts? Pages 223–32 in T. W. Hoekstra and M. Schak (eds.), *Arid Lands Management: Toward Ecological Sustainability.* Urbana: University of Illinois Press.

Pérez de Ribas, A. 1645. *Historia De Los Triunfos De Nuestra Santa Fé Entre Gentes Las Mas Bárbaras, Y Fieras Del Nuevo Orbe.* Madrid: A. de Paredes.

Perry, R. 1991. *Western Apache Heritage: People of the Mountain Corridor.* Austin: University of Texas Press.

Peterson, H. V. 1950. The problem of gullying in western valleys. Pages 407–34 in P. D. Trask (ed.), *Applied Sedimentation.* New York: John Wiley and Sons.

Pfefferkorn, I. 1949. *Sonora, a Description of the Province.* Trans. and ed. T. E. Treutlein. Albuquerque: University of New Mexico Press.

Philander, S. G. H. 1983. El Niño Southern Oscillation phenomena. *Nature* 302:295–301.

———. 1990. *El Niño, La Niña and the Southern Oscillation.* San Diego, Calif.: Academic Press.

Phillips, A., J. Marshall, and G. Monson. 1964. *The Birds of Arizona.* Tucson: University of Arizona Press.

Phillips, F. J. 1912. *Emory Oak in Southern Arizona.* USDA Forest Service Circular 201. Washington, D.C.

Phillips, W. S. 1963. *Vegetational Changes in Northern Great Plains.* University of Arizona Agricultural Experiment Station Report 214, Tucson.

Pierson, E. A., and R. M. Turner. 1998. An 85-year study of saguaro (*Carnegiea gigantea*) demography. *Ecology* 79:2676–93.

Polley, H. W., H. B. Johnson, and H. S. Mayeux. 1992. Carbon dioxide and water fluxes of C_3 annuals and C_3 and C_4 perennials at subambient CO_2 concentrations. *Functional Ecology* 6:693–703.

———. 1994. Increasing CO_2: Comparative responses of the C_4 grass *Schizachyrium* and grassland invader *Prosopis.* *Ecology* 75:976–88.

Pool, F. 1935. Reminiscences. Unpublished ms., Arizona Historical Society, Tucson.

Pope, D. P., J. H. Brock, and R. A. Backhaus. 1990. Vegetative propagation of key southwestern woody riparian species. *Desert Plants* 10(2):91–95.

Potter, R. L., J. L. Petersen, and D. N. Ueckert. 1984. Germination responses of *Opuntia* spp. to temperature, scarification, and other seed treatments. *Weed Science* 32:106–10.

Powell, H. M. T. 1931. *The Santa Fe Trail to California, 1849–1852.* San Francisco: Grabhorn Press.

Powell, J. W. 1962. *Report on the Lands of the Arid Region of the United States.* Ed. W. Stegner. Cambridge, Mass.: Belknap Press of Harvard University Press.

Pradeau, A. F. 1953. Nentvig's 'Description of Sonora.' *Mid-America* 35:81–90.

Priestley, H. I. 1916. *Jose de Galvez, Visitor-General of New Spain, 1765–1771.* Berkeley: University of California Press.

Pumpelly, R. 1870. *Across America and Asia.* New York: Leypoldt and Holt.

Pyke, C. B. 1972. *Some Meteorological Aspects of the Seasonal Distribution of Precipitation in the Western United States and Baja California.* Water Resources Center Contribution no. 139, UCAL-WRC-W-254. University of California, Los Angeles.

Pyne, S. J. 1982. *Fire in America.* Princeton, N.J.: Princeton University Press.

Quinn, W. H., V. T Neal, and S. E. Antunez de Mayolo. 1986. Preliminary report on El Niño occurrences over the past four and a half centuries. Oregon State University, Corvallis, College of Oceanography, Reference 86-16, NSF # ATM-85 15014.

Radding, C. 1997. *Wandering Peoples: Colonialism, Ethnic Spaces, and Ecological Frontiers in Northwestern Mexico, 1700–1850.* Durham, N.C.: Duke University Press.

Randerson, D. 1986. Mesoscale convective complex type storm over the desert Southwest. NOAA Technical Memorandum NWS WR-196, Western Region, Natl. Weather Service. Salt Lake City, Utah: U.S. Dept. Commerce.

Rasmusson, E. M. 1984. El Niño, the ocean/atmosphere connection. *Oceanus* 27:5–12.

Rea, A. M. 1997. *At the Desert's Green Edge: An Ethnobotany of the Gila River Pima*. Tucson: University of Arizona Press.

Reagan, A. B. 1924. Recent changes in the plateau region. *Science* 60:283–85.

Reap, R. M. 1986. Evaluation of cloud-to-ground lightning data from the western United States for the 1983–1984 summer seasons. *Journal of Climate and Applied Meteorology* 25:785–99.

Reff, D. T. 1991. *Disease, Depopulation, and Culture Change in Northwestern New Spain, 1518–1764*. Salt Lake City: University of Utah Press.

Reichenbacher, F. W., S. E. Franson, and D. E. Brown. 1998. *North American Biotic Communities*. Map, scale 1:10,000,000. Salt Lake City: University of Utah Press.

Report of the Attorney General. 1904. Washington, D.C.: GPO.

Report of the Governor of Arizona to the Secretary of the Interior. 1885. Washington, D.C.: GPO.

———. 1896. Washington, D.C.: GPO.

Report on Barracks and Hospitals with Descriptions of Military Posts. 1870. Circular no. 4, War Department, Surgeon General's Office. Washington, D.C.: GPO.

Reyes, A. M. de los. 1938. Memorial sobre las misiones de Sonora, 1772. *Boletín del Archivo General de la Nación* 9:276–320.

Reynolds, H. G. 1950. Relation of Merriam kangaroo rats to range vegetation in southern Arizona. *Ecology* 31:456–63.

———. 1958. The ecology of Merriam kangaroo rat (*Dipodomys merriami* Mearns) on grazing lands of southern Arizona. *Ecological Monographs* 28:111–127.

Reynolds, H. G., and J. W. Bohning. 1956. Effects of burning on a desert grass-shrub range in southern Arizona. *Ecology* 37:769–77.

Reynolds, H. G., and G. E. Glendening. 1949. Merriam kangaroo rat a factor in mesquite propagation on southern Arizona range lands. *Journal of Range Management* 2:193–97.

Reynolds, H. G., and S. C. Martin. 1968. *Managing Grass-shrub Cattle Ranges in the Southwest*. U.S. Department of Agriculture Handbook 162. Washington, D.C.

Rich, J. L. 1911. Recent stream trenching in the semi-arid portion of southwestern New Mexico, a result of removal of vegetation cover. *American Journal of Science* 32:237–45.

Ricketts, N. B. 1996. *The Mormon Battalion: U.S. Army of the West, 1846–1848*. Logan: Utah State University Press.

Rivera, R. L., and C. E. Freeman. 1979. The effects of some alternating temperatures on germination of creosotebush, *Larrea tridentata* (Zygophyllaceae). *Southwestern Naturalist* 24:711–14.

Roberts, N. 1989. *The Holocene, an Environmental History*. New York: Basil Blackwell.

Roden, G. I. 1966. A modern statistical analysis and documentation of historical temperature records in California, Oregon and Washington, 1821–1964. *Journal of Applied Meteorology* 5:3–24.

Roeske, R. H., M. E. Cooley, and B. N. Aldridge. 1978. *Floods of September 1970 in Arizona, Utah, Colorado, and New Mexico*. United States Geological Survey Water-Supply Paper 2052. Reston, Va.

Rogers, G. F. 1982. *Then and Now: A Photographic History of Vegetation Change in the Central Great Basin*. Salt Lake City: University of Utah Press.

———. 1986. Comparison of fire occurrence in desert and nondesert vegetation in Tonto National Forest, Arizona. *Madroño* 33:278–83.

Rogers, G. F., H. E. Malde, and R. M. Turner. 1984. *Bibliography of Repeat Photography for Evaluating Landscape Change*. Salt Lake City: University of Utah Press.

Rogers, G. F., and J. Steele. 1980. Sonoran Desert fire ecology. Pages 15–19 in M. A. Stokes and J. H. Dieterich (technical coordinators), *Proceedings of the Fire History Workshop, October 20–24, 1980, Tucson, Arizona*. USDA Forest Service General Technical Report RM-81. Rocky Mountain Forest and Range Experiment Station, Fort Collins, Colo.

Rogers, G. F., and M. K. Vint. 1987. Winter precipitation and fire in the Sonoran Desert. *Journal of Arid Environments* 13:47–52.

Ropelewski, C. F., and M. S. Halpert. 1986. North American precipitation and temperature patterns associated with El Niño–Southern Oscillation. *Monthly Weather Review* 114:2352–62.

Rothrock, J. T. 1875. *Preliminary and General Botanical Report, with Remarks upon the General Topography of the Region Traversed*. Annual Report of the Chief of Engineers, 1875. Washington, D.C.: GPO.

Rucks, M. G. 1984. Composition and trend of riparian vegetation on five perennial streams in southeastern Arizona. Pages 97–107 in R. E. Warner and K. M. Hendrix (eds.), *California Riparian Systems: Ecology, Conservation, and Productive Management*. Berkeley: University of California Press.

Russell, F. 1908. *The Pima Indians*. Annual Report of the Bureau of American Ethnology no. 26, 1904–5. Washington, D.C.: GPO.

Saarinen, T. F., V. R. Baker, R. Durrenberger, and T. Maddock, Jr. 1984. *The Tucson, Arizona, Flood of October 1983*. Washington, D.C.: National Academy Press.

Sanchini, P. J. 1981. Population structure and fecundity in *Quercus emoryi* and *Q. arizonica* in southeastern Arizona. Ph.D. diss., University of Colorado, Boulder.

Sauer, C. O. 1932. *The Road to Cibola*. Ibero-Americana, no. 3. Berkeley: University of California Press.

———. 1934. *The Distribution of Aboriginal Tribes and Languages in Northwestern Mexico*. Ibero-Americana, no. 5. Berkeley: University of California Press.

———. 1935. *Aboriginal Population of Northwestern Mexico*. Ibero-Americana, no. 10. Berkeley: University of California Press.

———. 1944. A geographic sketch of early man. *Geographical Review* 34:529–73.

———. 1956. The agency of man on the Earth. Pages 49–69 in W. L. Thomas et al. (eds.), *Man's Role in Changing the Face of the Earth*. Chicago: University of Chicago Press.

Sauer, C. O., and D. Brand. 1931. Prehistoric settlements of Sonora. *University of California Publications in Geography* 5:67–148.

Sayre, N. 1999. The cattle boom in southern Arizona: Towards a critical political ecology. *Journal of the Southwest* 41(2):239–71.

Schmidt, K. M., and R. H. Webb. 2001. Researchers consider U.S. Southwest's response to warmer, drier conditions. *EOS* 82:475, 478.

Schmutz, E. M., M. K. Sourabie, and D. A. Smith. 1985.

The Page Ranch story: Its vegetative history and management implications. *Desert Plants* 7:13–21.

Schrader, F. C. 1915. *Mineral Deposits of the Santa Rita and Patagonia Mountains, Arizona*. USGS Bulletin 582. Washington, D.C.

Schroeder, A. R. 1956. Fray Marcos de Niza, Coronado and the Yavapai. *New Mexico Historical Review* 31:24–37.

Schulman, E. 1956. *Dendroclimatic Changes in Semiarid America*. Tucson: University of Arizona Press.

Schumm, S. A., and R. F. Hadley. 1957. Arroyos and the semiarid cycle of erosion. *American Journal of Science* 255:161–74.

Schwennesen, A. T. 1917. *Ground Water in San Simon Valley, Arizona and New Mexico*. USGS Water Supply Paper 425A. Washington, D.C.

Sears, P. B. 1947. *Deserts on the March*. Norman: University of Oklahoma Press.

Secretaría de Recursos Hidráulicos. 1954. *Boletín Hidrológico*, no. 11. Mexico, D.F.

———. 1959. *Boletín Hidrológico*, no. 13. Mexico, D.F.

Sellers, W. D. 1960. Precipitation trends in Arizona and western New Mexico. Pages 81–94 in *Proceedings of the 28th Annual Snow Conference, Santa Fe*.

———. 1965. *Physical Climatology*. Chicago: University of Chicago Press.

Sellers, W. D., R. H. Hill, and M. Sanderson-Rae. 1985. *Arizona Climate*. Tucson: University of Arizona Press.

Seymour, D. J. 1989. The dynamics of Sobaipuri settlement in the eastern Pimería Alta. *Journal of the Southwest* 31(2):205–22.

Shantz, H. L. 1905. A study of the vegetation of the mesa region east of Pike's Peak: The *Bouteloua* formation. *Botanical Gazette* 42:16–47, 179–207.

———. 1924. Natural vegetation: Grassland and desert shrub. Pages 15–29 in *U.S. Department of Agriculture, Atlas of American Agriculture*. Washington, D.C.

Shantz, H. L., and R. L. Piemeisel. 1924. Indicator significance of the natural vegetation of the southwestern desert region. *Journal of Agricultural Research* 28:721–801.

Shantz, H. L., and B. L. Turner. 1958. *Vegetational Changes in Africa*. University of Arizona, College of Agriculture Report 169, Tucson.

Shapley, H. 1953. *Climatic Change: Evidence, Causes, and Effects*. Cambridge, Mass.: Harvard University Press.

Shaw, H. 1998. *Wood Plenty, Grass Good, Water None*. Chino Valley, Ariz.: Juniper Institute.

Sheldon, C. 1993. *The Wilderness of the Southwest*. Ed. N. B. Carmony and D. E. Brown. Salt Lake City: University of Utah Press.

Sheps, L. O. 1973. Survival of *Larrea tridentata* S. & M. seedlings in Death Valley National Monument, California. *Israel Journal of Botany* 22:8–17.

Sheridan, T. E. 1995. *Arizona: A History*. Tucson: University of Arizona Press.

Shmida, A., and T. L. Burgess. 1988. Plant growth-form strategies and vegetation types in arid environments. Pages 211–41 in N. J. A. Werger, P. J. M. Van der Aart, H. J. During, and J. T. A. Verhoeven (eds.), *Plant Form and Vegetation Structure*. The Hague: SPB Academic Publishing.

Shreve, F. 1910. The rate of establishment of the giant cactus. *Plant World* 13:235–40.

———. 1911a. Establishment behavior of the palo verde. *Plant World* 14:289–96.

———. 1911b. The influence of low temperatures on the distribution of the giant cactus. *Plant World* 14:136–46.

———. 1912. Cold air drainage. *Plant World* 15:110–15.

———. 1915. *The Vegetation of a Desert Mountain Range as Conditioned by Climatic Factors*. Carnegie Institution of Washington Publication no. 217. Washington, D.C.

———. 1917a. The establishment of desert perennials. *Journal of Ecology* 5:210–16.

———. 1917b. A map of the vegetation of the United States. *Geographical Review* 3:119–25.

———. 1922. Conditions indirectly affecting vertical distribution on desert mountains. *Ecology* 3:269–74.

———. 1925. Ecological aspects of the deserts of California. *Ecology* 6:93–103.

———. 1934a. Rainfall, runoff, and soil moisture under desert conditions. *Annals of the Association of American Geographers* 24:131–56.

———. 1934b. Vegetation of the northwestern coast of Mexico. *Bulletin of the Torrey Botanical Club* 61:373–80.

———. 1935. The longevity of cacti. *Cactus and Succulent Journal (U.S.)* 7:66–68.

———. 1939. Observations on the vegetation of Chihuahua. *Madroño* 5:1–13.

———. 1942a. The desert vegetation of North America. *Botanical Review* 8:195–246.

———. 1942b. Grassland and related vegetation in Northern Mexico. *Madroño* 6:190–98.

———. 1942c. The vegetation of Arizona. Pages 10–23 in T. H. Kearney and R. H. Peebles, *Flowering Plants and Ferns of Arizona*. U.S. Department of Agriculture Miscellaneous Publication 423. Washington, D.C.

———. 1944. Rainfall of northern Mexico. *Ecology* 25:105–11.

———. 1951. *Vegetation of the Sonoran Desert*. Carnegie Institution of Washington Publication no. 591. Washington, D.C.

———. 1964. Vegetation of the Sonoran Desert. Volume 1, Part I, in F. Shreve and I. L. Wiggins, *Vegetation and Flora of the Sonoran Desert*. Stanford, Calif.: Stanford University Press.

Shreve, F., and A. L. Hinckley. 1937. Thirty years of change in desert vegetation. *Ecology* 18:463–78.

Siegel, R. S., and J. H. Brock. 1990. Germination requirements of key southwestern woody riparian species. *Desert Plants* 10:3–8.

Simpson, L. B. 1952. *Exploitation of Land in Central Mexico in the Sixteenth Century*. Ibero-Americana, no. 36. Berkeley: University of California.

Sims, P. L., J. S. Singh, and W. K. Lauenroth. 1978. The structure and function of the ten western North American grasslands. I. Abiotic and vegetational characteristics. *Journal of Ecology* 66:251–85.

Sinclair, J. G. 1922. Temperatures of the soil and air in a desert. *Monthly Weather Review* 50:142–44.

Slauson, L. A. 2000. Pollination biology of two chiropterophilous Agaves in Arizona. *American Journal of Botany* 87:825–36.

Smith, D. A., and E. M. Schmutz. 1975. Vegetative changes on protected versus grazed desert grassland ranges. *Journal of Range Management* 28:453–58.

Smith, K. M., and R. H. Webb. 2001. Researchers consider U.S. Southwest's response to warmer, drier conditions. *EOS* 82(41):475, 478.

Smith, S. D., and P. S. Nobel. 1986. Deserts. Pages 13–62 in N. R. Baker and S. P. Long (eds.), *Photosynthesis in Contrasting Environments*. Amsterdam: Elsevier Publishers.

Smith, W. 1986. *The Effects of Eastern North Pacific Tropical Cyclones on the Southwestern United States*. NOAA Technical Memorandum NWS WR-197. U.S. Department of Commerce, National Oceanic and Atmospheric Administration, Washington, D.C.

Southwestern Stockman. passim. Willcox, Ariz.

Southwest Watershed Hydrology Studies Group. 1958. *Annual Progress Report*. Project Offices, Tucson and Tombstone, Ariz.

Spalding, V. M. 1909. *Distribution and Movements of Desert Plants*. Carnegie Institution of Washington Publication no. 113. Washington, D.C.

Spaulding, W. G. 1990. Vegetational and climatic development of the Mojave Desert: The last glacial maximum to the present. Pages 166–99 in J. L. Betancourt, T. R. Van Devender, and P. S. Martin (eds.), *Packrat Middens: The Last 40,000 Years of Biotic Change*. Tucson: University of Arizona Press.

Spicer, E. H. 1962. *Cycles of Conquest*. Tucson: University of Arizona Press.

Spring, J. 1902. With the regulars in Arizona in the sixties. *Washington National Tribune*, November 20.

———. 1903. Troublous days in Arizona. *Washington National Tribune*, July–October.

Sprugel, D. G. 1991. Disturbance, equilibrium, and environmental variability: What is "natural" vegetation in a changing environment. *Biological Conservation* 58:11–18.

Stamp, L. D. 1961. Some conclusions. Pages 379–88 in L. D. Stamp (ed.), *A History of Land Use in Arid Regions*. Paris: UNESCO.

Standley, P. C. 1920–26. *Trees and Shrubs of Mexico*. 5 parts. Washington, D.C.: GPO.

Steenbergh, W. F., and C. H. Lowe. 1969. Critical factors during the first years of life of the saguaro (*Cereus giganteus*) at Saguaro National Monument, Arizona. *Ecology* 50:825–34.

———. 1976. Ecology of the saguaro, part 1: The role of freezing weather in a warm-desert population. Pages 49–92 in *Research in the Parks*. National Park Service Symposium, Series 1. Washington, D.C.: GPO.

———. 1977. *Ecology of the Saguaro, Part 2: Reproduction, Germination, Establishment, Growth, and Survival of the Young Plant*. National Park Service Scientific Monograph Series no. 8. Washington, D.C.

———. 1983. *Ecology of the Saguaro, Part 3: Growth and Demography*. National Park Service Scientific Monograph Series no. 17. Washington, D.C.

Steinhart, P. 1994. *Two Eagles/Dos Aguilas: The Natural World of the United States–Mexico Borderlands*. Berkeley: University of California Press.

Stephens, H. G., and E. M. Shoemaker. 1987. *In the Footsteps of John Wesley Powell: An Album of Comparative Photographs of the Green and Colorado Rivers, 1871–72 and 1968*. Denver: Johnson Books.

Stevens, R. C. 1964. The Apache menace in Sonora. *Arizona and the West* 6:211–22.

Stewart, O. C. 1951. Burning and natural vegetation in the United States. *Geographical Review* 41:317–20.

———. 1956. Fire as the first great force employed by man. Pages 115–33 in W. L. Thomas et al. (eds.), *Man's Role in Changing the Face of the Earth*. Chicago: University of Chicago Press.

Stromberg, J. C. 1993. Riparian mesquite forests: A review of their ecology, threats, and recovery potential. *Journal of the Arizona-Nevada Academy of Science* 27:111–24.

Sutton, O. G. 1953. *Micrometeorology*. New York: McGraw-Hill.

Swain, C. H. 1893. Report on the mines known as the Old Mowry Mines. Unpublished ms., Arizona Historical Society, Tucson.

Swetnam, T. W. 1990. Fire history and climate in the southwestern United States. Pages 6–17 in J. S. Krammes (technical coordinator), *Effects of Fire Management of Southwestern Natural Resources*. USDA Forest Service General Technical Report RM-GTR-191. Rocky Mountain Forest and Range Experiment Station, Fort Collins, Colo.

Swetnam, T. W., C. D. Allen, and J. L. Betancourt. 1999. Applied historical ecology: Using the past to manage for the future. *Ecological Applications* 64:1189–206.

Swetnam, T. W., and C. H. Baisan. 1996a. Fire histories of montane forests in the Madrean borderlands. Pages 15–36 in P. F. Ffolliott and others (technical coordinators), *Effects of Fire on Madrean Province Ecosystems*. USDA Forest Service General Technical Report RM-GTR-289. Rocky Mountain Forest and Range Experiment Station, Fort Collins, Colo.

———. 1996b. Historical fire regime patterns in the southwestern United States since AD 1700. Pages 11–32 in C. D. Allen (ed.), *Fire Effects in Southwestern Forests, Proceedings of the Second La Mesa Fire Symposium, March 29–31, 1994, Los Alamos, New Mexico*. USDA Forest Service General Technical Report RM-GTR-286. Rocky Mountain Forest and Range Experiment Station, Fort Collins, Colo.

Swetnam, T. W., C. H. Baisan, A. C. Caprio, and P. M. Brown. 1992. Fire history in a Mexican oak-pine woodland and adjacent montane conifer gallery forest in southeastern Arizona. Pages 165–73 in P. F. Ffolliott and others (technical coordinators), *Ecology and Management of Oak and Associated Woodlands: Perspectives in the Southwestern United States and Northern Mexico*. USDA Forest Service General Technical Report RM-GTR-218. Rocky Mountain Forest and Range Experiment Station, Fort Collins, Colo.

Swetnam, T. W., C. H. Baisan, and J. M. Kaib. 2001. Forest fire histories of the Sky Islands of La Frontera. Pages 95–119 in G. L. Webster and C. J. Bahre (eds.), *Changing Plant Life of La Frontera: Observations of Vegetation in the United States/Mexico Borderlands*. Albuquerque: University of New Mexico Press.

Swetnam, T. W., and J. L. Betancourt. 1990. Fire/Southern Oscillation relations in the southwestern United States. *Science* 249:1017–20.

———. 1992. Temporal patterns of El Niño/Southern Oscillation—wildfire teleconnections in the southwestern United States. Pages 259–70 in H. F. Diaz and V. Markgraf (eds.), *El Niño: Historical and Paleoclimatic Aspects of the Southern Oscillation*. Cambridge, Engl.: University of Cambridge Press.

———. 1998. Mesoscale disturbance and ecological response to decadal climatic variability in the American Southwest. *Journal of Climate* 11:3128–47.

Swift, T. T. 1926. Date of channel trenching in the Southwest. *Science* 63:70–71.

Sykes, G. 1931. Rainfall investigations in Arizona and Sonora by means of long-period rain gauges. *Geographical Review* 21:229–33.

Szuter, S. R. 1991. *Hunting by Prehistoric Horticulturists in the American Southwest*. New York: Garland Press.

Talbot, W. J. 1961. Land utilization in the arid regions of southern Africa. Part I: South Africa. Pages 299–331 in L. D. Stamp (ed.), *A History of Land Use in Arid Regions*. Paris: UNESCO.

Tamarón y Romeral, P. 1937. *Demostración del Vastísimo Obispado de la Nueva Vizcaya, 1765*. Mexico, D.F.: Antigua Librería Robredo de José Porrua e Hijos.

Taylor, W. P. 1936. Some effects of animals on plants. *Scientific Monthly* 43:262–71.

Taylor, W. P., C. T. Vorhies, and P. B. Lister. 1935. The relation of jack rabbits to grazing in southern Arizona. *Journal of Forestry* 33:490–98.

Tevis, J. H. 1954. *Arizona in the 50's*. Albuquerque: University of New Mexico Press.

Thomas, A. B. 1932. *Forgotten Frontiers: A Study of the Spanish Indian Policy of Don Juan Bautista de Anza, Governor of New Mexico, 1777–1787*. Norman: University of Oklahoma Press.

———. 1933. A description of Sonora in 1772. *Arizona Historical Review* 5:302–7.

———. 1941. *Teodoro de Croix and the Northern Frontier of New Spain, 1776–1783*. Norman: University of Oklahoma Press.

Thompson, R. S., S. W. Hostetler, P. J. Bartlein, and K. H. Anderson. 1998. *A Strategy for Assessing Potential Future Changes in Climate, Hydrology, and Vegetation in the Western United States*. U.S. Geological Survey Circular 1153. Reston, Va.

Thornber, J. J. 1910. *The Grazing Ranges of Arizona*. University of Arizona Agricultural Experiment Station Bulletin 65, Tucson.

Thornthwaite, C. W. 1948. An approach toward a rational classification of climate. *Geographical Review* 38:55–94.

Thornthwaite, C. W., C. F. S. Sharpe, and E. F. Dosch. 1942. *Climate and Accelerated Erosion in the Arid and Semi-Arid Southwest, with Special Reference to the Polacca Wash Drainage Basin, Arizona*. USDA Technical Bulletin 808. Washington, D.C.

Thwaites, R. G. (ed.). 1905. *Expedition from Pittsburgh to the Rocky Mountains*. Vol. 17, Early Western Travels. Cleveland: Arthur H. Clark.

Tolman, C. F. 1909. The geology of the vicinity of the Tumamoc Hills. Pages 67–82 in V. M. Spalding, *Distribution and Movements of Desert Plants*. Carnegie Institution of Washington Publication no. 113. Washington, D.C.

The Tombstone, 1885. Tombstone, Ariz.

Tombstone Daily Epitaph. passim. Tombstone, Ariz.

Trewartha, G. T. 1954. *An Introduction to Climate*. New York: McGraw-Hill Book Co.

Tschirley, F. H. 1963. A physio-ecological study of jumping cholla (*Opuntia fulgida* Engelm.). Ph.D. diss., University of Arizona, Tucson.

Tschirley, F. H., and R. F. Wagle. 1964. Growth rate and population dynamics of jumping cholla (*Opuntia fulgida* Engelm.). *Journal of the Arizona Academy of Science* 3:67–71.

Tuan, Y. 1966. New Mexican gullies: A critical review and some recent observations. *Annals of the Association of American Geographers* 56(4): 573–97.

Turnage, W. V., and A. L. Hinckley. 1938. Freezing weather in relation to plant distribution in the Sonoran Desert. *Ecological Monographs* 8:529–50.

Turnage, W. V., and T. D. Mallery. 1941. *An Analysis of Rainfall in the Sonoran Desert and Adjacent Territory*. Carnegie Institution of Washington Publication no. 529. Washington, D.C.

Turner, R. M. 1963. Growth in four species of Sonoran Desert trees. *Ecology* 44:760–65.

———. 1974. *Quantitative and Historical Evidence of Vegetation Changes along the Upper Gila River, Arizona*. U.S. Geological Survey Professional Paper 655-H. Washington, D.C.: GPO.

———. 1990. Long-term vegetation change at a fully protected Sonoran Desert site. *Ecology* 71(2):464–77.

———. 1992. Long-term saguaro population studies at Saguaro National Monument. Pages 3–11 in C. P. Stone and E. S. Bellantoni (eds.), *Proceedings of the Symposium on Research in Saguaro National Monument*. Globe, Ariz.: Southwest Parks and Monuments Association.

———. 1994a. Great Basin Desertscrub. Pages 145–55 in D. E. Brown (ed.), *Biotic Communities: Southwestern United States and Northwestern Mexico*. Salt Lake City: University of Utah Press.

———. 1994b. Mohave Desertscrub. Pages 157–68 in D. E. Brown (ed.), *Biotic Communities: Southwestern United States and Northwestern Mexico*. Salt Lake City: University of Utah Press.

Turner, R. M., S. M. Alcorn, and G. Olin. 1969. Mortality of transplanted saguaro seedlings. *Ecology* 50 (5):835–44.

Turner, R. M., S. M. Alcorn, G. Olin, and J. A. Booth. 1966. The effect of shade, soil, and water on saguaro seedling establishment. *The Botanical Gazette* 127(2–3):95–102.

Turner, R. M., J. E. Bowers, and T. L. Burgess. 1995. *Sonoran Desert Plants: An Ecological Atlas*. Tucson: University of Arizona Press.

Turner, R. M., and D. E. Brown. 1994. Sonoran Desertscrub. Pages 181–221 in D. E. Brown (ed.), *Biotic Communities: Southwestern United States and Northwestern Mexico*. Salt Lake City: University of Utah Press.

Turner, R. M., and M. M. Karpiscak. 1980. *Recent Vegetation Changes along the Colorado River Between Glen Canyon Dam and Lake Mead, Arizona*. U.S. Geological Survey Professional Paper 1132. Washington, D.C.

Turner, R. M., H. A. Ochung', and J. B. Turner. 1998. *Kenya's Changing Landscape*. Tucson: University of Arizona Press.

Undreiner, G. J. 1947. Fray Marcos de Niza and his journey to Cibola. *The Americas* 3:415–86.

U.S. Bureau of the Census. 1872. Ninth Census of the United States: 1870.

———. 1883. Tenth Census of the United States: 1880.

———. 1895. Eleventh Census of the United States: 1890.

U.S. Congress, House. 1859. Executive Document 108, 35th Cong., 2d Sess. Washington, D.C.

U.S. Congress, Senate. 1852. Executive Document 121, 32d Cong., 1st Sess. Washington, D.C.

———. 1880. Executive Document 207, 46th Cong., 2d Sess. Washington, D.C.

———. 1898. *Report of the Boundary Commission upon the Survey and Re-Marking of the Boundary between the United States and Mexico West of the Rio Grande, 1891 to 1896*. Senate Document 247, 55th Cong., 2d Sess. Washington, D.C.

U.S. Department of Commerce, Weather Bureau. 1965. Substation history, Arizona. Washington, D.C.

Van Devender, T. R. 1990. Late quaternary vegetation and climate of the Sonoran Desert: United States and Mexico. Pages 134–63 in J. L. Betancourt, T. R. Van Devender, and P. S. Martin (eds.), *Packrat Middens: The Last 40,000 Years of Biotic Change*. Tucson: University of Arizona Press.

———. 1995. Desert grassland history: Changing climates, evolution, biogeography, and community dynamics. Pages 68–99 in M. P. McClaran and T. R. Van Devender (eds.), *The Desert Grassland*. Tucson: University of Arizona Press.

Van Devender, T. R., J. I. Mead, and A. M. Rea. 1991. Late Quaternary plants and vertebrates from Picacho Peak, Arizona. *Southwestern Naturalist* 36:302–14.

Van Devender, T. R., and W. G. Spaulding. 1986. Development of vegetation and climate in the southwestern United States. Pages 131–56 in S. G. Wells and D. R. Haragan (eds.), *Origin and Evolution of Deserts*. Albuquerque: University of New Mexico Press.

Vasek, F. C. 1979/80. Early successional stages in Mojave Desert scrub vegetation. *Israel Journal of Botany* 28: 133–48.

———. 1980. Creosote bush: Long-lived clones in the Mojave Desert. *American Journal of Botany* 67:246–55.

Velasco, J. F. 1850. *Noticias Estadísticas del Estado de Sonora*. Mexico, D.F.: Imprenta de Ignacio Cumplido.

Veryard, R. G. 1963. A review of studies on climatic fluctuations during the period of the meteorological record. Pages 3–16 in *Changes of Climate*. Paris: UNESCO.

Villanueva Diaz, José. 1996. Influence of land-use and climate on soils and forest structure in mountains of the southwestern United States and northern Mexico. Ph.D. diss., University of Arizona, Tucson.

Vivó, J. A., and J. C. Gómez. 1946. *Climatología de México*. Mexico, D.F.: Instituto Panamericano de Geografía Historia.

von Eschen, G. F. 1958. Climatic trends in New Mexico. *Weatherwise* 11:191–95.

Vorhies, C. T., and W. P. Taylor. 1933. *The Life Histories and Ecology of Jack Rabbits*, Lepus alleni *and* Lepus californicus *ssp., in Relation to Grazing in Arizona*. University of Arizona Agricultural Experiment Station Technical Bulletin Number 49, 467–587, Tucson.

Wagoner, J. J. 1951. Development of the cattle industry in southern Arizona, 1870's and 80's. *New Mexico Historical Review* 26:204–24.

———. 1952. *History of the Cattle Industry in Southern Arizona, 1540–1940*. University of Arizona Social Science Bulletin no. 20, Tucson.

Wallén, C. C. 1955. Some characteristics of precipitation in Mexico. *Geografiska Annaler* 37:51–85.

Wallmo, O. C. 1955. Vegetation of the Huachuca Mountains, Arizona. *American Midland Naturalist* 54:466–80.

Waters, M. R. 1985. Late Quaternary alluvial stratigraphy of Whitewater Draw, Arizona: Implications for regional correlation of fluvial deposits in the American Southwest. *Geology* 13:705–8.

———. 1988. Holocene alluvial geology and geoarchaeology of the San Xavier reach of the Santa Cruz River, Arizona. *Geological Society of America Bulletin* 100: 479–91.

———. 1992. *Principles of Geoarchaeology, a North American Perspective*. Tucson: University of Arizona Press.

Waters, M. R., and C. V. Haynes. 2001. Late Quaternary arroyo formation and climate change in the American Southwest. *Geology* 29(5):399–402.

Weaver, J. E. 1954. *North American Prairie*. Lincoln, Neb.: Johnsen Publishing Co.

Webb, R. H. 1985. Late Holocene flooding on the Escalante River, south-central Utah. Ph.D. diss., University of Arizona, Tucson.

———. 1996. *Grand Canyon: A Century of Change*. Tucson: University of Arizona Press.

Webb, R. H., and V. R. Baker. 1987. Changes in hydrologic conditions related to large floods on the Escalante River, south-central Utah. Pages 196–204 in V. Singh (ed.), *Regional Flood-Frequency Analysis*. Dordrecht, The Netherlands: D. Reidel Publishers.

Webb, R. H., and J. L. Betancourt. 1992. *Climatic Variability and Flood Frequency of the Santa Cruz River, Pima County, Arizona*. USGS Water-Supply Paper 2379. Reston, Va.

Webb, R. H., and J. E. Bowers. 1993. Changes in frost frequency and desert vegetation assemblages in Grand Canyon, Arizona. Pages 71–82 in K. T. Redmond (ed.), *Proceedings of the Ninth Annual Pacific Climate (PACLIM) Workshop, April 21–24, 1992*. California Department of Water Resources, Interagency Ecological Studies Program Technical Report 34. Sacramento, Calif.

Webb, R. H., S. S. Smith, and V. A. S. McCord. 1991. *Historic Channel Change of Kanab Creek, Southern Utah and Northern Arizona*. Monograph no. 9. Grand Canyon, Ariz.: Grand Canyon Natural History Association.

Webb, R. H., J. W. Steiger, and E. B. Newman. 1988. *The Response of Vegetation to Disturbance in Death Valley National Monument, California*. U.S. Geological Survey Bulletin no. 1793. Washington, D.C.

Webb, W. P. 1931. *The Great Plains*. New York: Ginn and Co.

———. 1957. The American West—perpetual mirage. *Harper's Magazine* 214:25–31.

Weber, D. 1992. *The Spanish Frontier in North America*. New Haven, Conn.: Yale University Press.

Weekly Arizonan, 1859. Tubac, Ariz.

Weltzin, J. E., S. Archer, and R. K. Heitschmidt. 1997. Small mammal regulation of vegetation structure in a temperate savanna. *Ecology* 78:751–63.

Weltzin, J. E., and G. R. McPherson. 1995. Potential effects of climate change on lower treelines in the southwestern United States. Pages 180–93 in L. F. De-Bano et al. (technical coordinators), *Biodiversity and Management of Madrean Archipelago: The Sky Islands of Southwestern United States and Northwestern Mexico*.

USDA Forest Service General Technical Report RM-GTR-264. Rocky Mountain Forest and Range Experiment Station, Fort Collins, Colo.

Went, F. W. 1949. Ecology of desert plants, II. The effect of rain and temperature on germination and growth. *Ecology* 30:1–13.

———. 1957. *The Experimental Control of Plant Growth.* Waltham, Mass.: Chronica Botanica.

Went, F. W., and M. Westergaard. 1949. Ecology of desert plants, III. Development of plants in the Death Valley National Monument, California. *Ecology* 30: 26–38.

West, R. C. 1949. *The Mining Community in Northern New Spain.* Ibero-Americana, no. 30. Berkeley: University of California Press.

White, L. D. 1969. Effects of a wildfire on several desert grassland shrub species. *Journal of Range Management* 22:284–85.

White, S. S. 1948. The vegetation and flora of the region of the Rio de Bavispe in northeastern Sonora, Mexico. *Lloydia* 11:229–302.

Whitfield, C. J., and H. L. Anderson. 1938. Secondary succession in the desert plains grassland. *Ecology* 19: 171–80.

Whitfield, C. J., and E. L. Beutner. 1938. Natural vegetation in the desert plains grassland. *Ecology* 19:26–37.

Whittaker, R. H., and W. A. Niering. 1964. Vegetation of the Santa Catalina Mountains, Arizona. I. Ecological classification and distribution of species. *Journal of the Arizona Academy of Science* 3:9–34.

———. 1968. Vegetation of the Santa Catalina Mountains, Arizona. IV. Limestone and acid soils. *Journal of Ecology* 56:523–44.

Whyte, R. O. 1963. The significance of climatic change for natural vegetation and agriculture. Pages 381–86 in *Changes of Climate.* Paris: UNESCO.

Wiggins, I. L. 1964. Flora of the Sonoran Desert. In F. Shreve and I. L. Wiggins, *Vegetation and Flora of the Sonoran Desert.* 2 vols. Stanford, Calif.: Stanford University Press.

Wilcox, D. 1981. The entry of the Athapaskans into the American Southwest: The problem today. Pages 213–56 in D. Wilcox and B. Masse (eds.), *The Protohistoric Period in the North American Southwest, AD 1540–1700.* Anthropological Research Papers, no. 24. Tempe: Arizona State University.

Willett, H. C. 1950. Temperature trends of the past century. Pages 195–206 in *Centenary Proceedings of the Royal Meteorological Society.* London: Meteorological Society.

Wilson, T. B. 2001. Nutrient dynamics and fire history in mesquite (*Prosopis* spp.)–dominated desert grasslands of the southwestern United States. Ph.D. diss., University of Arizona, Tucson.

Wilson, T. B., R. H. Webb, and T. L. Thompson. 2001. *Mechanisms of Range Expansion and Removal of Mesquite* (Prosopis *spp.) in Desert Grasslands in the Southwestern United States.* USDA Forest Service General Technical Report RMS-GTR-81. Rocky Mountain Research Station, Fort Collins, Colo.

Winn, F. 1926. The West Fork of the Gila River. *Science* 64:16–17.

Woolhiser, D. A., T. O. Keefer, and K. T. Redmond. 1993. Southern Oscillation effects of daily precipitation in the southwestern United States. *Water Resources Research* 29:1287–95.

Worchester, D. E. 1941. The beginnings of the Apache menace of the Southwest. *New Mexico Historical Review* 16:1–14.

Wright, H. A., and A. W. Bailey. 1982. *Fire Ecology.* New York: John Wiley and Sons.

Wright, R. A. 1982. Aspects of desertification in *Prosopis* dunelands of southern New Mexico, U.S.A. *Journal of Arid Environments* 5:277–84.

Wright, R. G., and S. C. Bunting. 1994. *The Landscapes of Craters of the Moon National Monument: An Evaluation of Environmental Changes.* Moscow: University of Idaho Press.

Wyllys, R. K. 1931. Padre Luis Velarde's *Relación* of Pimería Alta, 1716. *New Mexico Historical Review* 6:111–57.

Yang, T. W., and C. H. Lowe, Jr. 1956. Correlation of major vegetation climaxes with soil characteristics in the Sonoran Desert. *Science* 123:542.

Young, F. D. 1921. Nocturnal temperature inversions in Oregon and California. *Monthly Weather Review* 49: 138–48.

Zedler, P. H. 1981. Vegetation change in chaparral and desert communities in San Diego County, California. Pages 406–30 in D. C. West, H. H. Shugart, and D. B. Botkin (eds.), *Forest Succession.* New York: Springer-Verlag.

Zhang, Y., J. M. Wallace, and D. S. Battati. 1997. ENSO-like interdecadal variability: 1900–1993. *Journal of Climate* 10:1004–20.

Zohary, M. 1962. *Plant Life of Palestine.* New York: Ronald Press.

Zuñiga, I. 1835. *Rápida Ojeada al Estado de Sonora.* Mexico: Juan Ojeda.

About the Authors

Ray Turner has been studying the dynamics of desert vegetation in southern Arizona and northern Mexico since shortly after his arrival in Tucson in 1954. A graduate of the University of Utah (B.S. 1948: botany) and Washington State University (Ph.D. 1954: botany), he taught at the University of Arizona (1954–62) before joining the U.S. Geological Survey.

Turner's early interest in vegetation dynamics led to authorship, with Rod Hastings, of the original edition of *The Changing Mile* (University of Arizona Press, 1965). He is also author of publications describing changes in riparian vegetation along the Gila and Colorado Rivers as well as changes in permanent vegetation study plots at the Desert Laboratory, Tucson, Arizona, and in MacDougal Crater, Pinacate Preserve, Sonora, Mexico.

Retired (by all administrative accounts) since 1989, he has subsequently coauthored two books, *Sonoran Desert Plants: An Ecological Atlas* and *Kenya's Changing Landscape* (both published by the University of Arizona Press, 1995 and 1998, respectively), and articles describing changes through time in saguaro and foothill paloverde populations at the Desert Laboratory.

Robert Webb has worked on long-term changes in natural ecosystems of the southwestern United States since 1976. He has degrees in engineering (B.S., University of Redlands, 1978), environmental earth sciences (M.S., Stanford University, 1980), and geosciences (Ph.D., University of Arizona, 1985). Since 1985, he has been a research hydrologist with the U.S. Geological Survey in Tucson and an adjunct faculty member of the Departments of Geosciences and Hydrology and Water Resources at the University of Arizona.

Webb does interdisciplinary work merging history, climate change, desert vegetation ecology, hydrology, geomorphology, and Quaternary geology to attempt to paint a picture of long-term change in the desert regions of the United States and Mexico. His interests lie in quantifying whether long-term ecosystem changes are induced by human land-use practices or by environmental factors. Webb has authored or edited eight books, including *Grand Canyon, A Century of Change* (1996) and *Floods, Droughts, and Changing Climates* (with Michael Collier, 2002) for the University of Arizona Press. His current research deals with long-term changes in riparian vegetation in the Southwest, mortality and recruitment rates for common desert plants, carbon sequestration in the ecosystems of the Mojave and Sonoran Deserts, climatic effects on flood frequency and stream channels throughout the region, and geomorphology of debris flows in Grand Canyon and other canyon systems of the Colorado Plateau. Repeat photography is one of the common techniques used in all of these studies.

Janice E. Bowers, a botanist with the U.S. Geological Survey, has lived and worked in the *Changing Mile* region since graduating from the University of Arizona with a B.S. in botany in 1976. Her research interests include life history of woody plants, reproductive biology of cacti, plant diversity at the landscape scale, and dynamics of plant populations in relation to climatic variability. She is the author of *A Sense of Place: The Life and Work of Forrest Shreve* (University of Arizona Press, 1988) and other historical accounts of ecological research in the Sonoran Desert. With Raymond M. Turner and Tony L. Burgess, she wrote *Sonoran Desert Plants: An Ecological Atlas* (University of Arizona Press, 1995), an encyclopedia describing the ecology and distribution of more than 300 woody plants. Among her other books are several essay collections published by the University of Arizona Press and two wildflower guides.

James Rodney Hastings was born in 1923 in Hayden, Arizona, where he lived until he reached college age. He attended the University of Chicago, where work on a bachelor's degree was interrupted by a call from the U.S. Army. Returning to Chicago after three and a half years in the military, he completed the bachelor's degree in chemistry and English in 1948. In 1952, he received a second B.A. degree from the University of Arizona. Following a series of teaching jobs, he returned to his hometown as a high school teacher. During this period, he became active in a drive to eliminate tax inequities between the twin mining communities of Winkelman and Hayden. As an outgrowth of the taxation drive, Hayden was incorporated, and in 1957 Hastings became the town's first mayor. During his tenure as mayor, a latent interest in the region's history was rekindled, and he enrolled in a Ph.D. program in history at the University of Arizona, completing that degree in 1963.

Hastings's main interest—human impacts on the region's landscape during the colonial era—

took a new tack when he decided that the standard use of old written accounts for reconstructing the region's past could be significantly strengthened by augmenting those accounts with new photographs taken at the reoccupied positions of old landscape photographs. Aided by an appointment with the University of Arizona's new Institute of Atmospheric Physics, he initiated an ambitious rephotography effort. That effort culminated in publication of *The Changing Mile* (University of Arizona Press) in 1965, and ultimately led to the establishment of the Desert Laboratory's extensive repeat photography archive. Hastings's broad interest resulted in articles on arroyo cutting, climatology, saguaro growth, and plant geography, as well as history. At the time of his death in 1974, he was a professor in the Atmospheric Sciences Department, University of Arizona.

Index

149 (pl. 54), *156–157* (pl. 58), *190–191* (pl. 71), 248, *250* (fig. 7.1), 264. *See also* frost damage; temperature

catastrophic, 12

distribution, 12–13

frequency and intensity, 13–15, *14* (fig. 1.3)

prickly pear and, 253

frost damage, *56–57* (pl. 10), *124–125* (pl. 42), *128–129* (pl. 44), *134–135* (pl. 47), *138–139* (pl. 49), *142–143* (pl. 51), *148–149* (pl. 54), *152–153* (pl. 56), *156–157* (pl. 58), *180–181* (pl. 66), *190–191* (pl. 71), 248, 300n. 15

frost-free zone, 12–13

frost sensitivity, of brittlebush, 255

fuelwood, *38–39* (pl. 1), *40–41* (pl. 2), *62–63* (pl. 13), *64–65* (pl. 14), *102–103* (pl. 31), *114–115* (pl. 37), 295n. 3. *See also* woodcutting

G

Gadsden Purchase, 26, 29

gallery forest, 36, *136–137* (pl. 48), *188–189* (pl. 70), *190–191* (pl. 71), 262

Gálvez, José de, 22

general circulation pattern, 6–7

Gentry, H. S., 292n. 67

germination, 11–12

of brittlebush, 255

of Chihuahuan whitethorn, 253

of cottonwood, 254

of creosote bush, 12, 253, 304n. 118

of foothill paloverde, 250

of mesquite, 248

of prickly pear, 301n. 42

of saguaro, 252

of velvet mesquite, 300n. 8, 304n. 118

giant cactus, *200–201* (pl. 76)

Gila River Basin, 266

Gird Dam, *138–139* (pl. 49), *140–141* (pl. 50)

Glendening, G. E., 300n. 8

global teleconnections, 6–7

global warming, 263–266, *264* (fig. 8.2)

Golder Dam, *192–193* (pl. 72)

Government Draw, *132–133* (pl. 46), *134–135* (pl. 47)

grama, *82–83* (pl. 23), *116–117* (pl. 38), *200–201* (pl. 76)

black, 92, *128–129* (pl. 44), *142–143* (pl. 51), *152–153* (pl. 56), 268

blue, 92, *94–95* (pl. 27), *124–125* (pl. 42)

hairy, *124–125* (pl. 42)

Rothrock, *140–141* (pl. 50)

sideoats, *42–43* (pl. 3), *48–49* (pl. 6), *80–81* (pl. 22), *108–109* (pl. 34), *110–111* (pl. 35), *140–141* (pl. 50), *142–143* (pl. 51)

sprucetop, *108–109* (pl. 34), *128–129* (pl. 44)

Gran Desierto, *222–223* (pl. 87)

grasses, desert, 164. *See also names of species*

grassland, 16, 20, 28–29, *74–75* (pl. 19), *80–81* (pl. 22), *82–83* (pl. 23), *88–89* (pl. 26), 160, 262. *See also* Plains Grassland; Semidesert Grassland

cattle and, 258–259

in early accounts, 28–29

gray thorn, *82–83* (pl. 23), *102–103* (pl. 31), *112–113* (pl. 36), *122–123* (pl. 41), *152–153* (pl. 56), *156–157* (pl. 58), 160, *170–171* (pl. 61), *174–175* (pl. 63), *196–197* (pl. 74)

grazing, 23, *46–47* (pl. 5), *70–71* (pl. 17), *88–89* (pl. 26), *132–133* (pl. 46), *210–211* (pl. 81), 253–254, 261. *See also* overgrazing

and arroyo cutting, 32–33, 261

and vegetation change, 258–259, 276

grazing rights, 30

Great Basin Desert, 3

Gregory, H. E., 33–34

Griffiths, David, xiii, *116–117* (pl. 38)

Grosvenor Hills, *66–67* (pl. 15), *122–123* (pl. 41)

growing season, 264

growth habits, of desert plants, 162–163

Guajalote Peak, *86–87* (pl. 25), *108–109* (pl. 34)

Guaymas, Mexico, *242–243* (pl. 97), *244–245* (pl. 98)

Guevavi Canyon, *70–71* (pl. 17)

Guzmán, Nuño de, 21

H

Hacienda Santa Rita, *106–107* (pl. 33)

Hack, J. T., 32

hackberry

desert, *72–73* (pl. 18), *74–75* (pl. 19), *102–103* (pl. 31), *112–113* (pl. 36), *170–171* (pl. 61), *172–173* (pl. 62), *174–175* (pl. 63), *182–183* (pl. 67), *184–185* (pl. 68), *188–189* (pl. 70), *190–191* (pl. 71), *194–195* (pl. 73)

netleaf, *112–113* (pl. 36)

Hastings, J. R., xiv, 91, 299n. 11

Helvetia, *100–105* (pls. 30–32)

Hereford, R., 296n. 27

hibiscus

Coulter, *196–197* (pl. 74)

desert, *196–197* (pl. 74), *200–201* (pl. 76), *240–241* (pl. 96)

Hinckley, A. L., 13, 15

Hirschboeck, K. K., 7

hopbush, *116–117* (pl. 38), *188–189* (pl. 70), *194–195* (pl. 73)

Hornaday Mountains, *214–215* (pl. 83), *216–217* (pl. 84), *220–221* (pl. 86)

horse latitudes, 3

housing development, *70–71* (pl. 17), *76–77* (pl. 20), *100–101* (pl. 30), *110–111* (pl. 35), *120–121* (pl. 40), *126–127* (pl. 43), *128–129* (pl. 44), *130–131* (pl. 45), *144–145* (pl. 52), *168–169* (pl. 60), *184–185* (pl. 68), 274–276. *See also* urbanization

Howell, D. J., 301n. 70

Huachuca Mountains, *126–127* (pl. 43), *128–129* (pl. 44), *130–131* (pl. 45), *142–143* (pl. 51), *152–153* (pl. 56)

Hubbs, C. L., 304n. 112

Huerfano Butte, *100–101* (pl. 30)

Hull, H. M., 300n. 8

human impact, xv–xvi, *64–65* (pl. 14), 90, *242–243* (pl. 97), 250, 274. *See also* housing development; road building; urbanization

Huntington, E., 297n. 105

and precipitation change, 268–274

Texas honey, 300n. 2

velvet, *58–59* (pl. 11), *70–71* (pl. 17), *88–89* (pl. 26), *124–125* (pl. 42), *130–131* (pl. 45), *138–139* (pl. 49), *168–169* (pl. 60), *170–171* (pl. 61), *172–173* (pl. 62), *174–175* (pl. 63), *180–181* (pl. 66), *188–189* (pl. 70), *190–191* (pl. 71), 260, 268, 300n. 2, 304n. 118

western honey, *228–229* (pl. 90), 300n. 2

mesquite bosque, *122–123* (pl. 41), *132–133* (pl. 46)

mesquite invasion, xiii, 20, 28, *40–41* (pl. 2), *46–47* (pl. 5), *50–51* (pl. 7), *52–53* (pl. 8), *56–57* (pl. 10), *60–61* (pl. 12), *62–63* (pl. 13), *64–65* (pl. 14), *66–67* (pl. 15), *68–69* (pl. 16), *70–71* (pl. 17), *72–73* (pl. 18), *74–75* (pl. 19), *76–77* (pl. 20), *78–79* (pl. 21), *82–83* (pl. 23), *84–85* (pl. 24), *86–87* (pl. 25), *88–89* (pl. 26), 90–91, *96–97* (pl. 28), *102–103* (pl. 31), *104–105* (pl. 32), *108–109* (pl. 34), *116–117* (pl. 38), *120–121* (pl. 40), *126–127* (pl. 43), *128–129* (pl. 44), *134–135* (pl. 47), *136–137* (pl. 48), *146–147* (pl. 53), *148–149* (pl. 54), *150–151* (pl. 55), 160, 248, 257–259, 295n. 99, 296n. 33, 304n. 119

mesquite mortality, *80–81* (pl. 22), *84–85* (pl. 24), *98–99* (pl. 29), *148–149* (pl. 54), *150–151* (pl. 55), *152–153* (pl. 56), *156–157* (pl. 58)

meteorology, synoptic, of rainfall regimes, 5–6

Mexican crucillo, *102–103* (pl. 31), *104–105* (pl. 32), *110–111* (pl. 35), *112–113* (pl. 36), *146–147* (pl. 53), *170–171* (pl. 61), *174–175* (pl. 63)

Mexican devilweed, *138–139* (pl. 49)

Mexicans, 23–25

Mexican tea, *124–125* (pl. 42), *134–135* (pl. 47), *146–147* (pl. 53), *148–149* (pl. 54), *184–185* (pl. 68)

Mexico, camera stations in, *165* (fig. 6.3)

mills, *142–143* (pl. 51), 302n. 6

    Corbin Mill, *140–141* (pl. 50)

    Gird Mill, *140–141* (pl. 50), *142–143* (pl. 51)

Millville, *140–141* (pl. 50)

mines and mining, 23, *62–63* (pl. 13), *100–101* (pl. 30), *106–107* (pl. 33), *114–115* (pl. 37), *118–119* (pl. 39), *138–139* (pl. 49), *142–143* (pl. 51), *204–205* (pl. 78), 256–257. *See also* woodcutting

    Arizona Mining Company, *118–119* (pl. 39)

    Cerro Colorado Mine, *118–119* (pl. 39)

    El Plomo Mine, *38–43* (pls. 1–3)

    Hacienda Santa Rita mine, *106–107* (pl. 33)

    Helena Mine, *114–115* (pl. 37)

    Helvetia mining camp, *100–101* (pl. 30), *102–103* (pl. 31), *104–105* (pl. 32)

    Salero Mine, *62–63* (pl. 13)

    Santa Rita Mining Company, 29

missions, 21–22, *120–121* (pl. 40)

Mitchell, J. M., Jr., 303n. 96

Mojave Desert, 3, 5–6, 164

Monkey Canyon, *54–55* (pl. 9)

Monkey Spring, *58–59* (pl. 11)

Montezuma Peak, *126–127* (pl. 43)

Moore, R. C., 34

Mormon tea, *134–135* (pl. 47), *170–171* (pl. 61)

mortality, 256. *See also* mesquite mortality; oak mortality; saguaro decline

mortonia, *100–101* (pl. 30), *104–105* (pl. 32), *144–145* (pl. 52)

Mount Benedict, *76–77* (pl. 20)

Mount Fagan, *112–113* (pl. 36)

Mule Mountains, *154–155* (pl. 57)

N

Native Americans. *See* American Indians

natural gas pipeline, *56–57* (pl. 10)

Neilson, R. P., 292n. 52

Nentvig, J., 22

New Mexico, 266

Nichol, A. A., 299n. 32

Niza, Fray Marcos de, 21

Nogales, Arizona, *78–79* (pl. 21)

Nogales, Sonora, *64–65* (pl. 14)

nurse plants, saguaro and, *226–227* (pl. 89), 249–250, 300nn. 22, 23

O

oak establishment, *38–39* (pl. 1)

oak mortality, *44–45* (pl. 4), *60–61* (pl. 12), *64–65* (pl. 14), *66–67* (pl. 15), *68–69* (pl. 16), *70–71* (pl. 17), *72–73* (pl. 18), *84–85* (pl. 24), *86–87* (pl. 25), 90, *98–99* (pl. 29), *116–117* (pl. 38), *138–139* (pl. 49), 254, 257, 268–274, 302n. 8

oaks, *54–55* (pl. 9), *76–77* (pl. 20), *190–191* (pl. 71), 254, 260, 298n. 8

    Arizona white oak, 36, *48–49* (pl. 6), *50–51* (pl. 7), *98–99* (pl. 29), *102–103* (pl. 31)

    Emory oak, 36, *38–39* (pl. 1), *40–41* (pl. 2), *42–43* (pl. 3), *44–45* (pl. 4), *46–47* (pl. 5), *48–49* (pl. 6), *52–53* (pl. 8), *54–55* (pl. 9), *60–61* (pl. 12), *64–65* (pl. 14), *76–77* (pl. 20), *78–79* (pl. 21), *86–87* (pl. 25), *88–89* (pl. 26), 90, *100–101* (pl. 30), *102–103* (pl. 31), *138–139* (pl. 49), 260, 298n. 24

    Mexican blue oak, 36, *42–43* (pl. 3), *44–45* (pl. 4), *46–47* (pl. 5), *54–55* (pl. 9), *60–61* (pl. 12), *62–63* (pl. 13), *72–73* (pl. 18), *76–77* (pl. 20), *78–79* (pl. 21), *88–89* (pl. 26), *98–99* (pl. 29), *110–111* (pl. 35), 260

    pattern of change, 254

    Toumey oak, 36, *54–55* (pl. 9), *96–97* (pl. 28)

Oak Woodland, 16, 36, *38–89* (pls. 1–26), 90–91, *96–97* (pl. 28), *126–127* (pl. 43), *138–139* (pl. 49), 163, 248, 253–254, 256, *269* (figs. 8.7 and 8.8)

    changes in, 90, 298n. 32

    fire and, 259–260

    and precipitation change, 267–274

ocotillo, *46–47* (pl. 5), *56–57* (pl. 10), *60–61* (pl. 12), *62–63* (pl. 13), *66–67* (pl. 15), *68–69* (pl. 16), *72–73* (pl. 18), *74–75* (pl. 19), *76–77* (pl. 20), 90, *100–101* (pl. 30), *104–105* (pl. 32), *108–109* (pl. 34), *110–111* (pl. 35), *112–113* (pl. 36), *114–115* (pl. 37), *116–117* (pl. 38), *118–119* (pl. 39), *136–137* (pl. 48), *138–139* (pl. 49), *144–145* (pl. 52), 160, 162, 164, *178–179* (pl. 65), *180–181* (pl. 66), *190–*

ranches and ranching, 29, *52–53* (pl. 8)
    Appleton-Whittell Research Ranch, *94–95* (pl. 27)
    Babocomari ranch, 24
    Rancho Punta de Cirios, *240–241* (pl. 96)
    Salero Ranch, *38–39* (pl. 1)
    Yerba Buena Ranch, *82–83* (pl. 23)
ratany, *210–211* (pl. 81), 253
    range, *196–197* (pl. 74)
    white, *196–197* (pl. 74), *202–203* (pl. 77), *206–207*
        (pl. 79)
Reagan, A. B., 297n. 105
red brome, *176–177* (pl. 64)
Red Rock Canyon, *96–97* (pl. 28), *98–99* (pl. 29)
Redrock Creek, *98–99* (pl. 29)
Reeves, R. W., 33, 266
reproduction
    of jumping cholla, 254
    of mescal, 254
resource competition, oak-mesquite, 91
restoration of landscape, 261, 277
Rhus
    Rhus choriophylla, 58–59 (pl. 11), *100–101* (pl. 30)
    Rhus microphylla, 146–147 (pl. 53)
riparian forest, grassland and, 262
river decline, 262
rivers, early descriptions of, 26–28. *See also* San Pedro
    River; Santa Cruz River
road building, 33, *110–111* (pl. 35), *120–121* (pl. 40),
    122–123 (pl. 41), *178–179* (pl. 65), *186–187*
    (pl. 69), *196–197* (pl. 74), *212–213* (pl. 82), *232–*
    233 (pl. 92), 302n. 7
Robles Pass, *196–197* (pl. 74)
rock pickers, *212–213* (pl. 82)
rodents, 257–258, 268, 276–277
Roskruge, George, xiv, *58–59* (pl. 11), *66–67* (pl. 15),
    80–81 (pl. 22), *82–83* (pl. 23)
Roth, V. S., 301n. 70
Russell, F., 293n. 23
Russian thistle, *156–157* (pl. 58)

S

Sabino Canyon, *186–187* (pl. 69), *190–191* (pl. 71)
Sabino Creek, *188–189* (pl. 70)
sacaton, *94–95* (pl. 27), *96–97* (pl. 28), *112–113* (pl. 36),
    124–125 (pl. 42), *128–129* (pl. 44), *130–131*
    (pl. 45), *134–135* (pl. 47), *136–137* (pl. 48), *138–*
    139 (pl. 49), *140–141* (pl. 50), *146–147* (pl. 53),
    148–149 (pl. 54), *150–151* (pl. 55), *152–153*
    (pl. 56), *154–155* (pl. 57), *158–159* (pl. 59), 262
    alkali sacaton, *88–89* (pl. 26)
Safford Peak, *202–203* (pl. 77)
Sageretia wrightii, 102–103 (pl. 31)
saguaro, 11–12, 15, 161–164, 166, *168–169* (pl. 60),
    170–171 (pl. 61), *172–173* (pl. 62), *174–175*
    (pl. 63), *176–177* (pl. 64), *178–179* (pl. 65), *180–*
    181 (pl. 66), *184–185* (pl. 68), *186–187* (pl. 69),
    188–189 (pl. 70), *190–191* (pl. 71), *192–193*
    (pl. 72), *194–195* (pl. 73), *196–197* (pl. 74), *198–*
    199 (pl. 75), *200–201* (pl. 76), *202–203* (pl. 77),
    204–205 (pl. 78), *206–207* (pl. 79), *210–211*

(pl. 81), *216–217* (pl. 84), *218–219* (pl. 85), *220–*
    221 (pl. 86), *222–223* (pl. 87), *224–225* (pl. 88),
    226–227 (pl. 89), 246, *251* (fig. 7.3), 252, 257,
    300nn. 22, 24, 302n. 18
    and American Indian food gathering, 19–20,
        293nn. 19, 21
    pattern of change, 248–250
saguaro decline, xiii, *168–169* (pl. 60), *170–171* (pl. 61),
    172–173 (pl. 62), *174–175* (pl. 63), *176–177*
    (pl. 64), *182–183* (pl. 67), *184–185* (pl. 68), *186–*
    187 (pl. 69), *200–201* (pl. 76), *208–209* (pl. 80),
    224–225 (pl. 88), 246, 250, 300n. 17
Saguaro National Monument, *168–177* (pls. 60–64), 246
Saguaro National Park, *168–177* (pls. 60–64), *182–183*
    (pl. 67), 246, 276
saguaro repopulation, *176–177* (pl. 64), *184–185*
    (pl. 68), *198–199* (pl. 75), *214–215* (pl. 83)
saguaro study plot, *186–187* (pl. 69), *188–189* (pl. 70)
Saint David, *156–157* (pl. 58)
saltcedar, *154–155* (pl. 57)
San Bernardino, 24
San Cayetano Mountains, *66–67* (pl. 15), *82–83* (pl. 23),
    120–121 (pl. 40), *122–123* (pl. 41)
Sanford Butte, *68–69* (pl. 16)
San José de Sonoita, 24
San Pedro Riparian National Conservation Area, *128–*
    129 (pl. 44), *138–139* (pl. 49), *154–155* (pl. 57),
    262
San Pedro River, xiii, 9, 20, 26–28, 31–33, *124–125*
    (pl. 42), *128–129* (pl. 44), *132–133* (pl. 46), *134–*
    135 (pl. 47), *136–137* (pl. 48), *140–141* (pl. 50),
    142–143 (pl. 51), *152–153* (pl. 56), *154–155*
    (pl. 57), *156–157* (pl. 58), *158–159* (pl. 59), 258
    and vegetation change, 261–262
San Pedro Valley, 4, 29, 92, *93* (fig. 5.1), *124–159* (pls.
    42–59), 248, *250* (fig. 7.1), 258, 299n. 32
San Simon Valley, 4
Santa Catalina Mountains, *170–171* (pl. 61), *174–175*
    (pl. 63), *176–177* (pl. 64), *178–179* (pl. 65), *180–*
    181 (pl. 66), *184–185* (pl. 68), *186–187* (pl. 69),
    188–189 (pl. 70), *190–191* (pl. 71), *192–193*
    (pl. 72), *194–195* (pl. 73)
Santa Cruz River, 7, 9, 31, 33, *82–83* (pl. 23), *84–85*
    (pl. 24), *86–87* (pl. 25), *120–121* (pl. 40), *122–123*
    (pl. 41)
    and vegetation change, 262
Santa Cruz Valley, *80–81* (pl. 22), *82–83* (pl. 23), *108–*
    109 (pl. 34), 258
Santa Rita Experimental Range, *72–73* (pl. 18)
Santa Rita Mountains, 36, *44–45* (pl. 4), *48–49* (pl. 6),
    54–55 (pl. 9), *56–57* (pl. 10), *60–61* (pl. 12), *62–*
    63 (pl. 13), *72–75* (pls. 18–19), 92, *94–123* (pls.
    27–41), *182–183* (pl. 67)
satellite imagery, multispectral, xiv–xv
Sauer, C. O., 19–20, 34, 293n. 2
Schulman, E., 266
seasonal changes, plant response to, 161–162
seasonality, of precipitation, 268–274
secular trend, *vs.* unusual occurrence, 292n. 59
Sellers, W. D., 266–267
Semidesert Grassland, 16, 28, 92, 160, 163, 248, 253
    and fire, 20, 259–261
    and precipitation change, 267–274, *270* (figs. 8.9
        and 8.10)